A429433

3700

BEHAVIOR OF MARINE ANIMALS

Current Perspectives in Research

Volume 3: Cetaceans

BEHAVIOR OF MARINE ANIMALS
Current Perspectives in Research

Volume 1: Invertebrates
Volume 2: Vertebrates
Volume 3: Cetaceans

BEHAVIOR OF MARINE ANIMALS

Current Perspectives in Research

Volume 3: Cetaceans

Edited by
Howard E. Winn
Professor of Oceanography and Zoology
University of Rhode Island
Kingston, Rhode Island

and
Bori L. Olla
U. S. Department of Commerce
National Marine Fisheries Service
N. E. Fisheries Center
Sandy Hook Laboratory
Highlands, New Jersey

PLENUM PRESS • NEW YORK-LONDON

Library of Congress Cataloging in Publication Data

Winn, Howard Elliott, 1926-
　Behavior of marine animals.

　Includes bibliographies and indexes.
　CONTENTS: v. 1. Invertebrates. — v. 2. Vertebrates. — v. 3. Cetaceans.
　1. Marine fauna — Behavior. I. Olla, Bori, L., joint author. II. Title.
QL121W5.　　　591.5'2636　　　79-167675
ISBN 0-306-37573-7 (v. 3)

© 1979 Plenum Press, New York
A Division of Plenum Publishing Corporation
227 West 17th Street, New York, N.Y. 10011

All rights reserved

No part of this book may be reproduced, stored in a retrieval system,
or transmitted, in any form or by any means, electronic, mechanical,
photocopying, microfilming, recording, or otherwise, without written
permission from the Publisher

Printed in the United States of America

Measuring and studying an entrapped humpback whale (*Megaptera novaeangliae*)

CONTRIBUTORS

Ben Baxter	College of the Atlantic, Bar Harbor, Maine
Peter Beamish	Department of the Environment, Fisheries and Marine Service, Marine Ecology Laboratory, Bedford Institute of Oceanography, Dartmouth, Nova Scotia
Terje Benjaminsen	Institute of Marine Research, Bergen-Nordnes, Norway
Peter B. Best	Woodstock, Cape Town, South Africa
Oliver Brazier	Woods Hole Oceanographic Institute, Woods Hole, Massachusetts
David K. Caldwell	Biocommunication and Marine Mammal Research Facility, University of Florida, St. Augustine, Florida
Melba C. Caldwell	Biocommunication and Marine Mammal Research Facility, University of Florida, St. Augustine, Florida
Ivar Christensen	Institute of Marine Research, Bergen-Nordnes, Norway
William E. Evans	Hubbs/Sea World Research Institute, San Diego, California
Steven Katona	College of the Atlantic, Bar Harbor, Maine
Scott Kraus	College of the Atlantic, Bar Harbor, Maine
Stephen Leatherwood	Bioanalysis Group, Naval Ocean Systems Center, San Diego, California
G. Victor Morejohn	Department of Biological Sciences, San Jose State University and Moss Landing Marine Laboratories, Moss Landing, California
David W. Morgan	Aquatic Behavior Laboratory, Department of Biology, University of Notre Dame, Notre Dame, Indiana

Judy Perkins	Weston, Massachusetts
Paul J. Perkins	Department of Zoology and Graduate School of Oceanography, University of Rhode Island, Kingston, Rhode Island
Graham S. Saayman	Psychology Department, University of Cape Town, Rondebosch, C. P., South Africa
Algis G. Taruski	Graduate School of Oceanography, University of Rhode Island, Kingston, Rhode Island
Colin K. Tayler	Museum, Snake Park and Oceanarium, Port Elizabeth, South Africa
Thomas J. Thompson	Department of Zoology and Graduate School of Oceanography, University of Rhode Island, Kingston, Rhode Island
William A. Walker	Section of Mammalogy, Natural History Museum of Los Angeles County, Los Angeles, California
Hal Whitehead	Cambridge University, Cambridge, England
Howard E. Winn	Department of Zoology and Graduate School of Oceanography, University of Rhode Island, Kingston, Rhode Island

PREFACE

Four years ago we began soliciting articles for this volume from authors who were engaged in comprehensive research on whales. From the outset we decided not to limit the subject matter to behavior but to also include natural history. Much of what is known about the behavior of whales arose from studies whose principal aim was not behavior, much as it did for other animal groups before behavior was considered a distinct discipline. Thus in many of the articles behavior is closely intertwined with natural history and in others is completely overshadowed by a basic natural history approach.

Our aim was to have the articles contain a review of the literature and include research findings not previously published. For all intents and purposes this aim has been realized, albeit perhaps not in as balanced a fashion in terms of species or subject matter as was originally planned. Nevertheless, we believe the articles present a wide range of informative works with a myriad of approaches and techniques represented.

We are grateful to the contributors for their patience and understanding in awaiting publication, which has taken much longer than we originally expected. We are also grateful for the assistance of a number of people, especially Julie Fischer and Lois Winn for their editorial efforts, and Jill Grover, Carol Samet, and Lois Winn for their help in indexing.

Howard E. Winn
Bori L. Olla

INTRODUCTION

A decrease in the size of animal populations may be attributed to a number of factors, but the causes are often not easily assignable. In the case of certain species of whales at least, overexploitation is clearly the cause for reduced numbers. The efficiency of "man the hunter" as a predator has increased so rapidly that no vertebrate animal has developed a strategy to deal with it, least of all whales, which reproduce slowly. Therefore, preservation of these animals depends upon the predator establishing a strategy that will allow the prey to survive.

Implementation of conservation measures, to be of any value, must be based on a firm knowledge of the life history and habits of the animals. Unfortunately, as is true for most marine animals, too little is known to paint the total picture of whales' life habits. The degree of knowledge varies, running the gamut from anecdotal accounts in the natural environment to well documented behavioral patterns of smaller species observed in oceanaria.

Studying whales *in situ* presents special problems. Direct observation of animals that can move rapidly over long distances and in an environment that is often not easily accessible to human observers makes the acquisition of knowledge a painfully slow process, with information coming in small bits and pieces. Observations that seem trivial when considered within the context of other animals are oftentimes significant and of extreme value because the base of information for whales is so thin. As is the case with behavioral field studies in general, cause and effect relationships between physical parameters and particular habits depend almost entirely on correlations that may or may not be valid.

Indirect observations have been put to good use and much information has been gathered in this way. Acoustical monitoring has been one of the more important methods for observing whales indirectly, mainly because sound plays such an important role in their behavioral repertoire. But even with the voluminous amount of data gathered via acoustical monitoring, the meaning of most sounds is not yet well understood. One major reason for this is the difficulty involved in making direct behavioral observations synchronously with sonic recordings.

Much has been learned from studying those species that can be held within aquaria. However, one problem that arises from these studies is the difficulty of extrapolating behavior patterns from the aquarium to the sea with confidence. Results from animals held in captivity under conditions that may modify habits in often immeasurable ways make it incumbent upon the researcher to interpret his findings with a view to those behaviors which may transcend the aquarium.

It is obvious that no matter where and how it is studied, the whale requires the application of a wide range of innovative methodologies and techniques to answer the most basic questions. New techniques such as individual recognition based on natural markings, radio tagging, and obtaining skin samples to determine sex, along with increased interest and support, can be expected to yield some exciting results in the near future. The papers presented in this volume represent the direction of future studies and are but a small measure of what must yet be learned before real progress in understanding these animals can be made.

Howard E. Winn
Bori L. Olla

CONTENTS

Chapter 1
Some Recent Uses and Potentials of Radiotelemetry in Field Studies of Cetaceans

    Stephen Leatherwood and William E. Evans

I. Introduction	1
A. Historical Background	2
B. Transmitter and Receiver Development	4
II. Radio Tracking from Aircraft and Satellites	8
A. Hand-Held Receivers and Automatic Direction Finders	10
B. Fuselage-Mounted Antennae	10
C. Tests of Performance	11
D. Relocation and Data Collection	11
E. Gigi	14
III. Shore Station Monitoring	14
A. Radio Tracking Using Shore Stations	14
IV. Radiotelemetry and Coordinated Environmental Sampling	16
V. Multiple Animal Radio Tracks	17
VI. Methods for Attaching Radio Transmitter Packages	21
VII. Summary and Recommendations	28
References	29

Chapter 2
Identification of Humpback Whales by Fluke Photographs

    Steven Katona, Ben Baxter, Oliver Brazier, Scott Kraus, Judy Perkins, and Hal Whitehead

I. Introduction	33
II. Methods	34
III. Results	35
IV. Discussion	39
V. Summary	43
References	44

Chapter 3
The Natural History of Dall's Porpoise in the North Pacific Ocean

G. Victor Morejohn

I.	Introduction	45
II.	Taxonomic Considerations	46
III.	Distribution and Population Shifts	48
IV.	Predators	50
V.	Coloration	51
VI.	Parasites and Commensals	53
	A. Helminth Parasites	53
	B. Epizoic Commensals	55
VII.	Feeding Habits	56
VIII.	Morphology	57
	A. External Morphology	57
	B. Organ Weights	65
IX.	Reproduction	65
	A. Female Reproductive Tract	69
	B. Male Reproductive Tract	70
X.	Behavior	70
	A. Swimming	71
	B. Diving	75
	C. Play	75
	D. Agonistic Behavior	76
	E. Senses	77
XI.	Discussion and Conclusions	77
	References	81

Chapter 4
The Northern Right Whale Dolphin *Lissodelphis borealis* Peale in the Eastern North Pacific

Stephen Leatherwood and William A. Walker

I.	Introduction	85
II.	Distribution, Seasonal Movement, and Abundance	88
	A. Smithsonian Institution	100
	B. National Marine Fisheries Service (NMFS)	101
	C. University of California	102
	D. Naval Undersea Center	102
III.	Herd Sizes	107
IV.	Herd Configuration	109

Contents

- V. Interspecific Associations 111
- VI. Swimming Behavior 112
- VII. Sound Production 114
- VIII. Strandings 116
- IX. Coloration 120
  - A. Coloration of Adults 120
  - B. Coloration of Calves 121
  - C. Color Variation 123
- X. Reproduction 124
- XI. Food Habits 126
- XII. Parasitism 127
- XIII. Morphometrics 127
  - References 138

Chapter 5
The Natural History of the Bottlenose Whale, *Hyperoodon ampullatus* Forster

Terje Benjaminsen and Ivar Christensen

- I. Introduction 143
- II. Distribution 144
  - A. Geographic Distribution 144
  - B. Migration 145
  - C. Water Depth 148
  - D. Water Temperature 149
  - E. Relation to Ice 150
  - F. Segregation 150
- III. Behavior 152
  - A. Social Organization 152
  - B. Care-Giving 153
  - C. Diving 153
- IV. Feeding 154
- V. Reproduction 156
  - A. Sexual Maturity in Males 156
  - B. Sexual Maturity in Females 157
  - C. Pregnancy 158
  - D. Lactation 158
  - E. Reproductive Cycle 158
- VI. Age and Growth 158
- VII. Summary 160
  - References 162

Chapter 6

The Socioecology of Humpback Dolphins (*Sousa* sp.)

Graham S. Saayman and Colin K. Tayler

- I. Introduction ... 165
- II. The Study Area ... 167
  - A. Zone I ... 168
  - B. Zone II ... 168
- III. Methods ... 170
- IV. Statistical Analysis of Results ... 171
- V. The Dolphins ... 171
  - A. Calf I ... 172
  - B. Calf II ... 172
  - C. Juvenile ... 173
  - D. Grayback ... 173
  - E. Whitefin ... 174
- VI. Results ... 175
  - A. Range ... 175
  - B. Seasonal Occurrence ... 176
  - C. Social Structure ... 180
  - D. Birth Periodicity ... 189
  - E. Maintenance Activities and Social Behavior ... 191
  - F. Utilization of Habitat ... 197
  - G. Diurnal Activity Cycle ... 203
  - H. Influence of the Tidal Cycle upon Behavior ... 203
  - I. Interactions with Other Animals ... 207
- VII. Discussion ... 211
  - References ... 223

Chapter 7

Social Organization in Sperm Whales, *Physeter macrocephalus*

Peter B. Best

- I. Introduction ... 227
  - A. Review of Previous Work ... 227
  - B. Scope of the Present Work ... 231
- II. Composition of Population ... 233
- III. Composition of Groupings ... 237
  - A. All-Male Groups ... 237
  - B. Groups of Mixed Sexes ... 245
  - C. Effective Sex Ratio in the Population ... 254

IV. Ontogeny of Schooling Behavior 256
       A. Segregation of Males from Mixed Schools 256
       B. Segregation of Males to Higher Latitudes 260
    V. Geographical Distribution of Groupings 262
   VI. Migrations 271
  VII. Social Organization and Its Development 273
       A. Form of Social Organization in the Sperm Whale 273
       B. Evolution of Polygyny 276
 VIII. Implications for Management 280
       References 285

Chapter 8

Behavior and Significance of Entrapped Baleen Whales

Peter Beamish

    I. Introduction 291
   II. Ice Entrapment 291
       A. Causes 291
       B. Experimental Methods 293
       C. Results 293
       D. Discussion 298
  III. Net Entrapments 300
       A. Causes 300
       B. Experimental Methods 301
       C. Results 302
       D. Discussion 306
   IV. Techniques of Measuring, Tagging, and Release 307
    V. Significance and Summary 308
       References 309

Chapter 9

The Vocal and Behavioral Reactions of the Beluga, *Delphinapterus leucas*, **to Playback of Its Sounds**

David W. Morgan

    I. Introduction 311
   II. Materials and Methods 313
  III. Results 318
       A. Playback of Sounds Recorded from the Captive Animals ... 318
       B. Playback of Sounds Recorded from the Saguenay Herd 327

IV. Discussion 332
　A. Playback of Sounds Recorded from the Captive Animals 332
　B. Playback of Sounds Recorded from the Saguenay Herd 337
　References 342

Chapter 10

The Whistle Repertoire of the North Atlantic Pilot Whale (*Globicephala melaena*) and Its Relationship to Behavior and Environment

Algis G. Taruski

　I. Introduction 345
　II. Materials and Methods 346
　III. Results 349
　IV. Discussion 364
　V. Summary 367
　　References 367

Chapter 11

The Whistle of the Atlantic Bottlenosed Dolphin (*Tursiops truncatus*)—Ontogeny

Melba C. Caldwell and David K. Caldwell

　I. Introduction 369
　II. Literature Review 370
　III. Methods 370
　IV. Equipment 372
　V. Differences in Whistles by Age Class 372
　　A. Sound Loops 373
　　B. Durations 373
　　C. Frequency Modulations 373
　　D. Stereotypy 374
　VI. Analyses of Whistles of Individual Infants 376
　　A. Infant Number 1 376
　　B. Infant Number 2 380
　　C. Infant Number 3 381
　　D. Infant Number 4 382
　　E. Infant Number 5 383
　　F. Infant Number 6 386
　　G. Infant Number 7 389
　　H. Infant Number 8 391
　　I. Infant Number 9 391

Contents

J.	Infant Number 10	392
K.	Infant Number 11	394
L.	Infant Number 12	395
M.	Infant Number 13	396
N.	Infant Number 14	397
VII.	Summary and Discussion	397
	References	401

Chapter 12
Mysticete Sounds
Thomas J. Thompson, Howard E. Winn, and Paul J. Perkins

I.	Introduction	403
II.	Methods	405
III.	Species Accounts	405
	A. Mysticete Taxonomy	406
	B. Family Balaenidae	406
	C. Family Eschrichtiidae	410
	D. Family Balaenopteridae	412
IV.	Miscellaneous Sounds	421
V.	Discussion	424
	References	428

Index 433

Volume 3: Cetaceans

Chapter 1

SOME RECENT USES AND POTENTIALS OF RADIOTELEMETRY IN FIELD STUDIES OF CETACEANS

Stephen Leatherwood

Bioanalysis Group
Naval Ocean Systems Center
San Diego, California 92152

and

William E. Evans

Hubbs/Sea World Research Institute
1700 South Shores Road
San Diego, California 92109

I. INTRODUCTION

In 1874, Charles M. Scammon, noted whaling captain/naturalist, observed:

> It is hardly necessary to say that any person taking up the study of marine mammals, particularly the cetaceans, enters a difficult field of research, since the opportunities for observing the habits of these animals under favorable conditions are but rare and brief. My own experience has proved that observation for months and even years may be required before a single new fact in regard to their habits can be obtained. This has been particularly the case with the dolphins, while many of the characteristic actions of the whales are so secretly performed that years of ordinary observation may be insufficient for their discovery.

In the more than 100 years since that statement, cetaceans have continued to be among the most elusive and difficult to study of all wild animals. They inhabit most of the world's major oceans and seas and even some freshwater rivers and lakes. Wherever they occur, they are visible to the confounded, air-bound biologist primarily during the brief periods when they break the air–water interface to breathe and spend the rest of their lives, in some species greater than 95% of their time, below the surface and well out of view. Attempts to observe them from underwater (see Evans and Bastian, 1969; Norris et al., 1974) and aircraft (e.g., Nishiwaki, 1962; Leatherwood, 1974a, 1975) have provided some new insights but still fail to get at some of the most important aspects of how the animals move in and exploit their three-dimensional environment.

This being the case, it is not surprising that the recent history of field studies of cetaceans includes a series of largely "show-and-tell" naturalistic approaches and attempts to modify old and develop new techniques for field studies.

Of the recent approaches used, perhaps none has shown more significant progress or offers more promise than radiotelemetry. This chapter reviews the most important recent developments in radiotelemetric technology as it applies to cetaceans, summarizes some of the most important recent applications, adds previously unpublished data to suggest some potential uses, and suggests directions in which we feel new research should progress.

As used in this chapter, radio tracking implies simply the attachment of a transmitter to an animal and subsequent determination of its position as a function of time. In a broader sense, radiotelemetry implies a greater variety of uses, including transmission of data about the animal's environment (e.g., water temperature, salinity, dissolved oxygen), behavior (e.g., diving depth, swimming speed, sound production), or physiological state (e.g., heart rate, body temperature) as a function of time and location.

A. Historical Background

During the last 25 years, radiotelemetry has come into prominence as a tool in biological studies in general (e.g., Mackay, 1970; Galler et al, 1972; Schevill, 1974). The importance and potential of radiotelemetric instrumentation and techniques in the study of animal behavior were recognized in print as early as 1963 (Slater, 1963). In particular, some potentials for the use of telemetry in studies of aquatic animal communication were identified and discussed (Evans and Sutherland, 1963). But before many of the then available and subsequently developed techniques could be applied to studies of marine mammals in their natural environment, a number of problems in packaging, frequency allocations, antenna and receiver design, and methods of attaching the instrument packages had to be overcome. In the past 15 years, with a rapid acceleration in the last five years, the state of the art in radiotelemetric instrumentation and techniques, as in other remote sensing technology applicable to the study of cetaceans (and other marine mammals), has increased several orders of magnitude. Some of the more visionary concepts underlying much of this development, including uses of multichannel instrument packages with environmental and acoustic sensors and plans for development of systems to permit continual monitoring of instrumented cetaceans from shore stations via satellites, have been identified and discussed (Evans, 1970). Much of the most vital research and development effort has been directed toward these two important objectives. Despite significant progress in many areas, the state of the art continues to be embryonic and in need of significant further development.

The most important developments in radiotelemetry through 1973 as they apply to cetaceans were reviewed in detail by Norris et al. (1974). They may be summarized as follows:

Early attempts employed acoustic tracking devices developed for the study of fishes. Schultz and Pyle (1965) attempted to attach acoustic transmitters mounted on shallow harpoon heads to California gray whales (*Eschrichtius robustus*). Payne (1967, Rockefeller University, personal communication) similarly attempted to track humpback whales (*Megaptera novaeangliae*) using acoustic devices. In 1967–1968 one of us (Evans) tested the potential use of sonic transmitters attached by a suction cup to a captive *Tursiops truncatus* (unpublished data). None of these attempts met with any success. The primary problems identified were that (1) ranges obtainable were unacceptably short; (2) transducers, both transmitting and receiving, were inadequate; and, importantly for future approaches, (3) the projectors used frequencies that fell within the hearing ranges (e.g., see Johnson, 1966) of these highly acoustic animals. There were significant problems in all these cases with successful attachment and operation of the transmitters. But even if these technical problems had been overcome, it is highly questionable whether data obtained from these systems could have represented "normal" behavioral patterns for the tagged animals.

The next developments utilized radio transmitters. Schevill and Watkins (1966), for example, attempted to track unrestrained right whales by implanting radio beacons, attached to shallow darts, into the blubber. Although this early attempt was unsuccessful, it provided impetus to many of the developments which followed.

From 1967 through 1971, the Naval Undersea Center (now the Naval Ocean Systems Center), working largely with Ocean Applied Research Corporation of San Diego, California, progressed through numerous intermediate stages to develop a working radio-frequency transmitter for attachment to dolphins. These transmitters represented first-phase solution to the special problems of radio tracking marine mammals (Fig. 1). Their operation and construction are summarized by Martin et al. (1971). Because of the relatively large size of the animals to be tagged, weight and size of transmitters were not as restrictive as they were for small mammals. Some examples of the uses of this and the next generation of radio transmitters are discussed in Evans (1971).

These early transmitters were simple, submersible locator beacons, repackaged so they could be attached to a marine mammal (dolphin), and were intended primarily for tracking of the animals' movements. Radio frequencies in

Fig. 1. First prototype dolphin radio transmitter being tested, attached by a Teflon-coated pin to the dorsal fin.

the 11-m band were selected because of (1) their past history of being effective in transmitting from close to the sea surface; and (2) the availability of field-tested transmitters of appropriate size and transmission range. But because they employed a sea-water switch, operated by the link between the antenna tip and the transmitter housing, to transmit only at the surface, they also provided important data on respiration patterns and therefore, by inference, on diving behavior of the animals (see Fig.2).

B. Transmitter and Receiver Development

Second-generation instrumentation packages developed under this program added a pressure sensor and memory circuit to detect and store data on the maximum depth of each dive. These depths were coded to frequency and then transmitted on the carrier signal each time the animal surfaced. An example of the types of respiration and dive pattern data obtainable from these packages is shown in Fig. 3. These instrument packages were successfully deployed on common dolphins *Delphinus delphis* (Evans, 1971), pilot whales *Globicephala* sp., and killer whales *Orcinus orca* (Martin et al., 1971; Evans, 1974a). Subsequent designs added a channel to monitor the maximum water temperature at the animal's maximum dive depth. One of these packages was used in studying the behavior of a yearling California gray whale *Eschrichtius robustus* (Evans, 1974b).

Wuersig (1976), using equipment and techniques utilized by Evans (1971) and in consultation with the authors, radio-tagged 10 dusky dolphins (cited as dusky porpoises) *Lagenorhynchus obscurus* in Gulfo de San Jose, Argentina, and tracked them from shore and from rubber boat for an average of 6.5 days. Movement was monitored within a 750 sq km of the bay and an equal sized area of the adjacent open ocean. Some of the animals' positions and movements were determined using triangulation from two stations. The study revealed important information about movements, diving behavior, group behavior, and relationships to food sources of the species. In contrast to the patterns reported for *Delphinus* and *Stenella* (Evans, 1975; Perrin et al., 1976; Leatherwood and Ljungblad, 1979) the dusky dolphins seemed to dive during the day and remain close to the surface at night.

Between January 1975 and June 1976, Irvine et al. (1976) radio-tagged ten

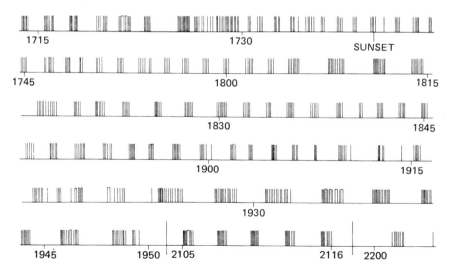

Fig. 2. Respiration pattern of an instrumented *Delphinus* as a function of time of day. Data inferred from transmission rate.

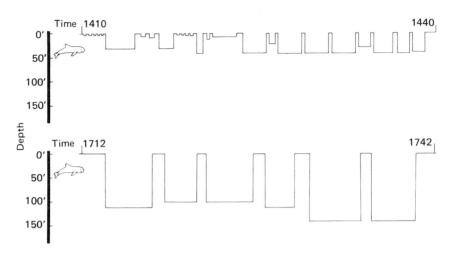

Fig. 3. Respiration pattern and diving depth pattern of a Pacific pilot whale.

Atlantic bottlenose dolphins with transmitters similar in design to those used by Evans (1971) and Wuersig (1976). The method of attachment was essentially the same except that the manufacturers' supplied saddle, which was too small for *Tursiops*, was replaced by a molded fiberglass saddle lined with open cell foam. Seven of the ten animals tagged were tracked for five days or more. Wuersig's average for *Lagenorhynchus obscurus* was 6.5 days and Evans's (1975) average for *Delphinus delphis* was less than 3 days, although Evans (1971) reported two relocations after 30 days.

Because of the inshore, coastal habitat of the animals they studied, Irvine *et al.* (1976) were able to accurately document many of the problems associated with attachment of radio transmitters to dorsal fins, of at least *Tursiops*. Tests on captive *Delphinus* (Evans, 1971) indicated that the perforation of the dorsal fin healed within 10 days and that the animals' swimming behavior was not adversely affected. However, the package used on the *Delphinus* was molded into a smooth hydrodynamic shape with the antenna in the midline anterior to the dorsal fin (see Fig. 1). These packages used by Wuersig were similar in design to those described by Evans (1971). Also similar were transmitters used by T. P. Dohl and K. S. Norris (1977, personal communication) Coastal Marine Laboratory, University of California, Santa Cruz, to tag and track *Delphinus, Lagenorhynchus,* and *Globicephala* off southern California and *Stenella* sp. off Hawaii and by Perrin *et al.* (1977) and Leatherwood and Ljungblad (1979) on *Stenella attenuata* in the eastern tropical Pacific. John Hall, Bureau of Sport Fisheries, Alaska (1976–1977), has been developing modifications for use on Dall's porpoise (*Phocoenoides dalli*).

Erickson (1976), utilizing killer whales captured in Puget Sound and donated by Sea World, Inc., to his University of Washington project, tested dorsal-fin-mounted VHF radio transmitters. The whales, captured at Budd Inlet at the southern end of Puget Sound, were fitted with radio transmitters while being held and observed in sea pens in the San Juan Islands.

Most of these packages were attached by means of a spring-loaded bolt placed through a nylon sleeve lining a catherized hole bored in the dorsal fin. The packs were equipped with soluble aluminum bolts to ensure that the packages were released from the animal, after some desired time period, leaving it unencumbered.

The package utilized on the gray whale (Evans, 1974b) was secured to the animal by surgical techniques with expected long life (Sweeney and Mattsson, 1974), although it too was equipped with corrosible studs for eventual release.

Since the completion of these efforts, numerous programs, some employing modified attachment mechanisms, transmitter housings, and tracking methods have been reported, all but one (Erickson's) utilizing the same basic radio frequency transmitters and receivers described above.

Norris and Gentry (1974) attached transmitters to three suckling California gray whales by means of expandable belly bands outfitted with corrosible links and tracked them for up to 5 hr. The three packs successfully released from the animals and were recovered from 5 to 17.5 hr later.

In a follow-up to this program, Norris et al. (1977) reported on more lengthy tracks of two mother-calf gray whale pairs within the lagoons of southern Baja California, Mexico, and up to 213 km south of the lagoons where they were tagged. Depth-of-dive sensors on one of the packs permitted definition of diurnal differences in diving behavior and respiration differences between quiescent and actively swimming calves.

Several converging programs have been attempting to develop radio transmitter tags which can be attached to cetaceans without capturing them. An example of this is the recently developed implantable whale beacon, based on an initial concept developed by Schevill and Watkins (1966) and developed in a more advanced stage by Ocean Applied Research Corporation in cooperation with Woods Hole Oceanographic Institution (Watkins), Naval Ocean Systems Center (NOSC), and Johns Hopkins University, with inputs from interested scientists from TINRO, Vladivostok, during the U.S./U.S.S.R. cooperative marine mammal program, as well as from National Marine Fisheries Service (NMFS).

Using a transmitter of this type, designed for delivery using a shotgun and intended for implantation in the blubber following the pattern of discovery tags (Ray and Wartzok, 1975), Ray and Wartzok (1976) tagged and successfully tracked a fin whale, *Balaenoptera physalus*, in the St. Lawrence River mouth for over 27 hr in August 1976.

During the same period and using much the same equipment, the NMFS Marine Mammal Division, Seattle, Washington, tagged three humpback whales, *Megaptera novaeangliae*, near Juneau, Alaska, and successfully tracked one of them intermittently for 6 days and 75 km (Tillman and Johnson, 1976). These experiments were successfully repeated in July 1977 with one track of approximately one week. A detailed history of the development and testing program which has resulted in an improved implantable whale tag has been recently prepared by Watkins and Schevill (1977). The implantable whale tag used by Ray and Wartzok (1976) and Tillman and Johnson (1976) was still very experimental in design. Tests conducted in Iceland in 1976 (Watkins and Schevill, 1977), using dead whales collected by the local shore whaling station, demonstrated serious problems with the transmitter power supply and tag penetration. The 1977 Iceland tests demonstrated that these problems had been solved, and that with the 1977 modifications the radio tag was ready for field trials on live balaenopterid whales. The tests conducted in eastern Alaska in July 1977 on *Megaptera* by the National Marine Fisheries Service (personal communication) employed the 1977 modified tags tested by Watkins in Iceland (Fig. 4a–c).

Some previously unpublished efforts by the authors (some in cooperation with NMFS) during the years 1971–1975 are reported below.

II. RADIO TRACKING FROM AIRCRAFT AND SATELLITES

One of the most important long-term objectives of developments in radiotelemetric instrumentation and techniques is the ultimate goal of monitoring movements, diving behavior, and selected environmental conditions by relays from tagged animals to satellites and from there to shore stations where data can be easily stored or quickly analyzed (Evans, 1971).

Research in support of the goal of satellite monitoring was recently identified as one of the most important problems for the cetacean research community (Anonymous, 1974). In response to this need, Ichihara (1972) described in detail some specific approaches toward location of instrumented whales, using satellites. His rather detailed treatment included a first look at the satellite receiver systems; relays and transmitter systems most likely to work; the instrumentation types, frequencies, and signal strengths required to work with an orbiting satellite; and some special problems with transmitter output, signal characteristics, and battery life. But the bulk of the treatment, and the only areas supported by new field data, dealt with the reliability of position fixes using available "electric wave" navigation systems (e.g., Decca, Loran, Omega, and two types of satellite navigation systems). During the summer of 1977, personnel from NMFS Southwest Fisheries Center, La Jolla, California, and NMFS Southeast Fisheries Center, National Space Technology Laboratories, Bay St. Louis, Mississippi, with cooperation from Sea World, Inc., tested the feasibility of transmitting from a swimming dolphin to the Nimbus 6 satellite. A preprototype VHF transmitter was attached by means of a harness to trained captive *Tursiops*. During the two weeks of testing it was documented that radio

Fig. 4. (a) Radio whale tag employed during the July 1977 tests: push rod and complete tag. (b) Modified radio whale tag system employed during the July 1977 tests. Gun was balanced with weight added to stock and forearm. (c) Three 1977 radio whale tags in finback carcass: Iceland. (From Watkins and Schevill, 1977; photos by Karen E. Moore, Woods Hole Oceanographic Institution.)

signals useful for determining position could be transmitted from the back of a free-swimming dolphin and received and processed by the Nimbus 6 satellite. Although most of the practical field problems remain unsolved, the feasibility of this approach has been demonstrated. Many of these problems can be solved by concentrating on improving this type of tracking technology using conventional aircraft.

Recent approaches by two long-term programs in radiotelemetric instrumentation and techniques (those of the Naval Undersea Center and the University of California at Santa Cruz) have recognized that if satellite monitoring is ever to be successful and routinely employed in cetacean research, there are a number of basic problems to be solved and research must progress through various intermediate stages. Most of the work to date has concentrated on adapting receiver systems for whale monitoring and data collection using low-altitude aircraft systems (1500 m). Results of some of those efforts are described below, in order of progressing difficulty.

A. Hand-Held Receivers and Automatic Direction Finders

On July 30, 1968, Evans (1971) tagged and released a male *Delphinus delphis* at approximately 33°20'N, 118°08'W off Catalina Island, California. During the next 8 days, he was able to relocate the animal twice from a Cessna 172 at 1000 feet using a hand-held radio receiver (OAR Finder's receiver). By July 31, the animal had moved to 33°19'N, 118°30'W, a straight-line distance of some 37 km. A week later, August 7, the same animal was at 33°30'N, 118°45'W, approximately 29 km straight-line distance out, on the opposite side of the island from the release site. Even this elementary tracking system could be used to relocate tagged animals in a limited area.

It was from this first attempt at radio tracking a fast moving delphinid at sea that it became obvious that a different, faster-responding receiving system was necessary. Although possible, it was extremely difficult to get an accurate directional fix with 2 sec of signal. This was approximately the amount of time the antennae on the *Delphinus* transmitter was exposed. It was decided that an automatic direction finder (ADF) specially designed to track cetaceans was needed. An ADF that could indicate relative direction $\pm 5°$ in any direction of the compass (360°) with a minimum 0.1 sec of signal was needed and not readily available. The instrument to solve this problem was designed and formulated under contract to the Navy by Ocean Applied Research Corp. The details of this design are discussed by Martin *et al.* (1971).

B. Fuselage-Mounted Antennae

The second step was to develop antenna systems specifically for aircraft. The first systems to be used in marine mammal tracking were described and illustrated by Evans and Leatherwood (1972) and by Evans (1974a). Basically, they consisted of the previously developed shipboard data recording and display instrumentation (Martin et al., 1971) with two modified antennae. The first, a fixed set of loops, was suspended from the aircraft belly about amidships oriented towards the bow of the aircraft (Fig. 5a,b) and resulted in a display of the animals' bearing relative to the aircraft orientation. The second was a motor-controlled sense antenna which could be lowered to the vertical position during flight and retracted for takeoffs, landings, and high-speed flights (Fig. 5b). The systems were configured for use on Naval S-2-type reconnaissance aircraft, but they can readily be adapted to other light aircraft.

C. Tests of Performance

Using this system, tests were conducted on May 23 and 25, 1972, to determine signal strengths detectable from a 250-MW, 27.585-mHz transmitter as a function of altitude and distance. In these tests, with the transmitter at the water surface to take advantage of the surface–ground–plane effect, the signal could be detected up to 43 km from 4500 m and 50 km from 1500 m. Directionality could be resolved at this altitude inside a 37- to 42-km distance from the two transmitters, respectively.

In subsequent tests with the package on a test animal, the transmission range increased to 65 km at 1800 m, at which time it was overpowered by shore interference. Maximum ranges with this combination of transmitter receiver systems probably approach 75 km under ideal conditions.

D. Relocation and Data Collection

On May 25 and 26, 1972, these tested systems were used to track an instrumented animal. At 1240 hr at approximately 32°20'N, 118°08'W, a 185-cm male Pacific common dolphin, *D. delphis* was captured, instrumented with the test transmitter with a depth-of-dive sensor, released, and tracked from shipboard until 0930 the following day. During the same period data were recorded from aircraft at 465 m for the periods 1248–1350 and 1410–1800 hr. The animal was relocated from aircraft at 1030 hr on the 26th and data recorded until 1500 hr. All the data on diving depths and cycles collected from aircraft during these periods were comparable to those obtained from shipboard. Furthermore, valuable observations of the tagged animals' free interaction with other members of the herd were possible. The tagged animal was relocated from aircraft June 15, 1973, at approximately 33°05'N, 118°35'W, nearly 75 km from the site where the last data were collected (Fig. 6).

Radio tags, for which only low-altitude aircraft receiving systems are

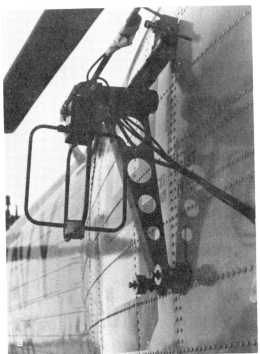

Fig. 5. (a) ADF loop antenna attached to side of Navy H-3 helicopter. (b) ADF loop and retractable sense antenna attached to belly of Navy S-3 ASW aircraft.

available, can be important in identifying specific stocks or populations of cetaceans. A simple example of their application in long-term studies of cetacean herd identification and movement can be illustrated by field exercises conducted off San Diego in 1972–1973.

Periodic aerial surveys (described in detail in Leatherwood, 1974a; Evans, 1975) were being used to conduct visual surveys for cetaceans. In support of this effort, a tagging cruise was conducted December 11–18, 1972, to mark individuals in some of the larger delphinid herds with static and radio tags. During the cruise, one animal from each of three herds was captured, radio tagged, and released. A 184.0-cm Pacific white-sided dolphin, *Lagenorhynchus obliquidens* (animal A) received a 27.575-mHz tag pulsing at the rate of 4 pulses/sec during time at the surface. Two Pacific common dolphins, *Delphinus delphis*, received similar tags, each transmitting on a different frequency and

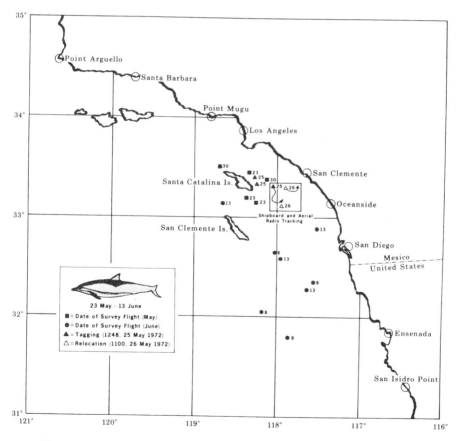

Fig. 6. Relocation of an instrumented *Delphinus* from an airborne radio tracking system.

producing different pulse rates. Assuming the animals remained with the same herds over long times, herds were identifiable by analysis of the signal format.

Each time herds of this species were located during flights in subsequent months, the receiving instrumentation in the aircraft was turned on and monitored sufficiently long for several respiration cycles of the animals to determine whether or not a radio-tagged animal was included in the group. On January 26, 1973, a group of an estimated 20 white-sided dolphins, located at 32°57'N, 117°18'W, was found to contain the radio-tagged animal A. On February 8, 1973, the same animal was found in the midst of an estimated 25 white-sided dolphins at 32°30'N, 117°17'W. In over 45 days, there was little net movement apparent for this particular group. Although this effort was intended only as a demonstration, the results do suggest a viable approach to coordinated studies of cetacean movements using coded radio tags and routine aircraft monitoring of a limited study area.

E. Gigi

In March, April, and May of 1973 the authors were able to successfully collect temperature and depth-of-dive information from an instrumented yearling California gray whale *(Eschrichtius robustus)*, named Gigi, released off southern California (Evans, 1974b). Despite damage to the antenna of the 1.0-W transmitter and aberrancies in the animal's swimming pattern, factors which combined to reduce the frequency of occurrence of transmission and transmitter performance, we were able to successfully relocate the whale and record up to 2 hr of data 14 days after its release (Fig. 7).

Although the results reported here are all limited, they do suggest progress towards high-altitude aircraft and eventual satellite monitoring of whales' movements and behavior. Norris *et al.* (1974) have pointed out that satellite receivers and transmitters are highly developed, although dedicated satellites specifically for wildlife assessment are needed before the necessary systems can be field tested and perfected.

III. SHORE STATION MONITORING

A. Radio Tracking Using Shore Stations

Another stated objective of much of the radiotelemetry research with cetaceans is the area of fixed-station monitoring of instrumented animals. Although some of those shore stations are ultimately slated for the satellite monitoring programs, all have intermediate and some will continue to have specialized use.

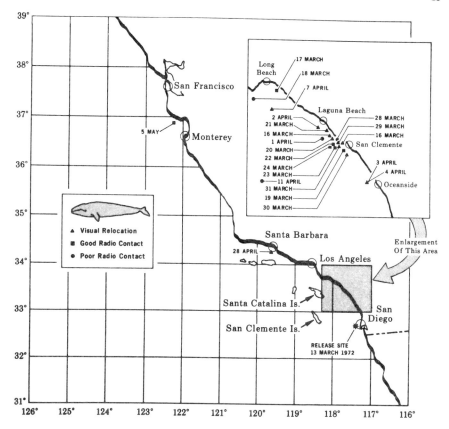

Fig. 7. Relocation of an instrumented gray whale from aircraft, ship, and mobile shore receiving stations.

For example, in some areas at least, Atlantic bottlenosed dolphins appear to have relatively restricted ranges of movement [e.g., Caldwell, 1955 (eastern Florida); Irvine and Wells, 1972 (Tampa Bay); Leatherwood and Platter, 1975 (Mississippi Sound)]. The situation may be true for bottlenosed dolphins (*Tursiops*) in other areas and for other species as well. Under those circumstances, if animals can be tagged, as described above, with transmitters coded by frequency and pulse characteristics to a particular individual or herd, and monitored from two or more shore stations, a continuous record of movements can be obtained from the triangulated determination of position at any given time.

Although detailed programs utilizing these techniques have not been conducted to date, several examples of the utility of even a single shore station have been demonstrated.

From January 1972 through December 1973, a monitoring station was maintained at the Southwest Fisheries Center of the National Marine Fisheries Service (La Jolla, California). In addition to loop and sense antennae, located atop the building some 150 m above sea level, and an ADF for monitoring position, the system included an 8-channel radio frequency scanner, which continually swept over the frequencies of interest, stopping on any frequency on which signals were received.

Using this system, signals were received in January and February 1973 from all three of the previously discussed animals tagged in December 1972.

That such systems can go mobile was demonstrated in 1972. Both a hand-held receiver, with compasses and polaris used from two separate positions to triangulate a position, and a modified instrument truck, equipped with standard shipboard tracking system for azmmith and signal strength, were used to track the gray whale Gigi (Evans, 1974*b*) during her movements along the southern California coast.

IV. RADIOTELEMETRY AND COORDINATED ENVIRONMENTAL SAMPLING

An example of a multiple method study of dolphin behavior and its environmental correlates, illustrated in Fig. 8, is described below.

As a portion of a detailed study on the biology and behavior of the Pacific common dolphin *Delphinus* cf. *D. delphis* in the northeastern Pacific (Evans, 1975), a radio track was conducted April 9–10, 1971. This particular study involved vessels of the U.S. Navy (*R/V Cape*) and California Department of Fish and Game (*M/V Alaska*). The approach was simply to radiotrack the animal, establish the depth to which it was diving, and make biological collections and environmental measurements at that depth. A 180-cm female *Delphinus* was captured, outfitted with a 27-mHz radio transmitter equipped to measure maximum depth of dive, and released at 1500 hr March 9, 1971. During the next 9 hr the animal's movements and diving behavior were continually monitored and recorded. Fathometer traces revealed concentrations of fishes, bottom contours, and density and movements of the light-sensitive deep scattering layer (DSL). Expendable bathythermographs examined changes in the vertical temperature structure of the water column. Measurements of sea-surface temperature gave continuing readings of changes in the horizontal temperature structure. Nansen bottles collected H_2O samples from depths to which the animal was diving and otter trawls collected organisms from that depth. During the short track, the animal displayed the nocturnal diving cycle described by Evans (1971) (Fig. 8). From 1500 to 1730 hr, dives were irregularly spaced and ranged in depth from 3 to 5 m. At 1730 hr the animal began making deeper dives, sometimes reaching 40 m. The fathometer traces during this period showed the DSL to be compressing and migrating towards the surface. No clear correlation

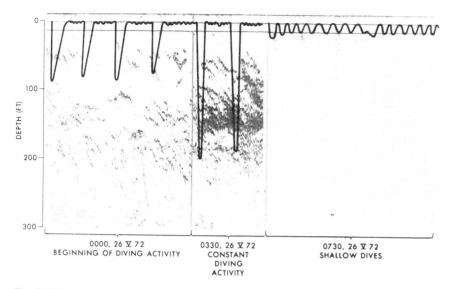

Fig. 8. Diving cycle of a *Delphinus* as a function of time of day and pressure of deep scattering layer (DSL).

of the onset of this behavior with sea-surface or vertical temperature structure could be determined. Once the depth of the deepest dives became constant at 40 m, the *M/V Alaska* made two trawls at that depth with an 11-m otter trawl, one at 1749–1809 hr, and one at 2235–2315 hr. These trawls collected a total of 100 bathylagids, 150 myctophids, 3 anchovies, 1 hake, 1 squid, and several euphausiids and sergestids. With the exception of one myctophid species caught, all have been found in the stomachs of *Delphinus* from the southern California continental borderland (Evans, 1975). It is assumed that the dives took advantage of the concentration of the organisms associated with the DSL towards the surface. Attempts to collect the radio-tagged animal to examine stomach contents to verify this speculation were unsuccessful. Nevertheless, the research does demonstrate an approach towards environmental monitoring intended to describe causal relationships between changes in animal behavior patterns and changes in environmental conditions.

V. MULTIPLE ANIMAL RADIO TRACKS

The criticism has been made that radio tracking studies involving a single animal which is extracted from a herd, outfitted with a radiotransmitter, released, and followed for a period of time may be giving an erroneous picture of the natural movement patterns and behavior of the species in general. We do not think this is so and have numerous bits of anecdotal evidence to suggest that after an initial period of adjustment to the tag, the animals move with their herds, behave in "typical" ways, and are essentially unaffected by the presence of the tag.

As an example of the tendency of a captured and radio-tagged animal to return to its own herd, we cite the case of a female *Delphinus* captured, tagged, released, and tracked for 24 hr (Evans, 1971). The animal used in this study was captured from a herd of an estimated 1000 individuals five miles off the east end of Catalina Island on February 12, 1969. The herd was milling and moving slowly to the northeast. The captive animal was transported away from the herd to an area approximately 28 km away, off the west end of Catalina Island, where she was released in an area with no delphinids in sight. The tagged animal proceeded slowly southeast following the coastline of the island. On two occasions she came in contact with small groups of *Delphinus* heading north. Although she joined with them for a short time, she shortly resumed her course southeast heading essentially to the general area from which she was captured. At approximately 1700 hr she joined a large herd (1000) and was seen in close company of four to five individuals. After joining, this aggregation started heading east toward Pt. Fermin. The tagged animal remained with this group until the track was terminated at 0900 hr on February 13, 1969. This is (a good bit of) circumstantial evidence, but the best way to see if the pattern is repeated and to examine some of the questions related to individual and group dependence on the herd, short-term recruitment, etc., is to tag a number of animals simultaneously and track them all over the same time period.

In November 1972, the National Marine Fisheries Service and personnel from NOSC conducted a joint operation from a chartered tuna purse seiner in the EASTROPAC. The purposes of the operation were (1) to examine the movement patterns and diving behavior of the major species of delphinids involved in the American tropical Pacific tuna fishery (Perrin, 1971), and (2) to test various methods for reducing the mortality of these species incident to fishing operations. One important portion of that work, as it related to the development and utilization of techniques for the study of the natural history and behavior of cetaceans, is described below.

In an attempt to test the effect or recapture of the same herd of porpoise on herd size, behavior, and time to recruit fish after a set, three *Stenella attenuata* from the same herd were radio tagged, one male (A) and two females (B, C). Once released, the objective was to track the herd and set on what was assumed to be the same herd five times. Each set was to be separated by 24 hr. The initial set was in the afternoon and the tagged animals were not released until almost sunset. The track is illustrated in Fig. 9. The chronology of events may be summarized as follows.

It was decided to make the first test the following day at 1000 hr and at 1000 hr on each consecutive day. After the release of the tagged animals, it became

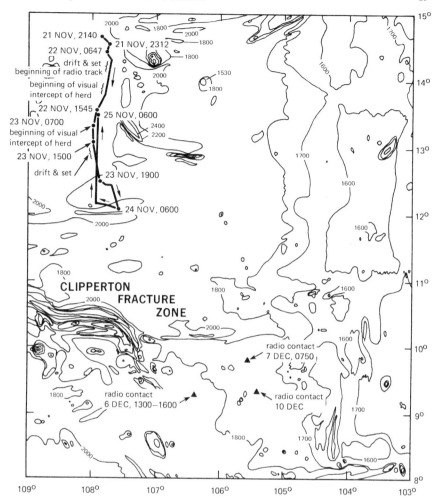

Fig. 9. Ship's track, and radio track and relocation of four instrumented *Stenella attenuata*.

increasingly obvious that the behavior of the male was quite different from that of the females. At first all animals appeared to stay on the same relative heading. After sunset the females began to move away from the male, and the decision was made to stay with the male. After 6 hr of tracking, the females separated from the male by an estimated 20 km, and the transmitted signal slowly faded. It was assumed the transmitters were out of range or had failed.

The vessel followed the male (A) until 1040 on November 22, when the net was set on the school containing him (set 3). One of the females (C) rejoined the male (A) at 0800 but separated from him again at 0930 before the set. An additional transmitter was placed on another male (D) during set 3. Also, a long-life, short-range transmitter was placed on another female (E). After the set, A, D, and E stayed together and were rejoined by C. Female B was not seen again following her departure with female C after initial release in set 2. A, D, E, and C were followed until 1036 (set 4) on November 23. During this period, the males (A and D) were segregated by some distance in the school from the females (C and E). Set 4 was made on the portion of the school containing A and D. They evaded capture by passing between the boat and the net skiff before completion of the net circle. After the set, A and D were followed until 2200 hr, when they separated. Male D was followed, with a faint signal in the background from A, until 1020 hr on November 24 (set 5). At this time, A and D had reconverged, but the set was made on the portion of the school containing D. Animal D again eluded capture in the seine. A and D rejoined after the set and were followed until 0700 hr on November 25, when the track was terminated because of an approaching storm. On December 6, 11 days after and approximately 285 miles distant from termination of the track, signals from one or both of the males (A and D) were picked up at 1255 hr and followed until 0750 hr on December 7, when strong signals were received from A in a school of less than 30 spotters. The school was not set on or further followed. Three days later, on December 10, at 0925 hr, signals were again received from A and D, and this time also from female C, which had not been heard from since separating from the males during set 4 on November 23. The males and the females were again segregated within the school. At the request of the cruise leader (W.E. Evans), the school was not set on and was not further followed.

Size and species composition of the school(s) containing the tagged animals changed during the course of the track, and fish were caught in each set. A large volume of data on respiration patterns and school structure was collected during the experiment.

The results of this multianimal track strongly support the utility of multianimal radio tracks in defining patterns which are otherwise very difficult to define.

Miscellaneous chart records from the tracks of the five spotted dolphins, *Stenella attenuata*, tracked during the 1971 cruise of the *Queen Mary* were analyzed in the following manner:

Assuming that each signal received represented a respiration, the respiration rate was determined as:

$$\bar{X} = \frac{\text{Number of respirations}}{\text{Number of minutes}}$$

Average dive times were computed as:

$$\bar{X} = \frac{(\text{Number of minutes} \times 60 \text{ sec/min}) - (\text{Number of respirations} - \bar{X}_2)}{(\text{Number of respirations} - 1)}$$

Average duration of respirations was computed as:
$$\bar{X}_2 = \frac{(\text{respiration}_1 + \ldots + \text{respiration}_i)}{\text{total respirations}}$$

And percent of time at the surface was computed as:
$$R_1 = \frac{\bar{X}_2}{\bar{X}_1 + \bar{X}_2} \times 100$$

The resultant values for samples of 7- to 18-min segments from the male *Stenella* (A and D) are shown in Table I. The combined average dive times for the two animals are shown in Fig. 10.

Animals spent from 80.3 to 94.2% of their time below the surface, a mean of about 86.5%. When the samples and hourly averages are analyzed, it appears that hourly averages for nighttime hours are generally below the over-all average, while those for daytime hours are generally above it. These data strongly suggest that spotter porpoise are primarily nighttime feeders. W.F. Perrin (personal communication; National Marine and Fisheries Service, SW Fisheries Center) concluded from comparative analysis of stomach contents of spinners and spotters that the two feed differently, without any clear picture of what isolates them, i.e., different times, depths, or selectivity for species.

Perrin *et al.* (1977) report from results of multiple tagging efforts that spotted dolphins moved an average of 2.6 km/hr or $\simeq$ 63 km/day. They further report that "during the fishing operation, the dolphin school on the average is chased before the net is set for about ½ hour at 19–28 km per hour and after release runs for 5.5–7.5 km before resuming normal activity."

Leatherwood and Ljungblad (1979) report on an approximately 11-hr track of a spotted dolphin, *Stenella attenuata,* in the eastern tropical Pacific in 1976. During the track the animal reached speeds in excess of 22 km/hr maintained speeds of greater than 19 km/hr for nearly 2 hr [as did a radio-tagged dolphin (spinner) which accompanied it], and exhibited three distinct diving patterns of the type described by Evans (1971) for *Delphinus delphis,* i.e., traveling, exploratory diving and feeding. These preliminary results corroborate the patterns observed in 1971 aboard the *M/V Queen Mary* and reported above and are not inconsistent with the range of values on movement reported by Perrin *et al.* (1977). Since the dolphins frequently move within the same general area of ocean for long periods, the point-to-point net movement figures from long-term tagging of various kinds may not clearly reflect distances actually traveled in short periods.

VI. METHODS FOR ATTACHING RADIO TRANSMITTER PACKAGES

The radio packages originally used in cetacean field studies were attached to the animal's dorsal fin by means of a nylon or Teflon bolt placed through a

Table I. Some Features of Dive Time and Respiration Rates of an Instrumented Free-Swimming Porpoise, *Stenella attenuata*

Animal	Time	Minutes	Respirations	$\bar{X}$	$\bar{X}_1$	$R_1{}^a$
A and D combined data						
	0250–0305	15	36	2.4	23.6	0.082
First day						
	0400–0407	7	15	2.1	27.8	0.070
	0945–1000	15	77	5.1	9.7	0.178
	1015–1030	15	87	5.8	8.3	0.202
	1030–1045	15	74	4.9	9.8	0.177
	1045–1100	15	89	5.9	8.1	0.206
	1100–1115	15	52	3.5	15.2	0.121
	1115–1130	15	26	1.7	32.5	0.061
	1130–1145	15	166	11.1	3.3	0.389
	1145–1200	15	20	1.3	45.2	0.044
	1200–1215	15	52	3.5	15.5	0.119
	1215–1230	15	75	5.0	10.0	0.174
	1230–1245	15	85	5.7	8.6	0.196
	1245–1300	15	101	6.7	6.9	0.233
	1300–1315	15	112	7.5	6.0	0.259
	1315–1330	15	59	3.9	13.4	0.135
	1330–1345	15	214	14.3	2.1	0.500
	1345–1400	15	278	18.5	1.1	0.656
	1400–1405	05	72	14.4	2.0	0.512
Second day						
	0600–0609	9	24	2.7	21.3	0.090
	0615–0630	15	60	4.0	13.1	0.138
	0630–0645	15	122	8.1	5.3	0.284
	0645–0700	15	67	4.5	11.5	0.154
	0700–0715	15	144	9.6	4.2	0.333
	0715–0730	15	169	11.3	3.2	0.396
	1403–1420	17	365	21.5	0.7	0.750
	2325–2340	15	64	4.3	12.2	0.147
	2340–2355	15	31	2.1	27.8	0.070
	2355–0010	15	43	3.5	19.3	0.098
Third day						
	0010–0025	15	31	2.1	27.8	0.070
	0025–0035	10	26	2.6	21.8	0.088
	1712–1727	15	44	8.8	18.8	0.101
	1727–1740	15	27	2.1	27.8	0.070
	1743–1758	15	52	3.5	15.5	0.119
	1758–1813	15	84	5.6	8.7	0.194
	1813–1828	15	29	5.3	9.4	0.183
	1828–1843	15	64	4.3	12.2	0.147
	1843–1848	5	23	4.6	11.4	0.156

(Continued)

Table I. *(Continued)*

Animal	Time	Minutes	Respirations	$\bar{X}$	$\bar{X}_1$	$R_1{}^a$
	1918–1933	15	55	3.7	14.5	0.127
	1933–1948	15	59	3.9	13.4	0.136
	1954–2010	16	26	1.6	36.2	0.055
	2010–2025	15	29	1.9	30.0	0.065
	2025–2035	10	16	1.6	37.8	0.053
	2050–2105	15	28	1.9	31.2	0.063

a Percent time at surface.

cauterized hole in the animal's dorsal fin (Evans, 1971). These bolts were spring-loaded and incorporated a bimetallic (brass and aluminum) construction nut which was designed to permit the package to come off the animal after some predetermined time period (Evans, 1971; Martin *et al.*, 1971).

The package was attached in a different way to Gigi, the captive yearling California gray whale, scheduled for release into the ocean after one year in captivity (Evans, 1974*b*). This package was attached surgically by means of four large polyvinyl-chloride-coated stainless-steel sutures (Sweeney and Mattsson, 1974) which held a base plate, contour-fitted to the animal's dorsal ridge. Threaded fasteners on the base plate matched to openings on the transmitter itself and, like the earlier release pins (Fig. 11), were equipped with bimetallic (aluminum and steel) links (Evans, 1974 *b*).

Numerous other instrument packages designed and tested during the period 1967 through 1972 for attachment to captive killer whales *Orcinus orca* and pilot whales *Globicephala* sp., trained to work untethered in the open sea and to return on command, utilized a variety of mountings on stretch belly-band systems (Bowers and Henderson, 1972).

Experiences with these previously described approaches converged in the development of radio packages for belly-band attachment to California gray whale calves. Several of these packages, illustrated and described in detail by Norris and Gentry (1974) and Norris *et al.* (1977), have been successfully attached to gray whales in Magdalena Bay, Baja California, Mexico; the whales have been subsequently tracked and the released packages have been recovered (Norris and Gentry, 1974; Norris *et al.*, 1977).

Another such package, designed for use in a study of diving behavior of the Pacific pilot whales off southern California, is illustrated in Fig. 12. The shaped foam on the backpack, pigmented for easy resighting, provides the flotation necessary for recovery. The stretch Leicra straps, and the extra holes in both ends of the Herculon sections permit adjustment of the package for attachment to an animal from 3 to 6 m in length. The smaller nylon straps at the base of the upper Herculon section and small D rings on their counterparts on the other side permit

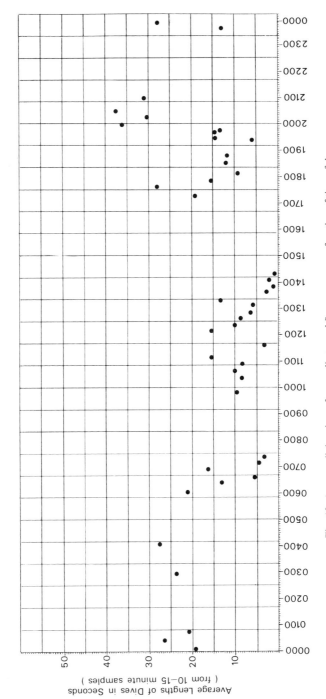

Fig. 10. Average diving time of two radio tagged *S. attenuata* as a function of time of day.

Fig. 11. Two-channel instrumentation package used to track a yearling California gray whale.

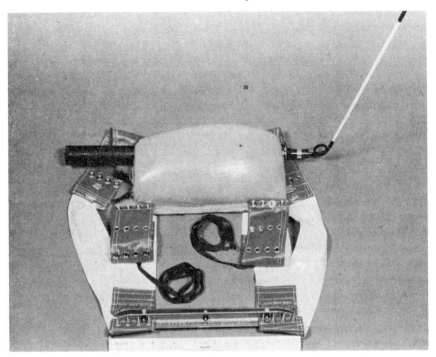

Fig. 12. A time-release instrumentation package designed for use on a pilot whale.

the package to be snugged up. The release pin is the bolt in the center of the belly attachment plate. When the soluble links in the center give way, release is effected by the springs in the outer two retainer pins. Strap- and harness-type attachments are appealing in many ways. It should be noted, however, that Bowers and Henderson (1972) determined that harness-type attachments, regardless of design, interfere significantly with the animal's locomotion. It is therefore suggested that this method of instrumenting cetaceans be considered only for short term (< 7 days) projects.

Instrument packages which can be recovered have figured importantly in recent developments. They can be outfitted with sensors for data not readily transmittable and with multichannel recorders for storage of those data.

Notwithstanding the fact that some of these systems have been successfully deployed on wild marine mammals, these radio transmitters need further development and testing. Furthermore, they are limited in that so far they can only be attached to captured animals. While this poses no great difficulty for many of the smaller cetaceans, the majority of the species of interest, including those species which pose current management problems by virtue of their exploitation in world whaling fisheries, are too large and cumbersome to be easily captured and outfitted with transmitters. There is also the added question of whether data obtained from a calf of a species so captured is necessarily reflective of patterns for the adults of the species. Recent evidence on the California gray whale, for instance, suggest that many gray whale females with calves participate in the migration from the lagoons late in the season, leaving in February, March, or early April, and that some may not participate in the migration at all (Leatherwood, 1974b).

For all these reasons, it is important that future radio-tag development concentrate on two major objectives: (1) methods of attachment without capture, and (2) development of attachment techniques which will allow equipment to stay on the host animal for extended periods of time (>12 months) without injury to the carrier. Development towards this goal is proceeding at a rapid rate. Since their development in the 1920s, the discovery tags have been one of the major tools used by biologists in the studies of whales (Rayner, 1940). Discovery tags are small stainless-steel bolts fired by means of a shotgun into the meat of a whale. The tags are stamped with information about the tagging agency and to whom they should be returned and, once recovered, can provide important information on the individual or species (Brown, 1962; Clark, 1962). But discovery tags are limited in that they are not visible on the living animals, and the whale must be killed before they can be of use.

The "team approach" technique used in developing and testing an implantable large-whale radio tag (Watkins and Schevill, 1977) could very well be a model for future programs involving cetacea and radiotelemetry. This tag can and should replace the discovery-type mark. It has been demonstrated that subcutaneous discovery marks can be tolerated by large whales with no apparent problem for periods of years. Data on tag-related mortality and tag loss rates are not available. The implantable whale tag is percutaneous, to allow exposure of the antennae. This presents problems of possibly high rejection rates due to the presence of a constant opening in the integument and resulting exposure to potential infection.

Erickson (1976) has experienced tag retention times of at least 5 months on the two *Orcinus* he has marked and tracked. In this case the tags were mounted on the anterior edge of the dorsal fin with surgically placed mount pins. Great care was taken, and the animals were observed and periodically examined for approximately a month before release. This technique approaches the optimum, but it is still not suitable for large-scale tagging programs, where the goal is obtaining movement data on individuals and groups of animals for months or years.

With the possible exception of the Erickson technique and the implantable whale tag, all the methods discussed are usable for short-term tagging only. Although Evans (1971, 1975) and others (e.g., Wuersig, 1976) did not directly observe package loss on pelagic species, Irvine *et al.* (1976) have most certainly demonstrated the strong possibility of not only tag loss but serious dorsal fin damage to *Tursiops*. This is probably partially due to the design of the attachment, but does place caution on wholesale long-term use.

VII. SUMMARY AND RECOMMENDATIONS

In the 15 years since some of the initial programs designed to use radio-telemetric techniques to study cetaceans (Evans and Sutherland, 1963) were conceived, considerable progress has been made. Thirteen species of cetacea have been radio tagged and successfully tracked (see Table II). Most of these programs have been concerned with the collection of locational data only; however, data such as depth of dive, and temperature at maximum depth have been telemetered.

The major problems have been logistic, especially costs involved with ship time and personnel. There have been and are problems with reliability of equipment in the field. Power supply size and antennae design, along with making electronics seaproof and "people"-proof have been other major problems. Radio tracking all animals is a tedious and time-consuming activity, and success seems to be a combination of prior preparation and luck. When dealing with cetaceans, which at the most spend less than 10% of their time at a location where radio transmission is possible, these restrictions are compounded.

There is no doubt that radiotelemetry, in terms of being a tool for conducting field studies of cetaceans, can be classified as a "scientific breakthrough." A word of caution is necessary, however: there are no "off-the-shelf"

Table II. Species of Cetacea Radio Tagged and Successfully Tracked

Species	Type of tag
Gray whale (*Eschrichtius robustus*)	Harness, and suture on back; 11-m band
Fin whale (*Balaenoptera physalus*)	Implantable tag; 11-m band
Humpback whale (*Megaptera novaeangliae*)	Implanted tag, 11-m band
Common dolphin (*Delphinus delphis*)	Dorsal fin bolt; 11-m band
Spotted porpoise (*Stenella attenuata*)	Dorsal fin bolt; 11-m band
Spinner porpoise (*Stenella longirostris*)	Dorsal fin bolt; 11-m band
Bottlenose porpoise (*Tursiops truncatus*)	Harness and dorsal fin; 11-m band
Pacific whiteside (*Lagenorhynchus obliquidens*)	Dorsal fin; 11-m band
Dusky dolphin (*Lagenorhynchus obscurus*)	Dorsal fin; 11-m band
Pilot whale (*Globicephala* cf. *G. scammoni*)	Harness and dorsal fin; 11-m band
Killer whale (*Orcinus orca*)	Harness and dorsal fin pin, 11-m, VHF
Dall's porpoise (*Phocoenoides dalli*)	Dorsal fin mount, VHF
Beluga (*Delphinapterus leucas*)[a]	VHF

[a] See Stasko, 1976.

systems available! Each study, each species is special, and the equipment needed must be designed with that in mind.

REFERENCES

Anonymous, 1974, Report of working group on biology and natural history, in: *The Whale Problem* (W.E. Schevill, ed.), pp. 5–10, Harvard University Press, Cambridge.

Bowers, C. A., and Henderson, R. S., 1972, Project Deep Ops: Deep Object Recovery with Pilot and Killer Whales, Naval Undersea Center Technical Publication No. 306 (November 1972), 86 pp.

Brown, S. G., 1962, A note on migration in fin whales, *Nor. Hvalfangst-Tid.* **51** (1): 13–16.

Caldwell, D. K., 1955, Evidence of home range in the bottlenosed dolphin, *J. Mammal.* **36**: 304–305.

Clark, R. 1962. Whale observation and whale marking off the cost of Chile in 1958, and from Ecuador towards and beyond the Galapagos Islands in 1959, *Nor. Hvalfangst-Tid.* **51**(7):265–287.

Erickson, A. W., 1976, Population Studies of Killer Whales *(Orcinus orca)* in the Pacific Northwest: A Radio Marking and Tracking Study of Killer Whales. Preliminary report on Marine Mammal Commission Contract MM5A012, 65 pp.

Evans, W. E., 1970, Uses of Advanced Space Technology in Upgrading the Future Study of Oceanology. American Inst. of Aeronautics and Astronautics Paper No. 70-1273, pp. 1–3.

Evans, W. E., 1971, *Orientation behavior of delphinids: Radio telemetric studies, Ann. N.Y. Acad. Sci.* **188**:142–160.

Evans, W. E., 1974a, Radiotelemetric studies of two species of small odontocete cetaceans, in: *The Whale Problem* (W. E. Schevill, ed.), pp. 385–394, Harvard University Press, Cambridge.

Evans, W. E., 1974b, Telemetering of temperature and depth information from a free ranging yearling California gray whale *Eschrichtus robustus*, *Mar. Fish. Rev.* **36**(4):52–58.

Evans, W. E., 1975, Distribution, Differentiation of Populations and Various Aspects of the Natural History of *Delphinus delphis* in the Northeastern Pacific, PhD dissertation, August 1975.

Evans, W. E., and Bastian, J., 1969, Marine mammal communication; social and ecological factors, in: *The Biology of Marine Mammals* (H. T. Anderson, ed.), pp. 425–475, Academic Press, New York.

Evans, W. E., and Leatherwood, J. S., 1972, Uses of an Instrumented Marine Mammal as an Oceanographic Survey Platform. Naval Undersea Center Technical Paper #331 (December 1972), 11 pp.

Evans, W. E., and Sutherland, W. W., 1963, Potential for telemetry in the study of aquatic animal communications, in: *Bio-Telemetry, the Use of Telemetry in Animal Behavior and Physiology in Relation to Ecological Problems* (L. E. Slater, ed.), pp. 217–224, Pergamon Press, New York.

Galler, S. R., Schmidt-Koenig, K., Jacobs, G. J., and Belleville, R. E. (eds.), 1972. *Animal Orientation and Navigation*, NASA Scientific and Technical Publication Office, Washington, D. C., 606 pp.

Ichihara, T., 1972, Tracking of whales using satellites, *Mon. J. Whales Res. Inst. of Tokyo*, 23pp.

Irvine, B. A., Kaufman, J. H., Scott, M. D., Wells, R. S., and Evans, W. E., 1976, A Study of the Activities and Movements of the Atlantic Bottlenosed Dolphin, *Tursiops truncatus*. Contract Dept. Marine Mammal Commission, #MM4AAC004, 34pp.

Irvine, A. B., and Wells, R. S., 1972, Results of attempts to tag Atlantic bottlenosed dolphins (*Tursiops truncatus*), *Cetology* **13**:1–5.

Johnson, C. S., 1966, Sound detection thresholds in marine mammals, in: *Marine Bioacoustics* (W.N. Tavolga, ed.), pp. 247–260, Pergamon Press, New York.

Leatherwood, J. S., 1974a, Aerial observations of migrating gray whales, *Eschrichtius robustus*, off southern California, 1969–1972, *Mar. Fish. Rev.* **36(4)**:45–59.

Leatherwood, J. S., 1974b, Behavioral interactions between gray whales, *Eschrichtius robustus*, and other marine mammals, *Mar. Fish. Rev.* **36(4)**:50–52.

Leatherwood, J. S., 1975, Observations of feeding behavior of bottlenosed dolphins (*Tursiops truncatus*) in the northern Gulf of Mexico and *Tursiops* cf. *T. gilli* off southern California, Baja California and Nayairit Mexico, *Mar. Fish. Rev.* **37(9)**:10–16.

Leatherwood, J. S., and Ljungblad, D. K. 1979. Some notes on nighttime swimming and diving behavior of a spotted dolphin, *Stenella attenuata*, based on radio telemetric data, Cetology (in press).

Leatherwood, J. S., and Platter, M. F., 1975, Aerial assessment of bottlenosed dolphins off Alabama, Mississippi and Louisiana, in: *Tursiops truncatus* Assessment Workshop. Final Report, U.S. Marine Mammal Commission, Contract MM5AC021 (D. B. Siniff and G. H. Waring, eds.), pp. 49–86, Rosensteil School of Marine and Atmospheric Science, University of Miami, Miami, Florida, 141 pp.

Leatherwood, S., Gilbert, J. W., and Chapman, D. G., 1978, An evaluation of some techniques for aerial censuses of bottlenosed dolphins, *J. Wildl. Manage.* **42**:239–250.

Martin, H. B., Evans, W. E., and Bowers, C. A., 1971, Methods for radio tracking marine mammals in the open sea. Proc. IEEE, Conf. on Eng. Ocean Environ., pp. 44–49.

Mackay, R. S., 1970, *Biomedical Telemetry*, 2nd ed., John Wiley, New York, 388pp.

Nishiwaki, M., 1962, Aerial photographs show sperm whales' interesting habits, *Nor. Hvalfangst-Tid.* **51(10)**:395–398.

Norris, K. S., Evans, W. E., and Ray, G. C., 1974, New tagging and tracking methods for the study of marine mammal biology and migration, in: *The Whale Problem* (W. E. Schevill, ed.), pp. 395–408, Harvard University Press, Cambridge.

Norris, K. S., and Gentry, R. L., 1974, Capture and harnessing of young California gray whales *Eschrichtius robustus*, *Mar. Fish. Rev.* **36(4)**:58–64.

Norris, K. S., Goodman, R. M., Villa-Ramirez, B., and Hubbs, L., 1977, Behavior of California gray whale, *Eschrichtius robustus* in southern Baja California, Mexico, *Fish. Bull.* **7(1)**:159–172.

Perrin, W. F., Evans, W. E., and Holts, D. B., 1977, Movements of Pelagic Dolphins (*Stenella* spp.) in the Eastern Tropical Pacific as Indicated by Results of Tagging, with Summary of Tagging Operations, 1969–1976, NMFS, Southwest Fisheries Center, Admin. Rept. No. LJ-77-6. 38 pp. (unpublished manuscript).

Ray, G. C., and Wartzok, D., 1975, Tests of an Implantable Beacon Transmitter for Use on Whales, Rept. to NMFS in compliance ESP No. E4. MMP No. 99, Johns Hopkins University, 8 pp. (unpublished manuscript).

Ray, G. C., and Wartzok, D., 1976, Radio-Tagging of Fin and Blue Whales, Rept. to NMFS in compliance, Marine Mammal Permit No. 134. Johns Hopkins University, 5 pp. (unpublished manuscript).

Rayner, G. W., 1940, Whale marking, progress and results to December, 1939, *Discovery Rep.* **19**:245–284.

Scammon, C. M., 1874, *The Marine Mammals of the N.W. Coast of North America; Together with an account of the N. American whale fishing*, J. H. Carmany and Co., San Francisco.

Schevill, W. E., 1974, *The Whale Problem, A Status Report*, Harvard University Press, Cambridge, 419 pp.

Schevill, W. E., and Watkins, W. A., 1966, Radio tagging of whales, *Woods Hole Oceanogr. Inst. Rep.* **66(17)**:1–15.

Schultz, J., and Pyle, C., 1965, Cat bites whale, *Yachting* **118(43)**.

Slater, L. E., 1963, *Bio-Telemetry*, MacMillan Co., New York, 372 pp.

Stasko, A. (ed.), 1976, Underwater Telemetry Tracking Aquatic Animals News-letter Vol. **6(2)**:33 pp. Environ. Sci. Div. Oak Ridge National Lab. Oak Ridge, TN 37830.

Sweeney, J. C., and Mattsson, J. L., 1974, Surgical attachment of a telemetry device to the dorsal ridge of a yearling California gray whale, *Eschrichtius robustus*, *Mar. Fish. Rev.* **36(4)**:20–22.

Tillman, M. F., and Johnson, J. H., 1976, Radio Tagging of Humpback Whales, Rept. to National Marine Fisheries Service, Permit No. 136, 4 pp. (unpublished manuscript).

Watkins, W. A., and Schevill, W. E., 1977, The Development and Testing of a Radio Whale Tag. *Woods Hole Oceanogr. Inst.* Woods Hole, Massachusetts, 38 pp. (unpublished manuscript).

Wuersig, B., 1976, Radio Tracking of Dusky Porpoise (*Lagenorhynchus obscurus*) in So. Atlantic, A Preliminary Analysis, Sci. Consult. on Marine Mammals, Bergen, Norway, FAO of UN. ACMRR/MM/SC/83, 20 pp.

Chapter 2

IDENTIFICATION OF HUMPBACK WHALES BY FLUKE PHOTOGRAPHS

Steven Katona, Ben Baxter, Oliver Brazier, Scott Kraus, Judy Perkins, and Hal Whitehead

College of the Atlantic, Bar Harbor, Maine 04609
Woods Hole Oceanographic Institute, Woods Hole, Massachusetts 02543
Cambridge University, Cambridge, England

I. INTRODUCTION

Among the large baleen whales, humpbacks are notable for the variety of different white and black patterns and slight morphological differences which can be observed between individuals (Lillie, 1915; Matthews, 1937; Pike, 1953; Schevill and Backus, 1960). Observations in the field and inspection of photographs suggest that variations in the shape of the dorsal fin or in the disposition of body scars can sometimes be used to identify individual humpback whales. It is primarily the pattern on the underside of the flukes which appears to us to provide the most positive discrimination between individual humpback whales. This pattern varies from nearly all white to nearly all black, but characteristically contains a variety of black or white patches, lines or streaks.

A typical humpback breathing sequence includes up to eight or ten breaths, separated by perhaps 15 or 20 sec, before a dive averaging 6 to 8 min. After the last breath of a series, the humpback often arches its back and tail, and if the whale dives deep, the flukes are usually lifted into the air. Fluke patterns may also be visible during lobtailing behavior. At these times, the flukes can be photographed and individual whales can be identified.

The authors of this chapter may be reached at the following addresses: Steven Katona, Ben Baxter, and Scott Kraus: College of the Atlantic, Bar Harbor, Maine 04609; Oliver Brazier: Woods Hole Oceanographic Institute, Woods Hole, Massachusetts 02543; Judy Perkins: 150 Chestnut Street, Weston, Massachusetts; Hal Whitehead: Cambridge University, Cambridge, England.

Recent estimates suggest that about 1200 (Mitchell, 1973b; Winn et al., 1975) humpback whales (*Megaptera novaeangliae*) inhabit the western North Atlantic Ocean, ranging from Greenland and north of Iceland, south to the West Indies and Venezuela (Leatherwood et al., 1976).

The annual migration cycle of humpback whales in the western North Atlantic Ocean was considered by Kellogg (1929) to include feeding during the warm months from New England northward, and winter calving in warm, southerly waters.

It is estimated that about 1000 humpbacks winter and calve on the Navidad, Silver, and Mouchoir banks, north of the Dominican Republic, and that others winter and calve near Puerto Rico, the Virgin Islands, and other islands of the lower Antilles (Townsend, 1935; Erdman, 1970; Winn et al., 1975). During March and April, humpbacks are found near Bermuda (Payne and McVay, 1971; Brazier, unpublished data). In the Gulf of Maine, humpbacks appear from April and May through November (Kellog, 1929; Schevill and Backus, 1960; Katona, 1976, and unpublished data; Katona et al., 1977), and in Newfoundland they occur from March or April through summer. Humpbacks have been reported to overwinter on Newfoundland's Grand Bank or other northerly regions (Kellogg, 1929; Williamson, 1961), but it is not likely that many do so (Winn et al., 1975). A native fishery from villages in Greenland takes about 10 humpbacks each summer (Mitchell, 1974) and a similar fishery in Bequia during the winter takes several per year (Burgess, 1970). The effect of those fisheries on the total humpback population is not known.

Comparison of fluke photographs can provide information on the movements of individual humpbacks and the lengths of time they remain in particular areas. It could therefore be used to investigate the existence of separate populations and the relative sizes of stocks of humpbacks. This is of considerable importance for the proper management of the species, and both Mitchell (1974) and Winn et al. (1975) have called attention to it. Over 160 humpbacks have been tagged in the North Atlantic during the past 10 years (Mitchell, 1970, 1973a, 1974; Christensen and Oritsland, 1974; Christensen, 1975; P. C. Beamish, personal communication), but only one recovery has been reported.

II. METHODS

Humpback whale photographs were obtained during the period from 1969 to 1976. Photographs from the Mount Desert Rock Lightstation, Maine (43° 58' N, 68° 06' W), were made from the lighthouse through a telescope and in nearby waters from small craft. All other photographs were made at sea. Although color transparency film was used in early work, we have found it both cheaper and better to use black and white film. Recent photographs were on Kodak Tri-X film exposed at ASA 1200 and developed in Acufine developer for 7 min at 20°C. Kodak Plus-X film was also used at ASA 320 and developed in the same way. The high exposure speed permitted use of a shutter speed of 1/1000 of a second in most circumstances, which was necessary to prevent blurring caused by the movement of long lenses. When necessary, we copied color slides or individual 16-mm motion-picture frames onto Pan-X film in order to make negatives for conversion to black and white format. All photographs were printed in 2 1/2-in. × 4-in. format for cataloging and comparison.

To produce the most useful photographs, the plane of the flukes should be nearly perpendicular to the camera since distortion of the pattern by parallax can be confusing. This can be achieved by studying the breathing pattern of an individual whale (or a group of whales), following them slowly in a boat, and preparing to take the photo from behind if the whale arches up the tailstock in preparation for its dive. Use of a motor-drive camera is helpful. Additional features of the whale which could be useful in identification, including dorsal fins, flippers, scars, and barnacle growths, are photographed when possible. Detailed field notes are important to prevent confusion. Behavioral events such as associations between whales, or occurrence of breaching, lobtailing, flipper slapping, or feeding should also be noted, since photographic comparisons may in time reveal individual behavior patterns. At sea, we have used lenses of focal lengths up to 300 mm to provide larger images and to allow us to photograph at greater distances from the whales. Motor vessels are most convenient for this work, since they are easier to maneuver than sailboats. Diesel engines seem to be less disturbing to the whales than gasoline engines.

III. RESULTS

The basic results of our work are these: (1) fluke patterns are apparently unique to individual whales; (2) photographs of flukes are useful for recognizing individual whales; and (3) comparisons of fluke photographs have allowed us to recognize individuals after long periods or after movement from one place to another.

We can distinguish all humpback whales for which we have obtained clear photographs of patterns on the ventral side of the fluke (total, 78 whales). Figures 1–8 demonstrate the variety of these markings. The most difficult flukes to distinguish have been those which are nearly all black. In those cases, particularly sharp and detailed photographs are needed, because discriminations depend largely on small differences in the position of small scars or scratches. The disposition of barnacles may be useful on occasion but is probably not reliable over long periods.

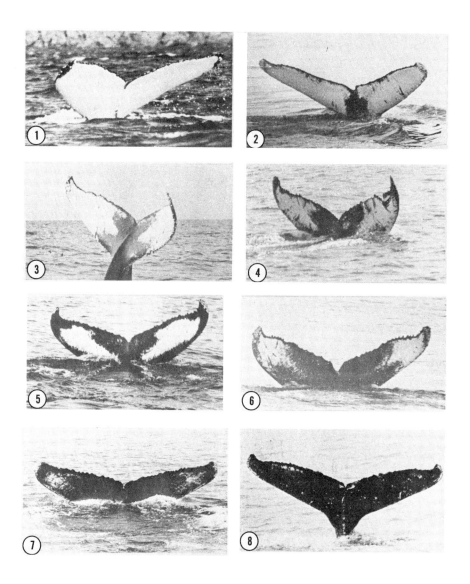

We have taken pictures of 43 whales from the Gulf of Maine (1974-1976, Steven Katona, Ben Baxter, Oliver Brazier), 13 whales from Bermuda (1969-1976, Oliver Brazier, Judy Perkins), and 22 whales from Newfoundland (1975-1976, Judy Perkins, Hal Whitehead). We have not found consistent differences between markings of whales from these areas. One whale which was photographed at Bermuda on April 15, 1976, matches the photograph of a whale taken at Mt. Desert Rock Lightstation on July 27-29, 1976 (Figs. 9, 10). We have not found any photographic matches between photographs from Newfoundland and those from either the Gulf of Maine or Bermuda. Furthermore, no matching photographs were obtained between the 12 humpbacks observed at Mt. Desert Rock during the period from July 23 to August 3, 1976, and the 31 humpback photographed at either Jeffreys Ledge or Stellwagen Bank between May 2 and November 13, 1976. Five different individuals were photographed on both Jeffreys Ledge and Stellwagen Bank.

Figures 11 and 12 show a whale which was photographed on October 1, 1974, at Brier Island, Nova Scotia (44° 15' N, 66° 17' W), which matches the photograph of a whale at Mt. Desert Rock on July 26, 1976. This appears to be the same whale, and since the fluke pattern did not change during this two-year period, these markings are potentially valuable natural tags for identifying individual whales over long periods. Photographs of another individual taken at Mt. Desert Rock in early September 1975 match photographs taken in the same location during the period July 19-29, 1976 (Figs. 23, 24), but the 1975 photograph is not of sufficient quality to reproduce here. In addition to suggesting the persistence of fluke patterns, these photographs also suggest that individual humpbacks may visit particular areas year after year. Figures 13 and 14 show a whale that was photographed alive on July 23, 1976, at the northern end of Jeffreys Ledge (43° 07' N, 70° 05'W), then dead 165 miles ENE on August 21, 1976, 10 miles WSW of Brier Island (44° 12' N, 66° 37.5' W). The poor

---

Figs. 1-8. Examples of the variety of white and black markings that are commonly found on the ventral surface of humpback whale flukes. These patterns vary from nearly all-white (#1) to nearly all-black (#8), with most falling somewhere between (#2-7).

Figure	Location	Date	Photographer
1	Newfoundland	7/21/75	JP/HW
2	Mount Desert Rock	7/15/76	BB
3	Jeffreys Ledge	7/23/76	SKK
4	Bermuda	4/15/76	OB
5	Jeffreys Ledge	9/10/76	OB
6	Mount Desert Rock	7/15/76	BB
7	Mount Desert Rock	7/29/76	SKK
8	Bermuda	1968	R. Payne

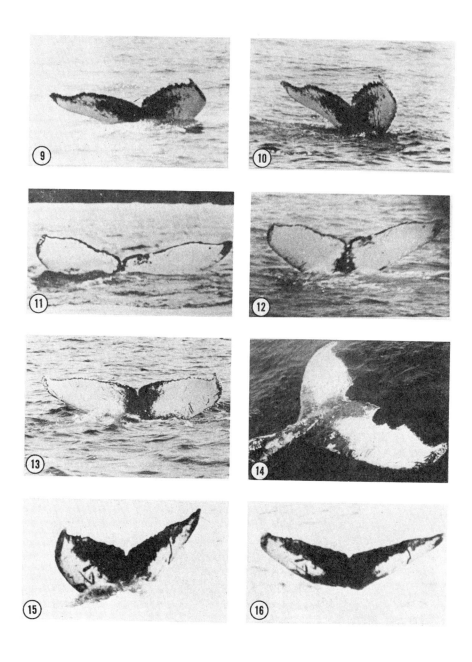

condition of this 46-ft female suggests that it had been dead for at least several weeks (during which time the carcass could have drifted a considerable distance). This provides evidence that death does not alter fluke markings and that they can be useful in identifying dead animals. The flukes of the carcass were attacked by sharks while observers were present. This whale carcass was also photographed (Ben Baxter) September 25, 1976, on the north shore of Long Island, Nova Scotia. At that time the fluke markings and skin were completely worn off, but the corpse was identified by photographically matching the shark bites on the posterior edge of the flukes.

Comparisons of fluke photographs taken at the same locations revealed that separate groups of humpback whales frequented Mt. Desert Rock for up to one month and remained in the Jeffreys Ledge area for up to three months. Twelve of the 31 whales seen at Jeffreys Ledge were seen on at least two separate photographic expeditions, demonstrating the feasibility of the method for recognition of individuals (Figs. 17–24). A humpback was photographed on August 2, 1976, off Cape Bonavista, and again a match was obtained with a photograph taken on August 25, 1976, about 30 miles away (Figs. 15 and 16).

IV. DISCUSSION

This evidence seems to indicate that each whale has unique fluke patterns. Comparison of the natural marks on these whales and the use of photographs to

Figs. 9–16. Examples of resightings of four different humpback whales by recognition of their characteristic fluke patterns. In each case the resighting occurred at a considerable distance from the initial sighting. Photos 11 and 12 were taken nearly two years apart. These photographs indicate the stability of identifiable markings after movement from one place to another (Figs. 9 and 10), over long periods of time (Figs. 11, 12, 15, and 16), and after death (Figs. 13 and 14).

Figure	Location	Date	Photographer
9	Bermuda	4/15/76	OB
10	Mount Desert Rock	7/27-8/2 1976	BB, SKK
11	Brier Island, N.S.	10/1/74	A. Davidson
12	Mount Desert Rock	7/26/76	BB
13	Jeffreys Ledge	7/23/76	SKK
14	Bay of Fundy 44°12'N, 66°37.5'W	8/21/76	J. Hain
15	Newfoundland Cape Bonavista Ledge 53°05'N, 48°45'N	8/2/76	K. Balcomb
16	Funk Island, Nfld. 53°00'N, 49°35'N	8/25/76	JP/HW

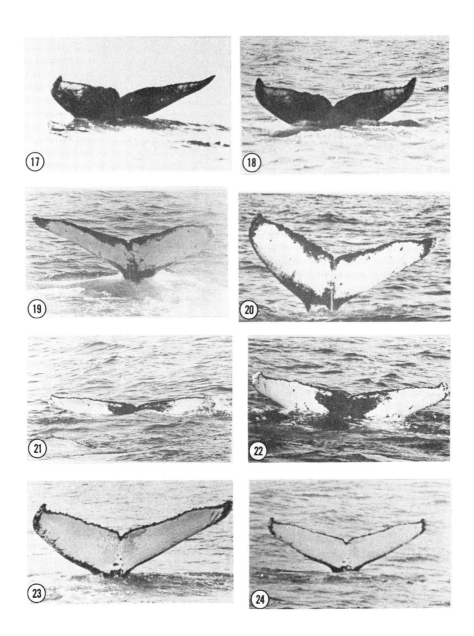

verify the details provide a convenient way to recognize individuals. In many ways, the exploitation of these natural markings appears to be superior to large-scale tagging programs: it is cheaper, not harmful to the animals, all the whales are marked, and the markings are highly visible from astern if the whale flukes up. It is, of course, possible that apparently identical photographs do not necessarily portray the same individual, but the amount of information contained in a fluke pattern is usually quite large, and the chance that the flukes of different animals would match exactly must be quite remote.

Slight morphological differences have also been used to distinguish individual killer whales (Bigg et al., unpublished manuscript), right whales (Payne, 1976), and California gray whales (Hatler and Darling, 1974; Darling, 1977).

The use of photographic recognition of fluke markings could contribute to an understanding of humpback populations. Comparison of photographs taken at all suspected breeding areas for a period of several years might reveal whether separate breeding populations exist. Comparison of these photographs with others taken throughout the northern portion of the species range might help to show whether such populations have different migratory routes or feeding ranges. This information will be important in future management decisions. Cooperative work by many people over long periods of time will be necessary in order to collect these photographs. We are trying to encourage the assembly of additional photographs of western North Atlantic humpbacks from colleagues and photographers, and we hope that anyone having photographs which could be useful for individual identification will contact one of us. If any hunting is carried out in any breeding or feeding areas, fluke photographs of captured animals should be obtained as a matter of course, since these photographs may constitute a match with earlier photos of the living whale. All photographs which are mentioned in this paper, along with all other photographs which were available

←

Figs. 17–24. Examples of pictures of four different humpback whales taken a few days apart in the same general area. These demonstrate the potential use of this technique as a method for documenting an individual whale's movements.

Figure	Location	Date	Photographer
17	Jeffreys Ledge	9/14/76	OB
18	Jeffreys Ledge	9/24/76	BB
19	Jeffreys Ledge	9/29/76	OB
20	Jeffreys Ledge	7/23/76	SKK
21	Jeffreys Ledge	10/26/76	OB
22	Jeffreys Ledge	10/30/76	OB
23	Mount Desert Rock	7/29/76	SKK
24	Mount Desert Rock	7/26/76	SKK

as of May 1977, have been assembled by Kraus and Katona (1977) into a catalog of individually identified humpback whales. Copies of that catalog have been deposited in libraries of selected museums, universities, and aquariums throughout the general western North Atlantic range of the humpback.

This paper is intended to be a summary of the techniques we are using and of our initial results. Since this is a continuing project, new information is accumulating which cannot be detailed here. However, it is interesting to note several recent results. During the summer of 1977, we rephotographed five individuals which were previously photographed in the Gulf of Maine during 1976. It is probably of significance that whales which swam together in 1976 were also photographed together in 1977 on two different occasions. In the first instance, two humpbacks which we saw together off Provincetown on May 1, 1976, were again photographed together on April 22, 1977 (Figs. 25–28), in the same area. Secondly, two whales which were seen at Mt. Desert Rock during July 1976 were photographed in a pod of four whales near Wooden Ball Island,

Figs. 25–28. Examples of resightings of two different humpback whales over a period of time. Both whales were photographed in the same general area in early 1976 and 1977.

Figure	Location	Date	Photographer
25	Off Provincetown	5/1/76	SKK
26	Off Provincetown	4/22/77	C. Ramsdell, K. Farrar
27	Off Provincetown	5/1/76	SKK
28	Off Provincetown	4/22/77	C. Ramsdell, K. Farrar

Maine (43° 51′ N, 68° 48′ W), about 30 miles west, on July 22, 1977. These results suggest that humpback whales may form relationships which last for a long time, and they also demonstrate the potential of this technique for investigating social relationships. Another whale which was photographed at Jeffreys Ledge in 1976 was seen at Mt. Desert Rock in 1977, in contrast to the results discussed above.

V. SUMMARY

Photographs of 78 humpback whales in the Gulf of Maine, off Bermuda, and off Newfoundland were compared and identified from photographs of naturally occurring patterns on the ventral side of the flukes. The tendency of this species to raise the flukes in the air before diving enables these markings to be photographed easily. Photographs are included to show the diversity of fluke markings and to indicate that patterns remain constant for at least two years and also persist after death. One whale apparently migrated from Bermuda to Maine during spring–summer of 1976. The potential use of the comparison of fluke photographs to study migrations, population size, and population structure of this species in the western North Atlantic is discussed.

ACKNOWLEDGMENTS

We are glad to thank the following people who have contributed photographs to this study: Ken Balcomb and George Nichols (*R.V. Regina Maris*, Ocean Research Education Society, Boston, Massachusetts); Bill Davidson, Keith Farrar, Jane Frick, Frank Gardner, and Jim Hain (*R. V. Westward*, Sea Education Association, Woods Hole, Massachusetts); Ed Lemire, Roger and Katy Payne, George Peabody, and Cathy Ramsdell. We are also grateful to the Southwest Harbor Group of the U.S. Coast Guard and to the following people for logistic support or boat time during the course of the study: Brian Burke, Peter Cavaney, Dr. and Mrs. H. C. Frick, Mr. T. S. Leland, Mr. and Mrs. Francis McAdoo, Francis W. Peabody, and Mr. and Mrs. Donald B. Straus. Lisa Karnofsky provided valuable help in the darkroom.

Portions of this research were supported by grants from the Carolyn Foundation (S.K.), The Mount Desert Oceanarium (B.B.), The Oceanic Society (J.P., H.W.), and the Whale Fund of the New York Zoological Society (O.B.). We thank Mrs. Marguerite Braymer for the loan of a Questar® telescope and the College of the Atlantic for sponsoring parts of this research. Finally, we are grateful to William A. Watkins and William E. Schevill for critical readings of this manuscript.

REFERENCES

Bigg, M. A., MacAskie, I. B., and Ellis, G., MS, Abundance and movements of killer whales off eastern and southern Vancouver Island with comments on management. Preliminary report. Unpublished MS, Arctic Biological Station, Fisheries Research Board of Canada.

Burgess, T. W., 1970, The whalers of Bequia, *Natl. Rev.* **22(23)**:629.

Christensen, I., 1975, Norwegian whale-marking in the northeastern Atlantic in 1975, Coun. Meet. Int. Coun. Explor. Sea (N:19), pp. 1–5.

Christensen, I., and Oritsland, T., 1974, Whales and seals marked in the Northeast Atlantic in 1974, Coun. Meet. Int. Coun. Explor. Sea **1974(9)**:1–7.

Darling, J., 1977, The Vancouver Island gray whales (Waters), *Vancouver Aquar.* **2(1)**:4–19.

Erdman, D. S., 1970, Marine mammals from Puerto Rico to Antigua, *J. Mammal.* **51**:636–639.

Hatler, D. F., and Darling, J. D., 1974, Recent observations of the gray whale in British Columbia, *Can. Field Nat.* **88**:449–459.

Katona, S. K., 1976, Whales in the Gulf of Maine: 1975. Report of the Gulf of Maine Whale Sighting Network, College of the Atlantic, Bar Harbor, Maine 04609, 45 pp.

Katona, S. K., Richardson, D. T., and Hazard, R., 1977, *A Field Guide to the Whales and Seals of the Gulf of Maine*, 2nd ed., College of the Atlantic, Bar Harbor, Maine 04609.

Kellogg, R., 1929, What is known of the migrations of some of the whalebone whales, *Annu. Rep. Smithsonian Inst.* **1928**(Publ. 2981): 467–494.

Kraus, S., and Katona, S. (eds.), 1977, Humpback Whales in the Western North Atlantic. A Catalogue of Identified Individuals, College of the Atlantic, Bar Harbor, Maine 04609.

Leatherwood, S., Caldwell, D. K., and Winn, H. E., 1976, Whales, Dolphins, and Porpoises of the Western North Atlantic. A Guide to Their Identification, NOAA Tech. Rept. NMFS CIRC-396, Seattle, Washington, available from U.S. Govt. Printing Office, Washington, D.C. 20402.

Lillie, D. G., 1915, British Museum of Natural History, British Antarctic Expedition of 1910 ("Terra Nova"), *Nat. Hist. Rep. Zool.* **1(3)**:85–124.

Matthews, L. H., 1937, The humpback whale, *Megaptera nodoso. Discovery Rep.* **17**:7–92.

Mitchell, E., 1970, Request for information on tagged whales in the North Atlantic, *J. Mammal.* **51(2)**:378.

Mitchell, E. D., 1973a, Draft Report on Humpback Whales taken under Special Scientific Permit by Eastern Canadian Land Stations, 1969–1971, Internat. Whaling Comm., London, 23rd Rept., App. IV, Annex M, pp. 138–154.

Mitchell, E. D., 1973b, The status of the world's whales, *Nature Can.* **2(4)**:9–25.

Mitchell, E. D., 1974, Present status of northwest Atlantic fin and other whale stocks, in: *The Whale Problem* (W. E. Schevill, ed.), pp. 108–161, Harvard University Press, Cambridge.

Payne, R., 1976, At home with right whales. *Nat. Geographic* **149(3)**:322–339.

Payne, R. S., and McVay, S., 1971, Songs of humpback whales, *Science* **173**:585–597.

Pike, G. C., 1953, Colour pattern of humpback whales from the coast of British Columbia, *J. Fish. Res. Bd. Can.* **10(6)**:320–325.

Schevill, W. E., and Backus, R. H., 1960, Daily patrol of a *Megaptera*, *J. Mammal.* **41(2)**:279–281.

Townsend, C. H., 1935, The distribution of certain whales as shown by logbook records of American whaleships, *Zoologica (N.Y.)* **19**:1–50.

Williamson, G. R., 1961, Winter sighting of a humpback suckling its calf on the Grand Bank of Newfoundland. *Nor. Hvalfangst-Tid.* **50**:335–336, 339–341.

Winn, H. E., Edel, R. K., and Taruski, A. G., 1975, Population estimate of the humpback whale (*Megaptera novaeangliae*) in the West Indies by visual and acoustic techniques, *J. Fish. Res. Bd. Can.* **32**:499–506.

Chapter 3

THE NATURAL HISTORY OF DALL'S PORPOISE IN THE NORTH PACIFIC OCEAN

G. Victor Morejohn

*Department of Biological Sciences
San Jose State University and
Moss Landing Marine Laboratories
Moss Landing, California 95039*

I. INTRODUCTION

Dall's porpoise, *Phocoenoides dalli* (True), is commonly encountered over the continental shelf of North America and in offshore and pelagic waters of the western and eastern North Pacific Ocean. This striking black and white porpoise is characterized by great speed and a cone-shaped splash it makes when surfacing to breathe. Because of its great speed in the wild, making it extremely difficult to capture alive, and its hyperactivity resulting in high mortality after capture, much important information remains to be disclosed of the basic components of its life cycle.

At our laboratories, situated at the head of the Monterey Submarine Canyon, the continental shelf drops off sharply only a few miles from shore to depths in excess of 1000 fathoms. Offshore and pelagic waters thus are close at hand, and we have been able to study this little known species of porpoise throughout the year. An intensive study of Dall's porpoise was undertaken to learn as much as possible about this species in its natural environment. This study was conducted in Monterey Bay for a period that extended over 30 months. My conclusions are based on observations of animals at sea and on specimens collected during this period as beach-cast dead animals or animals taken in the wild. Specimens were available for every month of the year. In this paper, much of the literature relevant to the natural history of this species is reviewed, and new information gathered during our long-term study is presented for the first time.

II. TAXONOMIC CONSIDERATIONS

True (1885) described the species *Phocoena dalli*. Later, Andrews (1911) erected the genus *Phocoenoides* to include *dalli* as well as a new species, *P. truei*, described in the same publication. The description of *P. truei* was based on only one specimen. After careful study of several specimens of *P. dalli* collected off North American waters and one fresh specimen collected in San Francisco Bay, California, Benson and Groody (1942, p. 50) concluded ". . . that the evidence available at present justifies the recognition of only one form. This involves regarding *P. truei* as a synonym of *P. dalli*." Color pattern and shape of the caudal region were the major criteria employed by Andrews (1911) to distinguish between the two taxa. Later studies by Cowan (1944) and Benson (1946) further supported the view that only one species is involved. Although much of the discussion by these authors was concerned with the great amount of individual variation in color patterns and intensity of black or gray pigments, only Nishiwaki (1967) has been able to comment on relative distribution of the two forms, stating that the *P. truei* form is geographically restricted off the coast of Sanriku, Japan (33° N), to the southern part of the Sea of Okhotsk (42° N). In consideration of their similarity in color, Rice and Scheffer (1968; p. 10) stated that *P. truei* ". . . may be a color phase of *P. dalli*." Wilke *et al.* (1953) considered that the two forms may, in fact, represent subspecies, inasmuch as they found a *"truei"* type fetus in the uterus of a *"dalli"* type. In an extensive study of color variation in the species, Houck (1976) is of the opinion that the two forms are colormorphs and do not represent geographical races. Without prior publication of subspecific names, Nishiwaki (1972, pp. 115–118) considered the two forms as subspecies (*P. dalli dalli* and *P. dalli truei*). Both forms reach 2.1 m in length, but whereas *P. d. dalli* reaches 145 kg in weight, *P. d. truei* has been recorded only up to 100 kg.

In the final analysis, nomenclatorial and systematic decisions will have to be made based on a larger series of specimens of both forms, especially of the *truei* type, with due consideration given to degree of intergradation between them. My use of the subspecies designations *P. d. dalli* and *P. d. truei*, therefore, is provisional pending further studies.

Although Nishiwaki (1972, p. 118) states that the two forms ". . . differ in proportional skull measurements, dental formulae, vertebral formulae . . . and adult size," the vertebral formulae which he gives under the description of each subspecies do not, in fact, differ. In coloration, the two species may be distinguished as follows (descriptions follow in part True, 1885; Andrews, 1911; Nishiwaki, 1972; and our own studies of *P. d. dalli*):

Phocoenoides dalli dalli (Fig. 1a). Slate gray to black with prominent white area extending from side to side ventrally across the thorax from an area posterior to the flippers approximately in line with the dorsal fin and upward approxi-

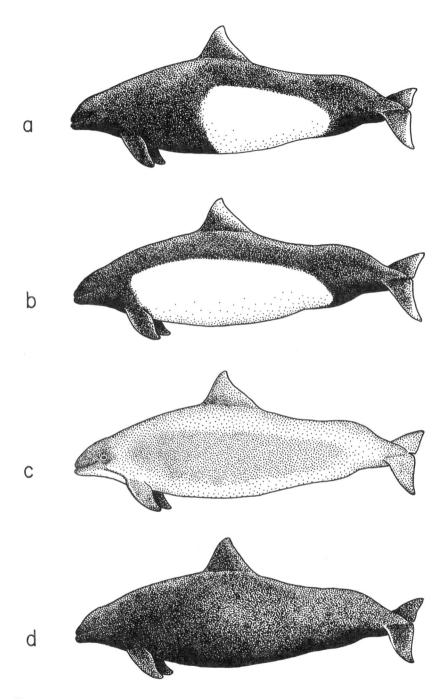

Fig. 1. Typical color patterns of adult *Phocoenoides*: (a) Dall's porpoise, *P.d. dalli*; (b) True's porpoise, *P.d. truei*; (c) "striped" *P.d. dalli*; (d) "black" *P.d. dalli*.

mately 2/3 the side of the body then continuing posterior to the anal region laterally. Color variations are described in another section.

Phocoenoides dalli truei (Fig.1b). Slate gray to black with prominent white area extending from side to side ventrally across the thorax between the bases of the flippers considerably in advance of the dorsal fin and upward extending anteriorly beyond the flipper bases and upward approximately 2/3 the side of the body and continuing posterior to the anal region laterally. Color variations are described in another section.

III. DISTRIBUTION AND POPULATION SHIFTS

According to Nishiwaki (1967), *P. dalli* is distributed from the northern Bering Sea north of the Aleutian chain from Cape Navarin southward across the Bering Sea to waters surrounding Japan in the western Pacific and to the eastern Pacific along the Alaskan, Canadian, and United States coastlines, southward to northern Baja California (32° N) (Fig. 2). They have been reported as far south as Bahia de Ballenas off central Baja California (Leatherwood *et al.*, 1972). A partial skull (SJSU #2526) was collected by the author on February 5, 1973, on the south shore of Scammon's Lagoon, Baja California. Seasonal movements in Japanese waters are well documented (Nishiwaki, 1966a). In the eastern North Pacific, Dall's porpoise is commonly seen within 100 miles of shore and usually in waters in excess of 100 fathoms deep (Yocum, 1946; Fiscus and Niggol, 1965;

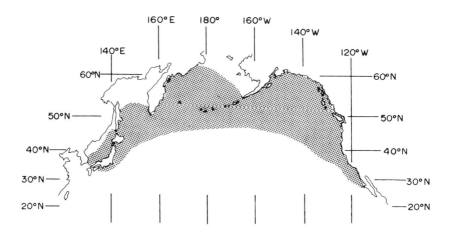

Fig. 2. Distribution of *P. dalli* in the North Pacific Ocean.

Table I. Percentages of Seasonal Sightings of Dall's Porpoises in Inshore Waters of British Columbia (48°30'N, 126°W to 55°30'N, 134°W) According to Size Groups[a]

Number in group	Winter, Dec–Feb (17 groups)	Spring, Mar–May (45 groups)	Summer, June–Aug (25 groups)	Fall, Sept–Nov (20 groups)	Total
1		7			3
2	18	4	20		9
3	18	11	8	10	11
4	18	16	20	10	16
5–9	40	29	36	30	32
10–14	6	18	16	30	17
15–29		9		5	5
30–49		4		10	6
50–100		2		5	2

[a] Modified from Pike and MacAskie, 1969, p. 29, Table 12.

Pike and MacAskie, 1969). It is seldom seen in shallow coastal waters; when it is thus seen, it is in channels or straits widely open at both ends with strong tidal currents (Cowan, 1944; Scheffer, 1949). The scattered reports of Dall's porpoise seen off Washington, Oregon, and California coasts (Cowan, 1944; Yocum, 1946; Lustig, 1948; Scheffer, 1949; Brown and Norris, 1956; Brownell, 1964; Fiscus and Niggol, 1965; Ridgway, 1966; Pike and MacAskie, 1969) can be used mainly for records of occurrence along its range of distribution but certainly not as evidence of migration.

In waters off Alaska and British Columbia, Dall's porpoise is sighted during most of the year (Cowan, 1944; Pike and MacAskie, 1969). Inshore sightings off British Columbia are shown in Table I. However, there are no sightings recorded in offshore waters of British Columbia during the winter months of December to February (Pike and MacAskie, 1969, p. 29, Table 12).

In considering the distribution of Dall's porpoise along the eastern Pacific seacoast of North America, Cowan (1944, p. 295) stated (probably due to paucity of sight records at that time) that ". . . occurrences south of Canadian waters are so few that they may be of an extralimital or vagrant nature." Observations and specimens collected off central California (Benson and Groody, 1942; Benson, 1946; Yocum, 1946; Brownell, 1964), the present study in Monterey Bay (Table II), and the extensive aerial and shipboard observations by personnel from the National Marine Fisheries Service (NMFS) pelagic fur seal cruises and NMFS albacore programs off southern California now provide sufficient evidence to indicate the common occurrence of Dall's porpoise throughout the year off California (Leatherwood and Fielding, 1974). During the months of April and May, these porpoises have been seen as far south as Turtle

Table II. Percentages of Seasonal Sightings of Dall's Porpoises in Waters of Monterey Bay, California (36° 50'N, 122°W), According to Size Groups

Number in group	Winter, Dec–Feb (17 groups)	Spring, Mar–May (18 groups)	Summer, June–Aug (41 groups)	Fall, Sept–Nov (19 groups)	Total
1			3	10	3
2	35	22	5	10	14
3	11	6	21	21	16
4	5	33	19	15	18
5–9	23	27	43	31	33
10–14	11	11	5	15	9
15–19	11	6	3		4

Bay, Baja, California (28° N, W.F. Perrin, observer). It may be said, therefore, that *P. d. dalli* occurs commonly as far south as 32° N with vagrant individuals moving as far south as 28° N.

Dall's porpoise has been observed nearly every month of the year throughout its range of distribution in the eastern North Pacific Ocean from Alaska to Baja California, yet there does not appear to be any strong suggestion of seasonal north–south movement of the entire population from one geographical area to another. The evidence provided by Brown and Norris (1956), Leatherwood *et al.* (1972), and Leatherwood and Fielding (1974) of shifts in subpopulations of Dall's porpoise north-south or inshore-offshore, qualify as examples of vertebrate migration as defined by Orr (1970). These shifts in subpopulations off southern California southward and inshore during the fall, especially in the region of the northern Channel Islands, show marked increases in numbers of individuals. Our studies (unpublished) of feeding habits of marine mammals in Monterey Bay have shown a high correlation of availability of prey species to marine mammal species diversity and density. Thus, the population shifts of Dall's porpoise off southern California (Leatherwood and Fielding, 1974) probably are largely related to increased availability of its preferred prey species in fall and winter.

IV. PREDATORS

The only known predator of Dall's porpoise other than man is the killer whale, *Orcinus orca*. Dall's porpoises apparently swim too rapidly for sharks to attack them in their range of distribution. The logbook records of predatory activities of killer whales in British Columbia waters (Pike and MacAskie, 1969,

p. 23, Table 7) list observations made of killer whales attacking schools of Dall's porpoises. A study by Rice (1968) of stomach contents of ten killer whales disclosed parts of Dall's porpoises that had been eaten by two of the whales. Killer whales are distributed along the entire coast of California, but their occurrence is sporadic, seasonal, and sparse compared to their relatively much greater abundance, and perhaps year-round residence, in waters off northern Washington and British Columbia (Bigg and Wolman, 1975).

On March 7, 1972, a male subadult Dall's porpoise was collected in Monterey Bay from a school of 12 porpoises. When the specimen was brought aboard, we saw long, raking tooth marks diagonally along both sides of the body that extended from the region posterior to the dorsal fin dorsally toward the caudal peduncle (Fig. 3). All the wounds were fresh, most exposing underlying muscle, still oozing pinkish lymphatic fluid. The predator was identified as a subadult male or adult female killer whale based on (1) the nature of the puncture wounds made by blunt, rounded teeth; (2) the distances between the probable apices of the teeth; and (3) the curvature of the maxillary and mandibular tooth rows. The killer whale attack appeared to be made from above.

V. COLORATION

The complexity of pigment intensity and pattern configuration in small odontocete cetaceans has interested many cetologists. Yablokov's (1963) study of color in cetaceans was primarily related to the role that color plays in the particular ecological setting of the cetacean species. The adaptive significance (or function) of delphinid pigmentation patterns was extensively considered by Mitchell (1970); and Perrin (1970, 1972) described the ontogeny and variation of color patterns of species of *Stenella* and other delphinids. Both Mitchell and Perrin agree that the most generalized and primitive pattern, using the counter-shading principle, is that which has a dark dorsal area that gradually shades into a lighter (often white) ventrum. Obliterative shading of this type essentially camouflages the animal from predator or prey from a side view. Mitchell defines this pattern as "saddled," and Perrin accepts it with modifications as his "cape system." Among the delphinids, *Stenella* cf. *S. longirostris* (Hawaiian and whitebelly spinners), juvenile *S. graffmani* (= *S. attenuata*), the several species of the genus *Tursiops, Phocoena phocoena*, and *P. sinus*, all typically are of the saddled or cape system pattern configuration. Other species of *Stenella* are born saddled, but ontogenetically assume other pigmentation patterns such as varying degrees of spotting and/or overlays. Mitchell (1970) interprets the "striped" pattern as found on *S. coeruleoalba* and other delphinids such as *L. obliquidens* as an evolutionary step up from the saddled configuration.

The striking black-and-white color pattern of Dall's and True's porpoises

Fig. 3. Dall's porpoise collected in Monterey Bay, California, with fresh killer whale tooth marks on both sides.

may be considered ". . . disruptive coloration, tending to break up the outline of the animal" (Mitchell, 1970, p. 729). With due consideration of the known predation on *Phocoenoides* by killer whales, this view cannot be contested. It appears, however, that the intensity of black pigment is at its highest only in adult male and female *Phocoenoides*. All subadults studied in the wild or as freshly collected specimens had slate-gray or grayish brown heads, and often these same colors were distributed along the flippers, caudal peduncle, and flukes. Other authors have noted similar variations (Benson and Groody, 1942; Cowan, 1944; Benson, 1946; Sheffer, 1949; Mizue and Yoshida, 1965; Morejohn et al., 1973). The variations noted in the gray or white frosting along the trailing edges of the dorsal fin and flukes (Brownell, 1964) is a maturational phenomenon showing wide individual variation. The greatest extent and lightness of the frosting is found on adults. The basic disruptive pattern configuration of light and dark is developed late *in utero* (Mizue and Yoshida, 1965, p. 27, Plate I, 2). However, other *in utero* color variants have been observed. Ridgway (1966) collected two pregnant females. One of the fetuses was illustrated in his publication (Ridgway, 1966, p. 107, Fig. 10). This fetus has the typical saddled pattern configuration of other delphinids, i.e., *Tursiops*, some *Phocoena*, some *Stenella*, etc. The other fetus (S.H. Ridgway, personal communication), was also saddled. The grayish-brown individuals observed in Monterey Bay, California, and described by Morejohn et al. (1973, p. 977) as ". . . resembling the harbor porpoise, *Phocoena phocoena*," were considered to be striped porpoises (Fig. 1c), as defined by Mitchell (1970), and not saddled. Thus, together with all-black adults and fetus described by Nishiwaki (1966b), the black female (C7) collected by us at Moss Landing (Fig. 1d), and the banded color variant (C137, a fetus) described by Morejohn et al. (1973), it would appear that several infrequently occurring color types occur in *Phocoenoides*. The banded variant and the saddled form have not been observed as subadults or adults (Fig. 4). The intensity of black pigment in the other color variations is probably related to age.

VI. PARASITES AND COMMENSALS

A. Helminth Parasites

Dall's porpoise serves as host to several helminth endoparasites and some ectocommensal diatoms. The helminth parasites have been reported from several parts of the body (Dailey, 1971, 1972):

1. Nasal Sinuses

Nasitrema dalli (Trematoda: Digenea); *Pharurus dalli* (Nematoda: Pseudaliinae).

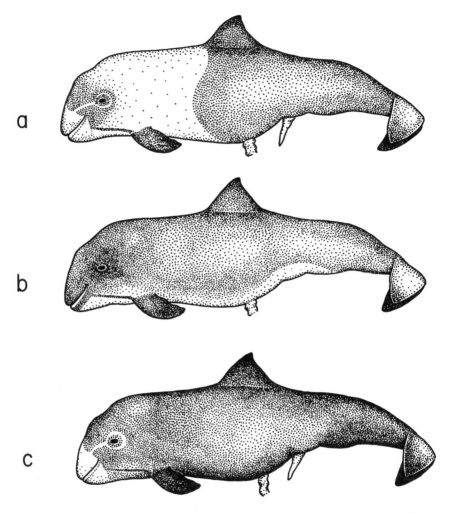

Fig. 4. Fetal color types of Dall's porpoise: (a) banded form which will develop the typical adult pattern before birth; (b) saddled or cape form which will develop into adult "striped" variant; (c) black form which will develop into adult "black" variant.

2. Lungs

Halocercus kirbyi (Nematoda: Pseudaliidae).

3. Hepatopancreatic Ducts

Campula oblonga (Trematoda: Digenea).

4. Stomach

Anisakis simplex (Nematoda: Hetercheilidae).

5. Mammary Glands

Placentonema sp.

6. Blubber of Genital Region

Phyllobothrium sp. (W. J. Houck, personal communication, from Dailey, 1972, p. 552).

B. Epizoic Commensals

An ectocommensal diatom fauna new to this porpoise and other small odontocetes was studied by Morejohn and Hansen (manuscript). Five forms of *Cocconeis ceticola* Nelson (1920) were found distributed on the epidermis posterior to the blowhole; on the trailing edges of the flippers, dorsal fin, and flukes; on the lateral sides of the caudal peduncle; and on both sides of the narrow prepeduncular keel posterior to the dorsal fin (Fig. 5).

Our studies (Morejohn and Hansen, manuscript) showed that *P. dalli dalli* was host to the five forms of the diatom *C. ceticola*. Two forms, *C. ceticola* and *C. ceticola* forma *ovalis*, were hosted in common with *P. phocoena, L. obliquidens,* and *Physeter catodon*, but the oceanic *S. coeruleoalba* hosted no diatoms at all. *C. ceticola* forma *constricta* and *C. ceticola* forma *subconstricta* have been described by Nemoto (1956, 1958) from sperm whales and baleen

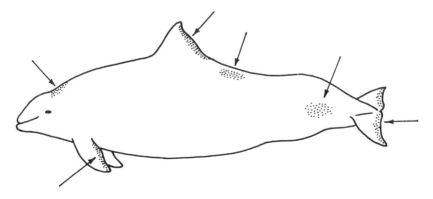

Fig. 5. Diagrammatic drawing of Dall's porpoise showing areas (stippled) of diatom attachment.

whales of the Bering Sea and the Antarctic region. These two forms of diatoms occurred on *Phocoenoides* collected in Monterey Bay, California (Table III) and may be evidence of population shifts of individual porpoises from northern waters. No copepods or barnacles have yet been reported to occur on this porpoise.

VII. FEEDING HABITS

Several investigators studying the feeding habits of *Phocoenoides* in the western and eastern North Pacific Ocean have reported fish, cephalopods, and crustaceans in the diet of this porpoise. Wilke *et al.* (1953) and Wilke and Nicholson (1958) found that most Dall's porpoises had taken mainly two species of squid (*Ommastrephes sloani pacificus* and *Watasenia scintillans*) and smaller amounts of lantern fish (Myctophids) and a deep water hake-like fish (*Laemonema morosum*) in Japanese waters. Mizue and Yoshida (1965) also reported that these porpoises ate squid, some fish, and shrimp from waters of the Bering Sea between eastern Kamchatka and north and south of the Aleutian Islands. From two animals taken in the Gulf of Alaska, Scheffer (1953) reported only capelin (*Mallotus villosus*). Further south off British Columbia and

Table III. Diatom Species Found on *Phocoenoides dalli dalli* and Two Other Small Delphinids Found in the Same Waters of Monterey Bay, California[a]

	Cetacean host		
Diatom species	*Phocoenoides dalli dalli*	*Phocoena phocoena*	*Lagenorhynchus obliguidens*
Cocconeis ceticola	X (9)[b]	X (4)	X (2)
C. ceticola f. *ovalis*	X (9)	X (4)	X (2)
C. ceticola f. *suborbicularis*	X (8)	X (2)	
C. ceticola f. *constricta*	X (8)		
C. ceticola f. *subconstricta*	X (2)	X (1)	X (1)
C. scutellum		X (1)	
C. costata var. *pacifica*		X (1)	
Nitzschia angularis		X (1)	
Navicula directa		X (1)	
Biddulphia aurita		X (1)	
Licomorpha abbreviata		X (1)	
Diploneis bombus		X (1)	
Amphora lineolata	X (1)		
Trachyneis aspera	X (1)	X (1)	
Stauroneis sp.		X (2)	

[a] Data source modified from Morejohn and Hansen (unpublished manuscript).
[b] Numbers in parentheses indicate cetacean sample size.

Washington, Pike and MacAskie (1969) reported two porpoises with squid in their stomachs and one with herring. Off Oregon and northern California, squid, hake (*Merluccius productus*), and jack mackeral (*Trachurus symmetricus*) were reported from Dall's porpoises (Fiscus and Niggol, 1965). In southern California waters, these porpoises were observed circling and perhaps feeding on anchovy (*Engraulis mordax*) and saury (*Cololabis saira*) by Brown and Norris (1956) and Norris and Prescott (1961). A specimen collected by Norris and Prescott contained hake, jack mackeral, and squid.

Our research efforts in Monterey Bay have attempted to document year-round feeding preferences of Dall's porpoise in the bay (Loeb, 1972). Since 1967 and through most of 1972, 38 porpoises have been collected. Stomachs were removed from 32 specimens, of which only 27 yielded food items (Table IV). Prey species were identified by use of fish otoliths (sagittae) and cephalopod upper and lower beaks. Food items eaten probably did not exceed 17 or 20 cm in total length. Based on the frequency of occurrence of food items throughout the year, it appears that hake (*M. productus*), herring (*C. harengus*), juvenile rockfish (*Sebastes* sp.), anchovy (*E. mordax*), and squid (*Loligo opalescens* and *Gonatus* sp.) are the preferred prey species, with the squid (*L. opalescens*) eaten throughout the year. It must be kept in mind, however, that the sample size is minimal and conclusions drawn from this small sample at best can only be considered tentative. An analysis of the percentage of cephalopod beaks and fish otoliths found in these porpoise stomachs throughout the year has been made by Loeb (1972). She also considered the relative availability of the food items to the porpoise throughout the year. Most of the year, the porpoises were seen in waters over 100 fathoms in depth, whereas during the winter months, they were seen in deeper water or in inshore waters of about 50 fathoms. This change in habitat preference seems to be a reflection of availability of prey species (Table III).

VIII. MORPHOLOGY

Most specimens used in this study were weighed, measured and dissected shortly after collection. Of those that were found beach-cast, dead, only three were not considered fresh enough for measurements or weights. Entire specimens were weighed on platform scales accurate to the nearest pound. These weights were later converted to kilogram equivalents. Internal organs were weighed on a beam balance accurate to the nearest gram. Osteological studies will be reported elsewhere.

A. External Morphology

This species is one of the few small odontocetes that lacks an extended rostrum and long mandibles—the beak. It is a characteristic shared with the

Table IV. Prey Species Eaten by *Phocoenoides dalli dalli* in Monterey Bay, California (36°45'N) throughout the Year[a]

Prey Species	Jan (1)[b]	Feb (2)	Mar (1)	Apr (1)	May (1)	June (6)	July (3)	Aug (5)	Sept (2)	Oct (2)	Nov (1)	Dec (2)
Fishes												
Clupea harengus	X	X	—	X	—	X	X	X	X	X	X	—
Engraulis mordax	X	—	X	—	—	X	—	X	—	—	—	X
Spirinchus starksi	—	—	—	—	—	—	—	—	—	X	X	—
Bathylagus stilbius	X	—	—	—	—	—	—	—	—	—	—	—
Myctophidae	—	X	—	—	—	—	—	—	—	—	—	—
Lamparyctus regalis	—	—	—	—	—	—	—	—	—	X	—	—
Merluccius productus	X	X	—	—	X	X	X	X	X	X	X	X
Chilara taylori	X	—	—	—	—	—	—	—	—	—	—	—
Zoarcidae	—	—	—	—	—	—	—	—	—	X	—	—
Macrouridae	—	—	—	—	—	—	—	—	—	X	—	—
Peprilus simillimus	X	—	—	—	—	—	—	—	—	—	—	—
Sebastes sp. (Juv.)	X	—	—	—	—	X	X	X	X	—	X	X
Anoplopoma fimbria	—	—	—	—	—	—	—	—	—	X	—	—
Liparis sp.	—	—	—	—	—	—	—	—	—	X	X	—
Citharichthys sordidus	—	—	—	X	—	—	—	—	—	—	—	—
Cephalopods												
Sepiolidae	X	—	—	—	—	—	—	X	—	—	—	—
Loligo opalescens	X	X	X	X	—	X	X	X	X	X	X	X
Enoploteuthidae	—	—	—	—	—	X	—	—	—	—	—	—
Ommastrephidae	—	X	—	—	—	—	—	—	—	—	—	—
Histioteuthidae	—	—	—	—	—	X	—	—	—	X	—	—
Octopoteuthidae	X	—	—	—	—	—	—	—	—	—	—	—
Gonatus sp.	X	X	X	X	—	—	X	—	—	X	—	—
Octopodidae	—	—	—	—	—	X	—	—	—	X	—	—
Octopus bimaculata	—	—	—	—	—	—	—	—	—	—	—	X
Octopus sp.	X	—	X	—	—	—	—	—	—	—	—	—
Unidentified	X	—	—	—	—	X	—	—	—	—	—	—
Crustacea: Decopoda	—	—	—	—	—	—	X	—	—	—	—	—

[a] Data on fishes and cephalopods, in part, modified from Loeb (1972).
[b] Number in parentheses after month represents number of stomachs examined.

related genera *Phocoena* and *Neophocaena*. In contrast to these relatives, *Phocoenoides* is a powerfully built animal with a relatively larger body muscle mass than other delphinids studied (Ridgway and Johnston, 1966; Ridgway, 1972, p. 651). The caudal peduncle is more deeply keeled in adult males than it is in other male delphinids. The head is proportionately much smaller than in other species (Mizue and Yoshida, 1965). The animal is apparently adapted for fast swimming speeds, useful in attaining great depths rapidly. Many of the prey species it feeds on are known to be deep-water forms. Morphometrics of external body features are shown in Table V for males and in Table VI for females. These measurements show that, for most of the characteristics considered, there is little difference between the sexes. However, there are a few characters which show tendencies for one sex to be larger or smaller. Adult males tend to be longer than females. Subadult males also tend to be longer than subadult females. Indications of male–female differences are seen in the measurements from tip of the snout to: (1) the center of the eye, (2) the external ear (external auditory meatus), (3) the anterior insertion of the flipper, (4) the tip of the dorsal fin, (5) the center of the genital slit, and (6) the measurement of the length of the genital slit. In all these measurements, the subadult and adult females are longer than the subadult and adult males. In the measurement from the tip of the snout to the center of the umbilical scar, both subadult male and female are similar, but the adult female tends to be longer in this measurement than the adult male. Subadult males tend to have flukes with a wider span than subadult females, but as adults, the females have the wider span. The genital and anal openings were previously shown to be significantly more widely separated in the male than in the female (Morejohn *et al.*, 1973). Other than the tendencies to differ between the sexes, there are no statistically significant differences to be found in the characters measured. On the average, subadult males and females weigh about the same. The mean weights for adults of both sexes indicate no difference. However, I believe that a larger sample using other capture methods would have yielded more males of the 200-kg category and perhaps females exceeding 123 kg.

In contrast to other fish and squid-eating porpoises, *Phocoenoides* has the smallest known teeth among the delphinids. In fact, the gums lateral to and between the teeth take on a serrated tooth shape and extend beyond the occlusal plane of the teeth in adult specimens. These intertooth gingival papillae have been termed "gum teeth" by Miller (1929). I have noted in all specimens examined the development of a "dental pad" at the tip of the upper jaw, as is found in all ruminant artiodactyls that have evolutionarily lost all or most of their upper incisors. In *Phocoenoides*, this dental pad has no teeth anterior to it, and the teeth that occur lateral to it are the smallest in the maxillary tooth row. The dental pad thus appears to be a specialization of this species adapting it to better grasp and swallow cephalopods. Many other cephalopod-eating cetaceans lack teeth in the upper jaws (e.g., *Kogia*), or in some instances, these teeth remain unerupted throughout life (as in *Physeter* and several ziphiids); if the teeth are

Table V. Morphometrics of Male *Phocoenoides dalli dalli* (in Millimeters) Collected in Monterey Bay, California[a]

	Subadults															
	C25	C117	C118	C121	C122	C128	C129	C131	C140	C142	C145	C153	C168	Mean	±SD	N
Total length	1850	1705	1870	1920	1770	1850	1780	1930	1760	1690	1790	1700	2025	1801.3	82.7	13
Snout to																
center of eye		200	199	212	200	210	170	190	170	195	190	198	190	176.6	53.1	12
angle of jaw	120	102	100	114	102	110	85	117	88	110	100	106	118	104.5	10.7	13
external ear		207	264	265	272		275	235	230	160	260	260	275	249.1	34.7	11
center blowhole	200	196	280	217	202	200	210	177	185	170	200	186	225	196.6	15.9	13
anterior insertion																
of flipper	310	272	310	284	280		230	255	260	270	280	274	325	275.0	22.9	12
tip dorsal fin	888	814	825	846	820	920	840	810	750	720	800	810	950	821.1	51.7	13
umbilicus		760	867	837	790		773	885	780	760	790	821		806.3	44.3	10
center genital slit	1065	940	1145	1097	1060	1150	1020	1160	1030	1000	1055	1039	1019	1060.0	64.1	13
center anus	1320	1020	1325	1297	1235	1360	1200	1360	1190	1160	1220	1159	1140	1229.7	100.1	13
Eye																
height			12	9	11		8	12	10	10	10	11	17	11.1	2.3	11
length		18	18	17	19	21	18	17	18	18	20	21	25	19.2	2.3	12
Genital slit																
length	92	68	50	118	70	100		85	75	65	60	71		77.6	19.6	11
Flipper																
anterior insertion																
to tip	200	195	233	220	220	225	210	215	200	210	207	220	195	211.5	11.9	13
axilla to tip	130	143	164	160	150	155	150	155	140	150	140	159	125	147.8	11.7	13
maximum width	82	90	100	97	100	100	100	150	85	100	90	105	90	99.2	16.8	13
Dorsal fin																
height	115	158	154	180	185	195	150	162	150	155	140	145	135	155.7	21.5	13
length base	335	320	366	400	320	400	300	320	270	300	240	340	350	327.8	46.0	13
Flukes																
tip to tip	332	493	484	555	485	540	460	525	450	430	460	470	470	473.4	55.7	13
Depth tail notch	21	30	32	30	38	38	25	32	30	30	30	30	20	29.7	5.3	13
Weight (kg)		72.6	81.7				81.7		52.2	39.2	86.7	76.3	145.3	79.5	31.2	8

Dall's Porpoise

	Adults													
	C119	C120	C124	C125	C126	C132	C134	C135	C149	C159	C151			
Total length	1880	1770	1864	2251	2012	1860	1998	1980	1960	2102	1896	1966.5	127.5	11
Snout to														
center of eye	217	175	215	185		200	200	170	215	224	202	199.4	17.7	10
angle of jaw	117	93	115	105	150	105	115	185	108	109	107	118.9	24.8	11
external ear	260	235	283		296	260	138	240	290		264	254.1	45.4	9
center blowhole	194	180	238	203	295	215	185	150	216	228	182	207.6	23.5	11
anterior insertion of flipper	275	250	254	251	273	280	290	260	270	282	297	275.6	21.6	11
tip dorsal fin	840	829	827	1000	950	860	835		855	944	790	882.5	71.3	10
umbilicus	830	830		908	920	820	940	910	868		728	861.5	66.7	9
center genital slit	1060	1070	1108	1328	1230	1035	1195	1300	1182	1281	1091	1165.4	91.4	11
center anus	1240	1265	1288	1542	1760	1275		1420	1364	1470	1287	1391.1	163.4	10
Eye														
height	7	11	7			12	13	13	12		10	10.6	2.4	8
length	20	17	20			18	25	19	19		19	19.6	2.4	8
Genital slit														
length	89	85	105	110	160	62	105	80	82		81	96.0	26.8	10
Flipper														
anterior insertion to tip	213	195	230	210	224	230	184	213	227	220	216	214.7	14.5	11
axilla to tip	177	147	170	155	148	169	152	140	156	158	146	156.2	11.5	11
maximum width	102	90	103	100	100	115	115	95	99	101	87	99.7	7.5	11
Dorsal fin														
height	165	135	168	165	178	167	195	170	158	159	159	165.4	14.6	11
length base	360	335	310	404	340	300	524	310	300	495	380	368.0	77.4	11
Flukes														
tip to tip	510	404	520	510	550	540	510	495	492	450	481	496.5	41.0	11
Depth tail notch	34	26	35	32	45	32	30	30	31	30	25	31.8	5.3	11
Weight (kg)	94.9	97.2	112.1	213.4	148.9	99.4	112.0	119.9	110.8	147.5	98.9	123.2	35.2	11

[a] Measurements were modified from Norris (1961).

Table VI. Morphometrics of Female *Phocoenoides dalli dalli* (in Millimeters) Collected in Monterey Bay, California [a]

	Subadults											Adults				
	C8	C31	C101	C102	C116	C123	C144	C150	Mean	±SD	N	C136	C160	Mean	±SD	N
Total length	1586	1868	1750	1600	1765	1730	1920	1808	1753.4	117.2	8	1850	2021	1935.5	120.9	2
Snout to																
center of eye	208	193	190	185	175	202	198	197	193.5	10.3	8	180	223	201.5	30.4	2
angle of jaw	122	91	90	92	89	107	110	102	100.4	11.9	8	100	120	110.0	14.1	2
external ear	270	245	235	240	236	248	—	261	247.9	13.2	7	230	291	260.5	43.1	2
center blowhole	221	199	215	175	155	195	162	193	189.4	23.7	8	200	216	208.0	11.3	2
anterior insertion																
of flipper	305	285	260	270	272	283	309	280	283.0	16.9	8	280	320	300.0	28.3	2
tip dorsal fin	880	898	875	820	790	830	880	906	859.9	41.4	8	820	972	896.0	107.5	2
umbilicus	764	877	750	715	841		805	818	795.7	56.1	7	900	910	905.0	7.1	2
center genital slit	1094	1170	1180	1035	1164	1110	1216	1138	1138.4	57.2	8	1250	1365	1307.5	81.3	2
center anus	1146	1257	1220	1105	1225	1180	1290	1224	1205.9	59.8	8	1320	1429	1374.5	77.0	2
Eye																
height	8	11	—	—	11	12	13	12	11.2	1.7	6	10	10	10	0.0	2
length	13	20			18	20	20	20	18.5	2.8	6	17	23	20.0	4.2	2

Dall's Porpoise

	1	2	3	4	5	6	7	8	Mean	SD	n	1	2	Mean	SD	n
Mammary slits																
right	8	19	—	22	15	18	13	24	17.0	5.5	7	21	28	24.5	4.9	2
left	8	17	—	22	15	19	15	26	17.4	5.7	7	24	28	26.0	2.8	2
Genital slit length	92	102	—	120	110	133	120	89	109.4	16.1	7	125	125	125.0	0.0	2
Flipper																
anterior insertion to tip	202	202	220	190	195	203	200	218	203.8	10.4	8	200	242	221.0	29.7	2
axilla to tip	145	135	155	150	145	158	150	168	150.8	9.9	8	135	171	153.0	25.4	2
maximum width	95	90	105	90	89	97	90	94	93.8	5.4	8	90	109	99.5	13.4	2
Dorsal fin																
height	136	150	160	140	150	150	142	160	148.5	8.8	8	165	179	172.0	9.8	2
length base	265	262	340	300	350	320	370	307	314.3	38.7	8	290	392	341.0	72.1	2
Flukes																
tip to tip	442	443	430	408	455	440	483	498	449.9	28.8	8	480	557	518.5	54.4	2
Depth tail notch	21	19	—	32	32	25	26	27	26.0	4.9	7	14	30	22.0	11.3	2
Weight (kg)	56.8	45.4	—	106.7	73.0	87.6	80.8	82.6	73.9	19.7	7	125.3	120.8	123.0	3.2	2

[a] Measurements were modified from Norris (1961).

Table VII. Organ Weights (grams) of Male and Female *Phocoenoides dalli*

Specimen no.	Body wt. (kg)	Heart	%[a]	Liver	%	Spleen	%	Kidneys	%	Dia-phragm	%	Pancreas	%
Males													
C117	72.6	739	1.02	1971	2.71	18	0.02	620	0.85	582	0.80	192	0.26
C118	81.7	876	1.07	2523	3.09	44	0.05	646	0.79	811	0.99	168	0.20
C119	94.9	913	0.96	2611	2.75	11	0.01	573	0.60	913	0.96	180	0.19
C120	97.2	864	0.90	3572	3.67	18	0.01	805	0.83	899	0.92	160	0.60
C126	213.4	1680	0.79	5284	2.48	—	—	1792	0.84	—	—	—	—
C128	148.9	1232	0.83	2856	1.92	—	—	896	0.60	1568	1.05	—	—
C129	81.7	708	0.87	1740	2.13	27	0.03	686	0.84	740	0.91	217	0.27
C170[b]	22.2	268	1.21	1042	4.69	6	0.03	108	0.49	200	0.90	76	0.34
Females													
C102	73.0	640	0.87	—	—	24	0.03	620	0.85	635	0.87	132	0.18
C116	87.6	881	1.00	2924	3.33	20	0.02	664	0.76	804	0.92	185	0.21
C123	80.8	672	0.83	2518	3.11			619	0.77	615	0.76	168	0.21

[a] % = organ weight/body weight × 100.
[b] This was a beach-cast juvenile specimen not used in other calculations of this study.

present, they are small and widely spaced at the tip of the rostrum (as in *Globicephala* and *Grampus*).

B. Organ Weights

Many studies have been made of the internal organs of small odontocetes. Genera on which organ weights have been studied include *Phocoena, Tursiops, Lagenorhynchus, Delphinus, Stenella, Globicephala, Grampus, Inia, Physeter,* and *Delphinapterus*. A recent review of pertinent literature is given in Perrin and Roberts (1972). Organ weights for male and female *Phocoenoides* are shown in Table VII.

Ridgway and Johnston (1966) and Ridgway (1966, 1972, p. 651) considered the heart of *Phocoenoides* to be extremely large compared to *Tursiops* and *Lagenorhynchus*. Their comparisons were based on only one male and three female specimens of *Phocoenoides* (body weight range: 80-122 kg; heart weight range: 1024.0–1302.5; relative to body weight: 1.31). Our sample of eleven specimens of both sexes of subadults and adults (body weight range: 72.6–213.4 kg; heart weight range: 640–1680 g; mean relative to body weight: 0.94) does not support this view. The value of 0.94 is closer to the value (0.85) for the relative heart weight of *Lagenorhynchus obliquidens* (Ridgway, 1972, p. 651, Table 10-24). Using some of the data provided by Perrin and Roberts (1972, pp. 20-21, Table 2) and some of the data provided by Gihr and Pilleri (1969, pp. 19-20, Tables 1 and 2; p. 36, Table 8; p. 38, Table 9), values were computed for most subadults and adults of *Stenella graffmani, S. styx,* and *Delphinus delphis*. Table VIII shows the interspecies comparisons with *Phocoenoides*. Comparison of the scatter plots for heart, liver, and kidneys of *Phocoenoides* (Figs. 6–8) with similar plots of the organs of the two species of *Stenella* (Perrin and Roberts, 1972, pp. 24–29) demonstrated similar degree of slope (organ weight relative to body weight) of the two genera *Stenella* and *Phocoenoides*.

The data for brain weights from Ridgway (1966, p. 110, Table 2) and that from Gihr and Pillari (1969) for brain weights of *S. styx* and *D. delphis,* shown in the comparison to organ weights in Table VIII, indicate a smaller brain and larger heart and liver in *Phocoenoides*.

Because of the lack of sufficient specimens of the juvenile category (between newborn and subadult), no attempt was made to determine the allometric or isometric nature of particular organs of *Phocoenoides*.

IX. REPRODUCTION

Since Scheffer (1949, p. 116) stated, "No one knows when or where the Dall porpoise gives birth to its young," a considerable amount of information

Table VIII. Interspecies Comparisons of Percent Body Weight for Several Organs of Some Subadults and Adults of Both Sexes of Four Small Odontocete Species [a]

Organ	N	Wt. range (g)	Mean weight ± SD	Mean % body wt
Delphinus delphis (body weight range: 40–59 kg; mean = 51.0 ± 6.2 SD, N = 7)				
Brain	7	660–824	761.1 ± 61.0	1.49
Heart	7	210–350	306.4 ± 47.1	0.60
Liver	7	1055–1435	1280.0 ± 127.0	2.51
Spleen	7	20–55	32.0 ± 11.7	0.06
Pancreas	7	96–210	142.3 ± 47.2	0.28
Kidneys	7	225–580	383.9 ± 113.1	0.75
Stenella styx (= *coeruleoalba*) (body weight range: 45–71 kg; mean = 59.8 ± 9.8 SD, N = 6)				
Brain	6	690–930	820.0 ± 79.1	1.37
Heart	6	350–525	433.0 ± 62.3	0.72
Liver	5	1300–1800	1486.0 ± 189.7	2.48
Spleen	5	35–55	46.0 ± 8.2	0.08
Pancreas	5	135–200	160.0 ± 32.2	0.28
Kidneys	5	335–495	397.0 ± 65.1	0.66
Stenella graffmani (= *attenuata*) (body weight range: 43–84 kg; mean = 60.5 ± 14.1 SD, N = 14)				
Brain				
Heart	14	179–308	272.2 ± 42.2	0.42
Liver	14	974–2195	1487.4 ± 378.1	2.46
Spleen	14	24–72	41.9 ± 15.1	0.07
Pancreas				
Kidneys	14	270–962	565.7 ± 195.8	0.94
Phocoenoides dalli (body weight range: 52–213.4 kg; mean = 99.0 ± 44.7 SD, N = 11)				
Brain	2	760.3–794.3[b]	777.3 ± 24.0	0.88
Heart	11	640–1680	927.5 ± 300.8	0.94
Liver	10	1740–5284	2900.0 ± 984.8	2.93
Spleen	8	11–44	22.9 ± 9.7	0.02
Pancreas	9	132–217	178.7 ± 25.6	0.18
Kidneys	11	573–1792	794.0 ± 346.0	0.80

[a] Data for *D. delphis* and *S. styx* (= *coeruleoalba*) taken from Gihr and Pilleri (1969); data for *S. graffmani* (= *attenuata*) taken from Perrin (1972).
[b] From Ridgway (1966, p. 110, Table 2).

has been amassed that aids in clarifying the issue. Earlier, Cowan (1944, p.296) gave an account of observations by Mr. A. Lyons of Hardy Bay, Vancouver Island, British Columbia. Mr. Lyons had frequently observed very young Dall's porpoises in Goletas Channel in late August. He had not observed them earlier than August 7, but on August 9, he made an unsuccessful attempt to capture what appeared to be a newborn calf. A near-term fetus was found in a female studied by Miller (1929) off Alaska near Sullivan Island, Lynn Canal. This female was collected in November 1926. The fetuses and newborn young examined by

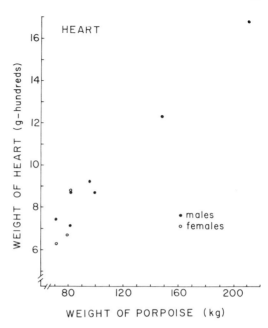

Fig. 6. Weight of heart relative to body weight for 7 male and 3 female Dall's porpoises.

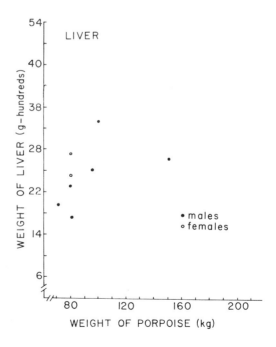

Fig. 7. Weight of liver relative to total body weight for 7 male and 2 female Dall's porpoises.

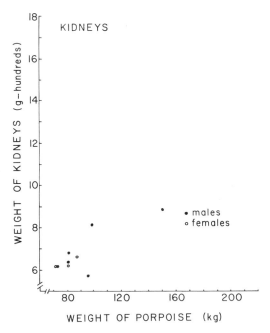

Fig. 8. Weight of kidneys relative to total body weight for 7 male and 3 female Dall's porpoises.

Mizue and Yoshida (1965) from the western Bering Sea led them to conclude that the calves are born between the middle of July and the first ten days of August. They estimated a range of 85–90 cm for a total length of a newborn calf.

Fetuses taken from female porpoises off California, in the Monterey Bay area and in the vicinity of the Channel Islands and San Diego, have been found in the winter and spring months (Table IX). Young porpoises, still in the late suckling stage, have been found freshly beach-cast, dead or dying, in February (1185 mm) and in July (1149 mm). The adult female (C160) found dead on the beach near Año Nuevo Point some 20 miles north of Monterey Bay on July 13, 1972, still had embryonic membranes extruding from the vaginal orifice and had obviously delivered its young a day or so earlier.

During the first year of growth, Mizue and Yoshida (1965) plotted data in support of a monthly rate of growth of 50 mm. Thus, the youngsters collected in February and July were about five and six months of age, born approximately six months apart. Young porpoises of this size range (115–120 cm) have been sighted at sea in Monterey Bay, both in January and August. Norris and Prescott (1961, p. 356) also observed young Dall's porpoise " . . . thought to be about three-quarters grown" off Catalina Island in southern California in the month of February. Therefore, based on the evidence from the literature and from our study of fetuses, newborn young, and small juveniles, parturition seems to take place throughout the year, at least from the eastern North Pacific waters of

Alaska to southern California. The assumptions made by Mizue and Yoshida (1965) of summer parturition in the western Bering Sea were based on only their gill-netted sample taken during July and August. A year-round study in Japanese and Bering Sea waters may yield results similar to those given in the present report.

A. Female Reproductive Tract

Among odontocete and mysticete cetaceans studied by myself and others, Dall's porpoise has been found to be unique in having almost no development of vaginal folds in the posterior walls of the vaginal vestibule (Morejohn and Baltz, 1972). The presence of these folds has been considered a character of ordinal significance and unique to the Cetacea (Harrison, 1949, 1969; Slijper, 1966). Another character of *Phocoenoides* which has been found to date only in mysticetes and not in odontocetes (Ohsumi, 1969) is the presence of a vaginal band in subadult female *Phocoenoides* (Morejohn and Baltz, 1972).

The largest developing follicles in the ovaries of subadult females ranged from 1.0 to 3.5 mm in diameter. In the pregnant female (C136), the *corpus luteum* of pregnancy was 22.0 mm in diameter, the longest *corpus albicans* measured 8.0 mm, and the largest primary follicle was 1.0 mm in diameter. The young of both females (C136, C160) had developed in the left uterine horn and only the left ovary had ovulated—a trait characteristic of many odontocetes. The two pregnant females studied by Ridgway (1966) also had sinistral ovulations and uterine implantations. The one female, examined by Ridgway (1966), which contained a term fetus had mammary glands full of thick brown milk. The mammary glands of female C160 were flaccid and contained watery brown fluid.

Table IX. Fetuses Taken from Female *Phocoenoides dalli dalli* in California Waters (SC = southern California, CC = central California) and Juveniles Found Dead or Dying on the Beach

Date collected	Length (cm)	Locality	Collector
April 21, 1965	87.6	Anacapa Island(SC)	S.H. Ridgway
January 8, 1965	40.7	Santa Barbara Channel near Santa Cruz Island(SC)	S.H. Ridgway
February 7, 1971	42.0	Monterey Bay(CC)	G.V. Morejohn
January 21, 1968	31.0	Between Santa Rosa and Santa Cruz Islands(SC)	B. Lenheim
March 22, 1952	45.0	La Jolla(SC)	C.L. Hubbs and C. Limbaugh
Juveniles:			
July 1971	114.9	Half Moon Bay(CC)	V. Loeb
February 5, 1974	118.5	Monterey Bay(CC)	D. Lewis

B. Male Reproductive Tract

The penis of male *Phocoenoides* appears to be typical of males of other delphinid species. The testes of six specimens ranged in length from 55 to 160 mm; width of testes ranged from 18 to 70 mm. Males with the largest testes were taken in January (C125) and in August (C135). However, these testes were flaccid and appeared in nonreproductive condition. It is suggested by Ridgway and Green (1967) that the smaller odontocetes along the Pacific coast of California (*Phocoena, Phocoenoides, Lagenorhynchus,* and *Delphinus*) have a reproductive peak from mid- to late summer. Our sample of adult male *Phocoenoides*, fetuses, and juvenile young does not necessarily support this view.

In summarizing the new information on reproduction in Dall's porpoise, we can now state that this species probably breeds and gives birth to young throughout the year in northeastern Pacific waters from Alaska to southern California.

X. BEHAVIOR

Of three individual *Phocoenoides* maintained in captivity, one died within two months, another lived for three months, and the third survived for 15 months after capture (Ridgway, 1966; Walker, 1975). They are considered to be one of the fastest of the smaller odontocetes and for this reason, extremely difficult to capture. Most individuals, when captured alive, die on the boat deck from capture stress. Contrasted to *Tursiops*, *Lagenorhynchus,* and *Globicephala*, which often can be observed at the bow of boats traveling at 4–8 knots, or *Delphinus* and *Lissodelphis* that can move along with a boat at 8–10 knots, *Phocoenoides* is not easily tempted to run with a boat for long unless the speed is 14–16 knots or faster. At such speeds, even in glassy water, observations of behavior are limited and, at best, of short duration. Our extended observations of this animal in the wild have allowed us to determine when it is most likely that Dall's porpoise will be in the bay and approximately where they will probably be. We have learned that their interest in running with a boat can be greatly extended from the usual five minutes up to a half hour. This is simply accomplished by varying speeds, running in figure-eights and circles, periodically coming to a gliding standstill then accelerating the motor and repeating our repertoire. In this manner, we have made many observations of different types of behavior.

Our observations were made from three vessels: *R/V Orca*, *R/V Amigo*, and *R/V Artemia*. The *R/V Orca* is a 28-ft inboard that produces no "rideable bow wave" as defined by Norris and Prescott (1961); the *R/V Amigo* (a charter vessel) is a 65-ft cabin cruiser converted with equipment and facilities for oceanographic studies and produces a rideable bow wave, but the bowsprit is some 8 ft above the sea surface. The *R/V Artemia* is a 48-ft U.S. Army tug also converted with equipment and facilities for oceanographic studies and used extensively in fish trawling operations. Its hull is of such construction that extensive "white water" is thrown ahead as it plows through the sea. It also produces a bow wave, but observations of porpoises below the bow are difficult to make because of the white water pushed forward. The *R/V Orca*, for reasons unknown to us, is more attractive to Dall's porpoise, and it is from this vessel that most of the observations were made.

A. Swimming

As is typical of cetaceans, Dall's porpoise swims by pumping up and down with its laterally displaced flukes. In compensation for the up-and-down strokes of the caudal peduncle, the head swings up and down as the animal moves through the water. As we approached the area where Dall's porpoise are often seen, usually along the 100-fathom contours of the Monterey submarine canyon, we would usually see the typical "cone-shaped" splash (Figs. 9 and 10) of the

Fig. 9. Cone-shaped splash of Dall's porpoise at play.

Fig. 10. Cone-shaped splash of striped variant of Dall's porpoise (above), and striped variant swimming about 0.5 m below the surface.

porpoises as they hurtled themselves through the water headed for our vessel. Often animals would appear close by seemingly from nowhere, race to our bow, criss-cross ahead with great swiftness, and then move off, perhaps occasionally splashing, then assume a slow, high, arched surfacing. This form of surfacing (Fig. 11) has been likened to a square box rolling over end-for-end (Brown and Norris, 1956). On many occasions, while trawling from the *R/V Artemia* at 2 or 3 knots in a porpoise area, we would sight them slowly surfacing in this manner. They were not seen to be moving in any one direction and surfaced to breathe every 40 or 50 sec. We made similar observations from the *R/V Orca* lying dead in the water and, when we started the engine and accelerated toward them at 12 to 14 knots, they disappeared from sight only to appear again wildly splashing toward us. We observed many at great distances (500–600 yards) at a right angle to our course direction (running at about 14 knots) and within 2–3 min they would be running with the vessel. On a flat sea, we can exceed 16 knots and, apparently, these porpoises are capable of greater speeds.

Most of our observations were made between 0900 and 1400 hr. Activities in the bay in mid- and late afternoon were frequently greatly reduced by strong winds and wave action. However, when weather permitted, we remained at sea throughout the afternoon. In the mornings and late afternoons, the porpoises appeared to be widely scattered and as we came into an area where they were scattered, splashes from several directions could be seen coming our way.

Fig. 11. Slow surfacing of Dall's porpoise during inactivity.

Members of a school apparently had gathered. They ran with the boat on all sides, usually three or four criss-crossing ahead of the boat, surfacing with great splashes every 15–25 sec. Generally, one individual was immediately before the boat (within a meter of the bow) or directly under the boat with its head projecting slightly beyond the bow and half a meter beneath the surface. This was a preferred position and individuals had to swim vigorously to maintain it. Periodically, the position had to be vacated in order for the porpoise to breathe and was immediately filled by another individual. One to four observers peering over the edge of the bow about 1.5 m from the closest porpoise recorded approximate sizes, coloration variations, scars, cuts, or other markings. Sometimes, as many as eight or ten porpoises were ahead of the vessel in an echelon formation, with longer porpoises of great girth running 2–3 m beneath the surface and the smaller ones vying with each other for the preferred position ahead of or under the bow. Only rarely did the larger individuals run with us for very long. These we assume were adults, the smaller ones subadults. Still smaller juvenile porpoises were sighted at several times during the year, but most of them appeared shy and either ran from the vessel or ran ahead some 10–20 m for a short time before veering off. [See Evans and Bastian (1969) for discussion of vertical and horizontal stratification of porpoise schools.]

From mid-morning to mid-afternoon, the porpoises were usually grouped closely and were seen more often surfacing slowly without splashing. At times, during midday runs, we could not entice these slowly surfacing porpoises to run with the boat. While drifting with the engine off among these porpoises, their periodic respirations could be seen and heard as they quietly surfaced. As noted earlier, they were not moving in any particular direction, but up and down in the water column. These animals may have been sleeping, based on their inactivity and their slow, shallow respiration, as contrasted to their explosive respirations when swimming rapidly before the bow. The observations made by Ridgway (1966, p. 105) on captive *Phocoenoides* indicated that they were "...hungrier and more active during the night-time feeding." Thus, based on their wide dispersal and activity in the mornings and late afternoons and their tight schooling and tendency not to run readily, if at all, with a vessel during midday, I am led to believe that they probably are mostly nocturnal feeders and rest during the greater part of the day.

Leaping from the water has been observed rarely in this porpoise; however, Yocum (1946) observed some leaping before a U.S. Naval vessel off the Farallon Islands west of San Francisco, California. On another occasion, after one was netted off Santa Barbara Island in southern California, Ridgway (1966) noted that it leapt from the water two or three feet, apparently trying to free itself from the net.

An interesting type of schooling behavior not observed in Monterey Bay is swimming in single file. This was observed by Norris and Prescott (1961) in San Pedro Channel in southern California waters. Four different schools were seen;

the first consisted of five individuals, the second of about 100 animals, and the other two consisted of 20 animals each. The first five were headed eastward; the others were proceeding on a southeasterly course. Individuals in each file were evenly spaced from the others in the same file about 100 ft apart.

B. Diving

The depths to which this species of porpoise dives during its feeding forays have not been measured by instruments attached to free-ranging animals. Depths to which it is thought to descend (>120 m) have been postulated by Ridgway (1966) and Ridgway and Johnston (1966). They based their estimates on several factors: (1) high daily food requirements to maintain body weight, (2) more massive skeletal musculature than other genera, (3) an extremely high blood-oxygen content, (4) a higher relative heart weight than other genera, and (5) a relatively thin blubber thickness. Coupled with these factors, they considered stomach contents of the *Phocoenoides* they had captured which revealed hake (*M. productus*), a deep-water fish that typically occurs below 120 m (about 67 fathoms). Our studies support this view and, additionally, other deep-water forms have been added to the prey species eaten (see Table IV) that are usually found at depths in excess of 100 fathoms, such as the liparids, zoarcids, and macrourids (Fitch and Lavenberg, 1968).

C. Play

Caldwell *et al.* (1966, p. 694) have defined "play" behavior as an action that ". . . apparently is performed for the sake of the activity itself, rather than for any observable effect it may have upon the environment, and behavior that cannot be shown to contribute immediately to any of the recognized internal drives that maintain life." This definition of play is adequate for the types of "play" behavior observed in this study. Dall's porpoises have been seen grasping lines of Nansen bottles and tugging at them. They have grasped lines of secchi disks and run off with them. Their running with fast-moving vessels I consider play behavior. At times, when porpoises were before the moving vessel, we would cut the engine to idle and glide along for a distance, the porpoises meanwhile also slowing down and turning sideways as if to look at the boat. Then, as we rapidly accelerated, they would lunge forward with great splashing and squealing to apparently resume their "play."

Subadult porpoises on several occasions were observed splashing about migrating gray whales, *Eschrichtius robustus*, and juveniles were seen playing about a feeding fin whale, *Balaenoptera physalus*.

D. Agonistic Behavior

Scott (1958) considered this type of behavior to include responses given when animals inflict injury or are in danger of being injured.

On two different occasions, we encountered three different schools in different areas of the bay. Each school had its own integrity and consisted of animals that we had previously seen together, recognizable by natural marks, sizes, and color variations. In each instance, one school ran for almost a mile with the boat and until we came upon one of the other schools. As the schools mingled, some porpoises stayed with the boat, but others milled about each other and one was seen to lie at the surface and list on its right side while another positioned itself at a perpendicular to the long axis of the body of the listed one and raised itself out of the water up to its pectoral fins with mouth open. The listed one rolled to expose its ventral surface, then sounded (Fig. 12). In relation to this behavior, Ridgway (1966, p. 107) described some of the difficulties encountered in attempting to train a captive male *Phocoenoides*. The animal seemed most responsive at night and least responsive during midday. Ridgway found that ". . . he showed a great deal of nervousness and irritability when being made to respond to something new. The irritation was manifested by the animal surfacing often, sometimes expelling air quite forcibly and listing a bit to one side." This listing behavior thus appears to be a response to a display of aggression on the part of another animal, in the latter case, a trainer.

At other times, a porpoise was seen to forcibly push another out of the preferred position before the bow. One larger animal swam next to another at the

Fig. 12. "Listing" behavior in Dall's porpoise. See text for description.

bow and was seen to "jaw clap" in the direction of the smaller animal and thereby displace it (John Hall, personal communication).

E. Senses

1. Hearing

Essentially nothing is known of the auditory capabilities of Dall's porpoise. The signals recorded by Evans (*in* Ridgway, 1966) and those recorded by Schevill *et al.* (1969) do not have the characteristic whistle typical of most delphinids studied. Only bursts of low-frequency clicks were noted that are comparable to those produced by *Phocoena phocoena*.

The external auditory meatus in two specimens studied by Norris and Prescott (1961) did not penetrate to the exterior. It penetrated the blubber and ended in the pigmented black layer beneath the epidermis. We have not noted imperforate external auditory meati in specimens used in this study; however, a female right whale dolphin, *Lissodelphis borealis* collected as a dead beach-cast specimen had no evidence of external auditory openings. This may be a variable characteristic among some delphinids.

2. Vision

The vision of *Phocoenoides* is probably as well developed as in other delphinids. We have seen them shy from sudden movements made by observers aboard our vessels. In repeated failures to catch them with stocking nets, they soon learn to swim lightly on their sides in order to watch the position of the net pole. Elsewhere in the swimming behavior section, I have described porpoises apparently watching us from the water.

3. Preanal Gland Pit

These glands open directly into the water and have been noted in sperm whales, *Physeter catodon*; the beluga, *Delphinapterus leucas* (Yablokov, 1961, as cited by Caldwell *et al.*, 1966); and the finless porpoise, *Neophocaena phocaenoides* (Nishiwaki and Kureha, 1975), and are considered important in chemical communication. Only one *Phocoenoides* specimen (C149) was recorded with a small opening 40 mm anterior to the center of the anus.

XI. DISCUSSION AND CONCLUSIONS

I have reviewed the taxonomic status of two species of *Phocoenoides* (*P. dalli* and *P. truei*) versus two subspecies of *P. dalli* (*P. d. dalli* and *P. d. truei*).

Dall's porpoise, ranges from the waters off Japan across the Bering Sea to waters off North America south to Baja California, Mexico; and True's porpoise is restricted to waters off the east coast of Japan.

Although seasonal movements of Dall's porpoise have been established in Japanese waters, a review of the pertinent literature for waters of the eastern Pacific Ocean mostly supports inshore and offshore movement at certain times of the year. Observations made throughout most of the year in their distribution range of the eastern Pacific demonstrate that they are present in these waters year-round.

Predation by the only predator known to feed on this porpoise, the killer whale, is reviewed, and a description of "one that got away" showing fresh killer whale tooth wounds is given. Probably related to killer whale predation on this animal is the black and white disruptive coloration found on both Dall's and True's porpoises. The striking black and white pattern may have adaptive significance.

I have briefly considered the ontogenetic aspects of coloration in other delphinid species. The primitive or ancestral obliterative shading (countershading) found in all species of adult *Tursiops*, some *Phocoena*, and some *Stenella* may also be seen in early fetuses of other porpoises such as in newborn young or juveniles of some *Stenella* (see Evans and Bastian, 1969, p. 467, Fig. 15 of mother and juvenile *S. plagiodon*). Adult color pattern and/or markings are formed as overlays on this ancestral color pattern, referred to as "saddled" by Mitchell (1970) or "cape" by Perrin (1972).

A close study of fetal color descriptions, photographs of fetuses, and close inspection of actual fetuses of *Phocoenoides* of several ages has established the fact that there are several color patterns in evidence during embryonic development. A black fetus was described by Nishiwaki (1966b, p. 201, Fig. 6) that was taken from an all-black female. The photograph shows a light area from the gape to the tip of the rostrum and lower jaw and a light line running posteriorly that connects with a light ring around the eye (called "bridle" stripe by Mitchell, 1970, p. 719). The fetal color variant described by Morejohn *et al*. (1973, p. 979, Fig. 2) has the same bridle stripe. A term fetus reported by Ridgway (1966, p. 107, Fig. 10) had a typical saddled or cape color pattern. The other fetus mentioned in the same report was described to me later by Ridgway (personal communication) and was also of the saddled pattern. Mitchell (1970) and Perrin (1972) have both demonstrated how adult spotted or striped porpoises developed from juvenile saddled or cape patterns. Thus, in consideration of the marked color intensity changes that take place during development in some species, it should not be surprising to develop an all-black adult from the fetus described by Nishiwaki (1966b) or the striped adult (Fig. 10) described and illustrated by Morejohn *et al*. (1973) from the fetus type shown by Ridgway (1966). No fetal variant of the type described by Morejohn *et al*. (1973) has been seen as an adult in the wild and therefore, because the smallest fetuses of *Phocoenoides* (31.0–45.0 cm in total length; see Table IX) are all of the type color variant described by Morejohn *et al*. (1973), I can only conclude that this color variant gradually changes in pattern so that in late-term fetuses, the relative position of light and dark areas of the adult are fixed. (See Mizue and Yoshida, 1965, p. 27, Plate I, Fig. 2 for photo of normally marked term fetus.) A term fetus collected by C.L. Hubbs is also normally marked.

The individual variations observed in the extent and distribution of white frosting on the trailing edges of the dorsal fin and flukes are to be considered phenomena of the aging process. Younger subadults and juveniles generally have darkly pigmented dorsal fin and flukes.

A review of the literature shows that trematodes, cestodes, and nematodes are typically found as the helminth endoparasitic fauna of *Phocoenoides*. Ectocommensal diatoms have also been found on the skin film of this species. Five forms of the diatom *Cocconeis ceticola*, some usually found on mysticetes and large odontocetes, were found on various parts of the porpoise's body, flippers, and flukes. Some of these diatoms were found to be hosted in common with other odontocetes. Two forms of these diatoms were previously described only from cetaceans occurring in the Arctic and Antarctic. Their occurrence on Dall's porpoises may be evidence of wide distributional shifts of certain individual porpoises.

Throughout their range of distribution, Dall's porpoise is known to feed on cephalopods and fishes, most of which are notably deep-water forms. Analysis of stomach contents of porpoises taken in this study has shown that hake, herring, juvenile rockfishes, anchovies, and cephalopods are taken as prey species through most of the year in Monterey Bay, California. Deep-water liparids, zoarcids, and macrourids taken occasionally are evidence that the porpoises dive to depths in excess of 100 fathoms. Approximately ten different species of cephalopods were eaten during the year. The one species of squid most frequently taken throughout the year was *Loligo opalescens*.

A morphometric study was made of Dall's porpoise which involved 19 measurements of external proportions and weights of several internal organs. There were no statistically significant differences between males and females. A development of the internal tip of the lower jaw to form a dental pad is considered an adaptation to predation on cephalopods. A comparison of relative weights of internal organs was made with several small delphinid species for which these data were available. Dall's porpoise was found to have the smallest brain and the largest heart and liver relative to body weight.

Based on the age (size) and time of year that fetuses were collected, and the evidence of parturition and occurrence of juvenile porpoises from Alaska to southern California through most of the calendar year, it was concluded that Dall's porpoise breeds throughout the year along its entire range of distribution in the eastern Pacific Ocean. Although adult males were collected throughout the year, those with the largest testes dimensions were taken in mid-winter and mid-summer and were in a nonreproductive state.

Behaviorally, Dall's porpoise was not observed to school in any particular manner when undisturbed by moving vessels; however, single-file formations were described off southern California. The animals are capable of great speed and can exceed 16 knots. Play behavior when running with boats is largely dependent on the speed of the boat. Varying speeds to slow (2–3 knots) or fast (12–14 knots) and changing directions in the same area were found to prolong play behavior in this porpoise. Few adults ran with the boats; when they did, they ran ahead of the bow deep in the water or off to the side and not for long. The subadults ran with the boats far more frequently. Dall's porpoise is believed to be inactive throughout most of the day, sleeping intermittently, and resuming active feeding in late afternoon and through the night and early morning. Open-mounthed displays of aggression on the part of porpoises to members of other schools resulted in a lateral display of the body to the aggressor, accompanied by a slight body list away from the aggressor and followed by immediate dispersal. Exercise of dominance privileges was noted when larger individuals forcibly displaced smaller ones at the preferred positions before the vessel's bow. In one instance, jaw clapping was noticed to accomplish the same response. Little work has been done on hearing and phonation in this species. To date, only low-frequency clicks have been recorded.

In view of the studies brought together in this account, I would hope that the areas of this contribution that show need for more research will be given consideration in development of future research programs.

ACKNOWLEDGMENTS

A study of this nature cannot be accomplished by one individual without considerable assistance from many others. My associates, colleagues, and former students over the past few years have aided greatly with observations at sea, assistance with beach-cast specimens, and many hours of processing specimens. I gratefully acknowledge the help given by D. Varoujean, D.M. Baltz, D. Lewis, V.J. Loeb, and L. Talent. J.S. Leatherwood provided aerial and shipboard observation data; J.E. Fitch, L. Pinkas, J.L. Cross, and R. Evans helped greatly with identification of fish otoliths and cephalopod beaks; and W.F. Perrin, D.W. Rice, and C.L. Hubbs provided field notes and fetuses in their care. Both W.F. Perrin and S. Leatherwood critically read the manuscript and provided many useful comments.

REFERENCES

Andrews, R. C., 1911, A new porpoise from Japan, *Am. Mus. Nat. Hist.* **3**:31-52.
Benson, S.B., 1946, Further notes on the Dall porpoise, *J. Mammal.* **27(4)**:368-373.
Benson, S.B., and Groody, T.C., 1942, Notes on the Dall porpoise *(Phocoenoides dalli), J. Mammal* **37(3)**:311-323.
Bigg, M.A., and Wolman, A.A., 1975, Live-capture killer whale *(Orcinus orca)* fishery, British Columbia and Washington, 1962-73, *Fish. Res. Bd. Can.*, Special Issue — Review of Biology and Fisheries for Smaller Cetaceans, **32**:1213-1221.
Brown, D. H., and Norris, K. S., 1956, Observations of captive and wild cetaceans, *J. Mammal.* **37**:311-326.
Brownell, R.L., Jr., 1964, Observations of Odontocetes in central California waters, *Nor. Halfangst-Tid.* **53(3)**:60-66.
Caldwell, D.K., Caldwell, M.C., and Rice, D.W., 1966, Behavior of the sperm whale, *Physeter catodon* L., in: *Whales, Dolphins and Porpoises* (K. S. Norris, ed.), pp. 1677-1717, University of California Press, Berkeley.
Cowan, I. M., 1944, The Dall porpoise *(Phocoenoides dalli* True) of the northern Pacific Ocean, *J. Mammal.* **25(3)**:295-306.
Dailey, M. D., 1971, Distribution of helminths in the Dall porpoise *(Phocoenoides dalli* True), *J. Parasitol.* **57**:1348.
Dailey, M. D., 1972, A checklist of marine mammal parasites, in: *Mammals of the Sea, Biology and Medicine* (S. H. Ridgway, ed.), pp. 528-589, Charles C. Thomas, Springfield, Illinois.
Evans, W. E., and Bastian, J., 1969, Marine mammal communications: Social and ecological factors, in: *The Biology of Marine Mammals* (H. T. Anderson, ed.), pp. 425-475, Academic Press, New York.
Fiscus, C. H., and Niggol, K., 1965. Observation of Cetaceans off California, Oregon and Washington, U.S. Fish Wildlife Service Spec. Sci. Report — Fisheries No. 498, 27 pp.
Fitch, J. E., and Lavenberg, R. J., 1968, *Deep Water Fishes of California,* 179 pp., University of California Press, Berkeley.
Ghir, M., and Pilleri, G., 1969, On the anatomy and biometry of *Stenella styx* Gray and *Delphinus delphis* L. (Cetacea, Delphinidae) of the western Mediterranean, in: *Investigations on Cetacea* (G. Pilleri, ed.), Vol. 1, pp. 15-16, Institute of Brain Anatomy, University of Berne, Switzerland.
Harrison, R. J., 1949, Observations on the female reproductive tract of the ca'aing whale, *J. Anat.* **83**:238-253.
Harrison, R. J., 1969, Reproduction and reproductive organs, in: *The Biology of Marine Mammals* (H. Andersen, ed.), pp. 253-348, Academic Press, New York.
Houck, W. J., 1976, The taxonomic status of the species of the porpoise genus *Phocoenoides*, Report of the Advisory Committee on Marine Resources Research, pp. 1-13, FAO, Bergen, Norway.
Leatherwood, S., and Fielding, M. R. 1974, A Summary of Distribution and Movements of Dall Porpoises *Phocoenoides dalli* off Southern California and Baja California, Working paper #42, FAO, United Nations, ACMRR meeting Dec. 16-19, 1974, La Jolla, Calif.
Leatherwood, S., Evans, W. E., and Rice, D. W., 1972, The Whales, Dolphins and Porpoises of the Eastern North Pacific; A Guide to Their Identification in the Water, Naval Undersea Center Tech. Publ. No. 282, pp. 1-175.
Loeb, V. J., 1972, A Study of the Distribution and Feeding Habits of the Dall Porpoise in Monterey Bay, California pp. 1-62. MA thesis, San Jose State University, California.
Lustig, B. L., 1948, Sight records of Dall porpoises off the Channel Islands, California, *J. Mammal.* **29(2)**:183.

Miller, G. S., Jr., 1929, The gums of the porpoise *Phocoenoides dalli* (True), *Publ. No. 2771, Proc. U.S. Natl. Mus.* **79**:1–4.

Mitchell, E., 1970, Pigmentation pattern evolution in delphinid cetaceans: An essay in adaptive coloration, *Can. J. Zool.* **48**:717–740.

Mizue, L., and Yoshida, K., 1965, On the porpoises caught by the salmon fishing gill-net in Bering Sea and the north Pacific Ocean, *Bull. Fac. Fish., Nagasaki Univ.*, No. 19:1–36.

Morejohn, G. V., and Baltz, D. M., 1972, On the reproductive tract of the female Dall porpoise, *J. Mammal.* **53(3)**:606–608.

Morejohn, G. V., and Hansen, J. C., Epizoic diatom flora on small odontocete cetaceans (unpublished manuscript).

Morejohn, G. V., Loeb, V., and Baltz, D. M., 1973, Coloration and sexual dimorphism in the Dall porpoise, *J. Mammal.* **54(4)**:977–982.

Nemoto, T., 1956, On the diatoms of the skin film of whales in the northern Pacific, *Sci. Rep. Whales Res. Inst.* **11**:90–132.

Nemoto, T., 1958, *Cocconeis* diatoms infested on whales in the Antarctic, *Sci. Rep. Whales Res. Inst.* **13**:185–191.

Nishiwaki, M., 1966a, Distribution and migration of marine mammals in the north Pacific area, Symposium No. 4, 11th Pacific Sci. Congress, pp. 40–41.

Nishiwaki, M., 1966b, A discussion of rarities among the smaller cetaceans caught in Japanese waters, in: *Whales, Dolphins and Porpoises* (K. S. Norris, ed.), pp. 192–204, University of California Press, Berkeley.

Nishiwaki, M., 1967, Distribution and migration of marine mammals in the north Pacific area, *Bull. Ocean Res. Univ. Tokyo* **1**:1–64.

Nishiwaki, M., 1972, General biology, in: *Mammals of the Sea, Biology and Medicine* (S. H. Ridgway, ed.), pp. 3–204, Charles C. Thomas, Springfield, Illinois.

Nishiwaki, M., and Kureha, K., 1975, Strange organ in the anal region of the finless porpoise, *Sci. Rep. Whales Res. Inst.* **27**: 139–140.

Norris, K. S., 1961, Committee on Marine Mammals. Standardized methods for measuring and recording data on the smaller cetaceans, *J. Mammal.* **42(4)**:471–476.

Norris, K. S., and Prescott, J. H., 1961, Observations on Pacific cetaceans of Californian and Mexican waters, *Univ. Calif. Publ. Zool.* **63(4)**:291–402.

Ohsumi, S., 1969, Occurrence and rupture of vaginal band in the fin, sei and blue whales, *Sci. Rep. Whales Res. Inst.* **21**:85–94.

Orr, R. T., 1970, *Animals in Migration*, London: Collier–Macmillan, 303 pp.

Perrin, W. F., 1970, Color pattern of the eastern Pacific spotted porpoise *Stenella graffmani lönnberg* (Cetacea, Delphinidae). *Zoologica, N.Y.* **54**:135–142.

Perrin, W. F., 1972, Color patterns of spinner porpoises (*Stenella* cf. *S. longirostris*) of the eastern Pacific and Hawaii, with comments on delphinid pigmentation, *Fish. Bull.* **70**:983–1003.

Perrin, W. F., and Roberts, E. L., 1972, Organ weights of non-captive porpoise (*Stenella* spp.) *Bull. South. Calif. Acad. Sci.* **71(1)**:19–32.

Pike, G. C., and MacAskie, I. B., 1969, Marine mammals of British Columbia, *Fish. Res. Bd. Can. Bull.* No. 171:1–54.

Rice, D. W., 1968, Stomach contents and feeding behavior of killer whales in the eastern north Pacific, *Nor. Hvalfangst-Tid.* **57(2)**:97–109.

Rice, D. W., and Scheffer, V. B., 1968, A List of the Marine Mammals of the World, U.S. Fish Wildlife Service Spec. Sci. Report—Fisheries No. 579:1–16.

Ridgway, S. H., 1966, Dall porpoise, *Phocoenoides dalli* (True): Observations in captivity and at sea, *Nor. Hvalfangst-Tid.* **55(5)**:97–109.

Ridgway, S. H., 1972, Homeostasis in the marine environment, in: *Mammals of the Sea, Biology and Medicine* (S. H. Ridgway, ed.), pp. 590–747, Charles C. Thomas, Springfield, Illinois.

Ridgway, S. H., and Green, R., 1967, Evidence for a sexual rhythm in male porpoises, *Nor. Hvalfangst-Tid.* **57(1)**:1–8.

Ridgway, S. G., and Johnston, D. G., 1966, Blood oxygen and ecology of porpoises of three genera, *Science* **151**(3709):456–457.

Scheffer, V. B., 1949, The Dall porpoise (*Phocoenoides dalli*) in Alaska, *J. Mammal.* **30**(2):116–121.

Scheffer, V. B., 1953, Measurements and stomach contents of eleven delphinids from the northeast Pacific, *Murrelet* **34**:27–30.

Schevill, W. E., Watkins, W. A., and Ray, C., 1969, Click structure in the porpoise, *Phocoena phocoena, J. Mammal.* **50**(4):721–728.

Scott, J. P., 1958, *Animal Behavior*, 281 pp., University of Chicago Press, Chicago.

Slijper, E. J., 1966, Functional morphology of the reproductive system in Cetacea, in: *Whales, Dolphins and Porpoises* (K. S. Norris, ed.), pp. 277–319, University of California Press, Berkeley.

True, F. W., 1885, On a new species of porpoise (*Phocoena dalli*) from Alaska, *Proc. U.S. Natl. Mus.* **8**:95–98.

Walker, W. A., 1975, Review of the live-capture fishery for smaller cetaceans taken in southern California waters for public display, 1966–73, *Fish. Res. Bd. Can.*, Special Issue — Review of Biology and Fisheries for Smaller Cetaceans, **32**:1197–1211.

Wilke, F., and Nicholson, A. J., 1958, Food of porpoises in the waters off Japan, *J. Mammal.* **39**(3):441–443.

Wilke, F., Taniwaki, T., and Kuroda, N., 1953, *Phocoenoides* and *Lagenorhynchus* in Japan with notes on hunting, *J. Mammal.* **34**(4):488–497.

Yablokov, A. V., 1961, The "sense of smell" of marine mammals, *Tr. Soveshch. Ikhtiol. Kom. Akad. Nauk S.S.S.R.* **12**:87–93 (original not seen by author).

Yablokov, A. V., 1963, O typakh okraski kitoobraxnydh, *Byull. Mosk. O. Ispyt. Prir. Otd. Biol.* **68**(6):27–41. (Types of colour of the Cetacea, Bull. Moscow Soc. Nat. Biol. Dep. Fish. Res. Board Transl. Ser. No. 1239).

Yocum, C. F., 1946, Notes on the Dall porpoise off California, *J. Mammal.* **24**(4):364.

Chapter 4

THE NORTHERN RIGHT WHALE DOLPHIN
Lissodelphis borealis PEALE IN THE
EASTERN NORTH PACIFIC

Stephen Leatherwood

Bioanalysis Group
Naval Ocean Systems Center
San Diego, California 92152

and

William A. Walker

Section of Mammalogy
Natural History Museum of Los Angeles County
Los Angeles, California 90007

I. INTRODUCTION

The northern right whale dolphin, *Lissodelphis borealis* (Peale), is one of the least known delphinids of the eastern Pacific. Since originally described (Peale, 1848), there have been only 32 additional records from the eastern Pacific added to the literature (Table I). These published accounts represent 16 sightings at sea, ten beach strandings, and six collections. Very little has been published on the biology, behavior, and distribution of this species.

This paper presents new, unpublished data gathered during the period from 1956 through 1976. Although these data do not permit an exhaustive treatment of the species, we consider it important to bring the status of knowledge of this species up to date.

Table I. Published Records of *Lissodelphis borealis* in the Eastern North Pacific

Source	Date	Location	Remarks
Peale (1848)	1848	46°06′50″N, 134°05′W	Type specimen, 122.0 cm collected by Peale
True (1889)	October 1868	200 mi off Cape Mendocino, Calif.	246.4-cm male collected by Dall
Scheffer and Slipp (1948)	October 15, 1939	Cohassett Beach, Wash.	Beach stranding, 213.0-cm male
	March 11, 1940	Ocean Beach, Copalis, Wash.	Beach stranding, 207.8-cm male
McGary and Graham (1960)	August 18, 1958	42°47′N, 175°08′E	Animal entangled in net
Norris and Prescott (1961)	June 12, 1948	Dillon Beach, San Francisco, Calif.	Beach stranding, 203.2-cm male collected by Orr
	Winter 1957	Between Anacapa and Santa Barbara Islands, Calif.	Sighting
	January 25, 1960	8 mi S Long Point, Catalina Island, Calif.	Sighting 90 animals
	January 27, 1960	Off E end Catalina Island, Calif.	Sighting 20 animals
	February 17, 1960	4 mi E Long Point, Catalina Island, Calif.	Sighting
	April 18, 1960	4 mi NNW Long Point, Catalina Island, Calif.	Sighting approx 60 animals
	May 5, 1960	San Pedro Channel, Calif.	Sighting 30–32 animals
	November 7, 1960	Manhattan Beach, Los Angeles County, Calif.	Beach stranding
Brownell (1964)	July 14, 1963	37°43′N, 123°40′W	Sighting two groups, 25 animals

The Northern Right Whale Dolphin

Reference	Date	Location	Remarks
Fiscus and Niggol (1965)	March 13, 1958	38°30'N, 124°15'W	Sighting (several animals)
	February 3, 1959	35°17'N, 121°44'W	Sighting 100 ± animals
	February 4, 1959	35°12'N, 122°04'W	Sighting 100 ± animals
	February 5, 1959	35°01'N, 121°33'W	Sighting 100 ± animals
	January 18, 1961	35°17'N, 122°06'W	Sighting 25 animals
	February 10, 1961	35°27'N, 121°48'W	Sighting 20 animals
Daugherty (1965)	1956	Torrey Pines, San Diego, Calif.	Beach stranding
Fitch and Brownell (1968)	November 8, 1967	Imperial Beach, Calif.	Beach stranding collected by Hubbs; stomach content analysis
Pike and MacAskie (1969)	July 2, 1959	50°N, 145°W	Sighting 2 animals
Wick (1969)	October 6, 1967	Cape Kiwanda, Ore.	Beach stranding, 160.0-cm female
Guiget and Schick (1970)	February 13, 1970	48°23'N, 126°52'W	203.2-cm female collected by MacAskie
Roest (1970)	July 14, 1966	Pismo Beach, Calif.	Beach stranding, 156.0-cm male
Harrison et al. (1972)	February 29, 1969	Catalina Channel, Los Angeles County, Calif.	212.0-cm male captured alive at sea; reproductive comments on 3 male specimens
Leatherwood et al. (1972)	1968–1971	Off San Clemente Island, Calif.	Sighting(s)
Schroeder et al. (1973)	1970–1973	Los Angeles County, Calif.	Parasitism of 2 beach-stranded specimens
Leatherwood (1974a)	January 19, 1972	32°05'N, 118°10'W	Sighting 1000 animals
Walker (1975)	February 27, 1972	Catalina Island, Los Angeles County, Calif.	217.0-cm male captured alive at sea; comments on biology and behavior

II. DISTRIBUTION, SEASONAL MOVEMENT, AND ABUNDANCE

The ranges, seasonal movements, and numbers of *L. borealis* are still rather poorly known. In the western North Pacific, the species has been reported to occur as far south as Cape Nojima (Nishiwaki, 1966), the northern half of Honshu Island (Kasuya, 1971), and Cape Inubo, Japan (Nishiwaki, 1972); and as far north as the southern Kuril Islands (Klumov, 1959) and Paramushir Island, approximately 51°N (Sleptosov, 1952). Within that broad range, Okada and Hanaoka (1940) reported that the species was "relatively" common in the northern Sea of Japan; Ohsumi (1972) reports that *L. borealis* is "very common off the Pacific coast of Japan, especially northwards"; and based on the taking of five specimens during experimental purse operations off the Kurils, Klumov (1959) concluded that the species could be commercially fished. *L. borealis* has historically been fished off much of the Japanese coast (Gilmore, 1951; Hawley, 1958–1960; Ohsumi, 1972). The history of the fishery for the species is summarized by Mitchell (1975).

Nishiwaki (1966) reported that from their population centers off Japan the "distribution of *L. borealis* gradually faded E.N.E.," although where they go when they leave Japanese waters is unknown. He added that there were no reports of *L. borealis* from Japanese fishing boats working the Bering Sea. Kasuya (1971) reported that, like *Phocoenoides dalli*, *L. borealis* off Japan are colder-water species and occur off the Pacific coasts, to about 35°N, only between September and June. No other details of the species' movements in the western North Pacific have been described.

In the eastern north Pacific, *L. borealis* was first known from the type specimen collected 200 miles off northern Oregon (Peale, 1848). This and all other published records of sightings of live *L. borealis* in the eastern North Pacific are summarized in Table I. Several points are of interest regarding those records. Scammon (1874) reported that right whale dolphins occurred from San Diego as far north as the Bering Sea. Most of the subsequent general publications summarizing the distribution of cetaceans in the North Pacific continued to report this as its range (Miller and Kellogg, 1955; Hall and Kelson, 1959; Daugherty, 1965; Hershkowitz, 1966; Walker, 1968). The northernmost published record is that of two animals observed off British Columbia at 50°N, 145°W (Pike and MacAskie, 1969). The southernmost record is that from approximately 32°05'N, 118°10'W (Leatherwood, 1974a). Although there are records from British Columbia (one sighting), Washington (four collections), and Oregon (one collection), by far the majority of the eastern North Pacific reports are from California (14 sightings and seven collections).

In Table II are listed 175 previously unpublished sightings of *L. borealis*. The approximate locations of all these sightings are plotted by month in Figs. 1–4. Some of these records are simply incidental observations made by marine

Table II. Previously Unpublished Sightings of *Lissodelphis borealis* (1961–1976)

No.	Date	Location Latitude (°N)	Location Longitude (°W)	Number of animals	Source of data[a]
1	January 12, 1961	33°33.8'	118°26.8'	?	5f
2	April 10, 1964	42°44'	124°52'	100+	2a
3	October 23, 1964	49°48'	154°51'	5 ± 1	1b
4	April 16, 1965	37°53'	124°17'	10+	2a
5	April 16, 1965	37°52'	124°17'	10	5a
6	January 21, 1966	37°38'	123°06'	100	2a
7	January 24, 1966	36°23'	122°34'	1	2a
8	February 2, 1966	37°38'	123°06'	100	1b
9	February 2, 1966	33°05'	121°00'	200±	2a
10	February 2, 1966	30°00'	120°49'	40±	2a
11	February 23, 1966	38°00'	124°00'	300±	2a
12	February 26, 1966	40°00'	124°32'	2	2a
13	February 26, 1966	40°10'	124°40'	100±	2a
14	March 24, 1966	33°05'	121°00'	200	1b
15	March 23, 1967	32°21'	119°26'	1	3
16	March 23, 1967	32°04'	119°41'	6	3
17	March 23, 1967	32°00'	119°47'	3	3
18	April 3, 1967	34°35'	126°21'	1250	3
19	April 3, 1967	34°29'	126°22'	150	3
20	June 30, 1967	32°59'	119°59'	20	3
21	September 16, 1967	38°43'	124°47'	100±	2a
22	September 17, 1967	37°48'	123°39'	25±	2a
23	September 17, 1967	37°51'	123°36'	11 ± 1	2a
24	September 17, 1967	37°53'	123°33'	250 ± 50	2a
25	September 17, 1967	37°47'	123°39'	25	1b
26	September 17, 1967	37°51'	123°35'	10	1b

(Continued)

Table II. *(Continued)*

No.	Date	Location Latitude (°N)	Longitude (°W)	Number of animals	Source of data[a]
27	September 18, 1967	37°19′	123°42′	3	2a
28	October 17, 1967	35°00′	122°43′	35	3
29	October 17, 1967	35°00′	122°58′	100	3
30	October 31, 1967	35°34′	122°09′	200	3
31	October 31, 1967	35°45′	122°15′	25	3
32	November 16, 1967	35°03′	123°14′	150	3
33	November 16, 1967	35°01′	123°07′	100	3
34	January 4, 1968	34°47′	121°39′	800	3
35	March 28, 1968	32°15′	120°05′	500	3
36	April 7, 1968	33°01′	118°33′	200	1a
37	April 7, 1968	32°36′	118°50′	25	1a
38	July 31, 1968	38°55′	124°55′	~500	1b
39	December 31, 1968	33°55′	119°55′	50	1b
40	February 7, 1969	33°49′	118°42′	1	5d
41	March 9, 1969	32°40′	118°25′	1	2a
42	February 4, 1970	32°35′	118°05′	?	1b
43	February 13, 1970	48°23′	126°52′	?	
44	August 20, 1970	32°08′	121°03′	55 ± 5	2b
45	April 7, 1971	32°54′	117°15′	1	1a
46	April 7, 1971	33°27′	118°40′	150	1a
47	May 7, 1971	37°52′	124°58′	150	1a
48	October 8, 1971	35°55′	124°45′	125	1a
49	November 8, 1971	32°34′	118°42′	70 ± 10	1a
50	December 14, 1971	32°08′	118°30′	30–35	1a
51	December 14, 1971	32°09′	118°30′	500 ± 50	1a
52	December 30, 1971	32°37′	117°55′	125 ± 25	1a

The Northern Right Whale Dolphin

53	December 30, 1971	32°39'	118°11'	125 ± 25	1b
54	December 30, 1971	32°32'	118°19'	7–10	1a
55	December 30, 1971	32°41'	118°28'	80 ± 10	1a
56	December 30, 1971	32°36'	118°18'	20+	1a
57	December 30, 1971	32°29'	118°36'	20	1a
58	December 30, 1971	32°31'	118°02'	8 ± 2	1a
59	December 30, 1971	32°27'	118°26'	80 ± 15	1a
60	December 30, 1971	32°28'	117°33'	150+	1a
61	January 7, 1972	32°24'	117°58'	8–10	1a
62	January 7, 1972	32°20'	117°56'	9	1a
63	January 7, 1972	32°13'	117°56'	1000+	1a
64	January 7, 1972	32°17'	118°31'	30	1a
65	January 7, 1972	32°12'	117°58'	45	1a
66	January 7, 1972	32°12'	117°56'	10	1a
67	January 7, 1972	32°47'	119°21'	2000+	1a
68	January 7, 1972	32°49'	119°23'	400	1a
69	January 7, 1972	32°57'	119°16'	5	1a
70	January 7, 1972	32°57'	119°23'	10	1a
71	January 25, 1972	31°02'	119°16'	100 ± 20	1a
72	February 3, 1972	31°30'	119°00'	450	1a
73	February 3, 1972	31°33'	118°26'	30–35	1a
74	February 3, 1972	31°43'	118°28'	35	1a
75	February 8, 1972	32°05'	118°27'	10	1a
76	February 8, 1972	32°08'	118°28'	8	1a
77	March 2, 1972	32°40'	119°20'	20	1a
78	March 2, 1972	33°07'	118°44'	20	1a
79	March 2, 1972	33°18'	118°45'	150–175	1a
80	March 2, 1972	33°20'	118°43'	200+	1a
81	March 5, 1972	32°28'	118°49'	75–100	1b
82	April 6, 1972	32°53'	118°25'	1	1a
83	April 6, 1972	33°01'	118°27'	1	1a
84	April 6, 1972	32°55'	118°26'	6	1a
85	April 6, 1972	32°55'	118°27'	1	1a

(Continued)

Table II. *(Continued)*

No.	Date	Location Latitude (°N)	Longitude (°W)	Number of animals	Source of data[a]
86	April 6, 1972	34°27'	118°30'	45	1a
87	April 27, 1972	34°27'	120°39'	45 ± 5	1a
88	April 27, 1972	35°23'	121°27'	200	1a
89	April 27, 1972	35°28'	121°12'	200+	1a
90	May 2, 1972	32°42'	119°09'	~200	1a
91	June 30, 1972	31°30'	121°00'	25	2b
92	July 30, 1972	31°30'	121°00'	25	2b
93	August 27, 1972	34°25'	122°05'	3	2b
94	September 21, 1972	37°30'	122°30'	~50	5g
95	October 17, 1972	32°55'	118°27'	150	1a
96	January 23, 1973	33°10'	118°25'	12	1a
97	January 24, 1973	36°37'	122°16'	12	5h
98	January 24, 1973	32°5'	117°56'	40	1a
99	February 6, 1973	32°58'	118°50'	275 ± 25	1a
100	February 8, 1973	32°58'	118°52'	1200	1a
101	February 17, 1973	29°24'	117°05'	5	2b
102	March 6, 1973	32°31'	118°02'	200	1a
103	March 14, 1973	36°57'	122°21'	18 ± 2	4
104	March 19, 1973	33°11'	121°41'	6	1b
105	April 6, 1973	33°30'	118°14'	2	1b
106	April 6, 1973	33°28'	118°28'	1	1b
107	April 6, 1973	33°28'	118°16'	400	1b
108	April 6, 1973	33°30'	118°30'	40	1b
109	October 12, 1973	39°24'	124°05'	7 ± 1	1b
110	December 23, 1973	30°31'	116°42'	35 ± 5	2b
111	January 9, 1974	32°29'	117°23'	35 ± 5	1a

The Northern Right Whale Dolphin

112	January 10, 1974	32°41'	118°32'	125–150	1a
113	January 10, 1974	32°47'	118°18'	45 ± 5	1a
114	January 11, 1974	33°01'	118°26'	50	1a
115	January 12, 1974	33°10'	118°45'	7	1a
116	January 12, 1974	33°12'	119°05'	30	1a
117	January 13, 1974	33°03'	118°28'	85–100	1a
118	January 16, 1974	33°01'	118°29'	25	1a
119	January 17, 1974	32°47'	118°16'	50	1a
120	January 29, 1974	32°45'	117°55'	12	5a
121	April 21, 1974	32°58'	118°31'	1	1b
122	April 23, 1974	32°53'	118°21'	2	1b
123	September 1, 1974	44°53'	125°16'	65 ± 10	2b
124	September 20, 1974	46°56'	125°42'	40 ± 10	2b
125	September 24, 1974	45°26'	125°32'	4 ± 2	2b
126	September 26, 1974	40°23'	125°11'	6 ± 1	2b
127	September 28, 1974	37°00'	124°31'	50 ± 15	2b
128	October 3, 1974	44°32'	124°59'	7 ± 3	2b
129	October 3, 1974	44°20'	124°59'	7 ± 2	2b
130	October 4, 1974	38°48'	124°00'	4	2b
131	October 4, 1974	37°47'	124°00'	6 ± 1	2b
132	October 4, 1974	37°55'	124°00'	6	2b
133	October 21, 1974	44°24'	124°59'	4	2b
134	October 21, 1974	44°25'	124°59'	6	2b
135	January 23, 1975	33°10'	118°25'	12	5c
136	February 5, 1975	32°54'	118°35'	55 ± 5	1a
137	May 22, 1975	33°06'	118°56'	1000+	1a
138	May 27, 1975	33°45'	119°44'	1	4
139	May 27, 1975	33°47'	119°47'	150	4
140	August 31, 1975	41°58'	126°30'	250	2b
141	September 30, 1975	38°18'	125°40'	30 ± 5	2b
142	October 1, 1975	37°14'	124°57'	50 ± 10	2b
143	October 2, 1975	36°55'	124°10'	40 ± 10	2b
144	October 2, 1975	36°30'	123°30'	40 ± 10	2b

(Continued)

Table II. *(Continued)*

No.	Date	Location Latitude (°N)	Location Longitude (°W)	Number of animals	Source of data[a]
145	October 23, 1975	44°50'	125°10'	6 ± 2	2b
146	November 12, 1975	33°24'	120°23'	25	4
147	November 12, 1975	32°44'	119°29'	25	4
148	November 13, 1975	33°23'	119°45'	20	4
149	December 16, 1975	32°21'	118°03'	18	1a
150	December 16, 1975	33°07'	118°05'	47 ± 1	1a
151	December 16, 1975	33°20'	118°15'	1	4
152	December 17, 1975	33°14'	118°08'	24	1a
153	December 17, 1975	32°57'	119°00'	20	4
154	December 17, 1975	32°54'	119°00'	4	4
155	December 17, 1975	32°47'	119°00'	500	4
156	December 17, 1975	32°20'	118°52'	17	4
157	January 7, 1976	33°25'	119°22'	100	4
158	January 8, 1976	32°56'	118°47'	275	4
159	January 8, 1976	32°50'	117°42'	200	4
160	January 9, 1976	32°52'	118°05'	20	4
161	January 10, 1976	33°26'	117°53'	30	4
162	February 11, 1976	33°25'	119°24'	50	4
163	February 11, 1976	33°18'	118°09'	35	1a
164	February 13, 1976	32°42'	118°35'	20	4
165	February 14, 1976	33°19'	118°50'	16	4
166	February 14, 1976	33°13'	118°56'	25	4
167	February 14, 1976	33°02'	119°09°	100	4
168	February 14, 1976	32°20'	119°20'	30	4
169	February 15, 1976	32°52'	117°45'	15	4
170	March 23, 1976	33°33'	118°34'	130	1a

171	March 23, 1976	32°32'	118°33'	4	1a
172	March 23, 1976	33°19'	118°27'	2	1a
173	March 23, 1976	33°11'	117°50'	30	1a
174	March 24, 1976	33°02'	118°06'	200 ± 20	1a
175	March 24, 1976	32°49'	117°39'	150 ± 20	1a

[a] Sources of data: 1. Naval Undersea Center, San Diego, Calif: a. Aerial Survey Records; b. Ship survey records. 2. National Marine Fisheries Service: a. Northwest Fisheries Center, Marine Mammal Division, Seattle, Washington (C.H. Fiscus); b. Southwest Fisheries Center, La Jolla, Calif. (W.F. Perrin). 3. U.S. National Museum, Smithsonian Institution, Pacific Pelagic Bird Research Program (R.L. Brownell and R.L. Delong). 4. University of California, Santa Cruz, Calif. (K.S. Norris, T. Dohl, J.D. Hall, L. Hobbs). 5. Incidental Records of: a. Scripps Institution of Oceanography (C.L. Hubbs); b. San Diego Museum of Natural History (R.M. Gilmore); c. United States Coast Guard, Sea Surface Temperature surveys (J.D. Hall); d. Marineland of the Pacific; e. British Columbia Provincial Museum (C. Guiguet); f. Lockheed Corporation (W.E. Evans); g. San Jose State College (G.V. Morejohn); h. U.S. Naval Post Graduate School, Monterey; i. Humboldt State College (W.J. Houck).

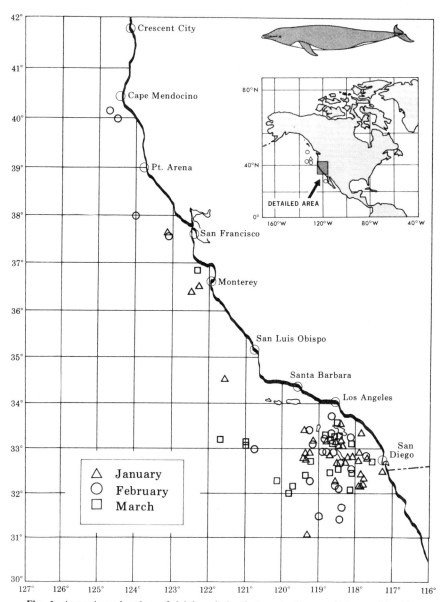

Fig. 1. Approximate locations of sightings during first quarter (January, February, March).

The Northern Right Whale Dolphin 97

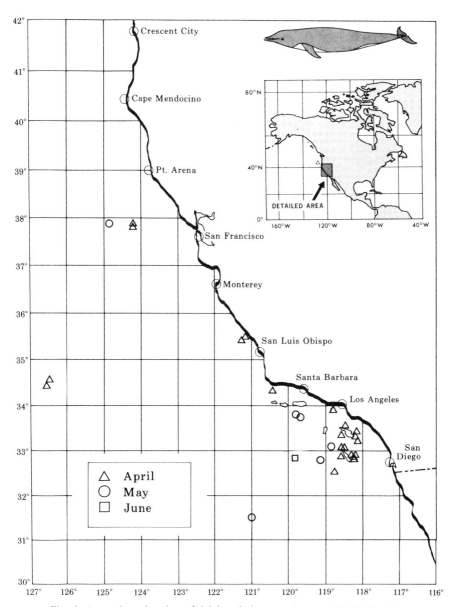

Fig. 2. Approximate locations of sightings during second quarter (April, May, June).

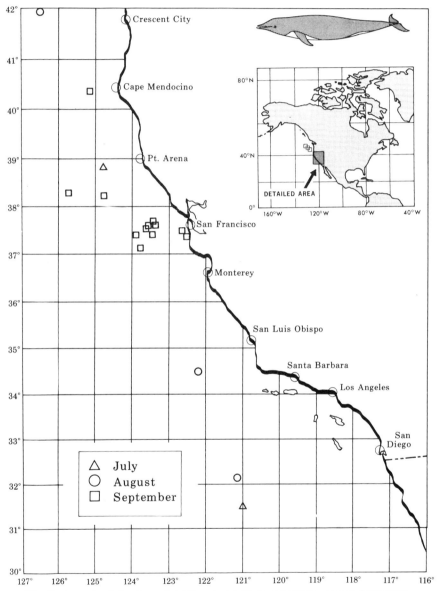

Fig. 3. Approximate locations of sightings during third quarter (July, August, September).

The Northern Right Whale Dolphin

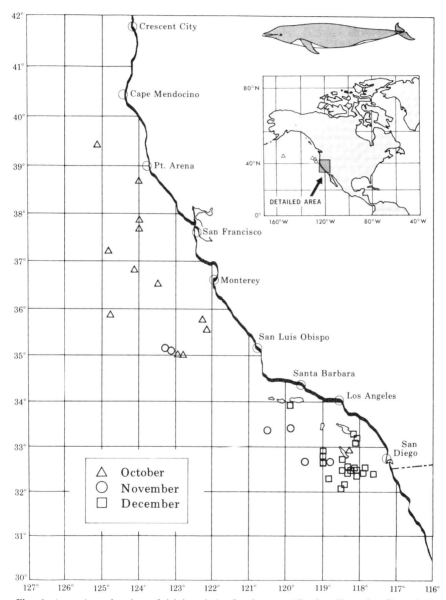

Fig. 4. Approximate locations of sightings during fourth quarter (October, November, December).

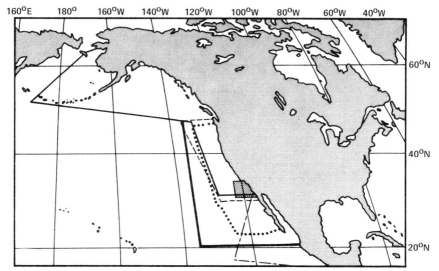

Fig. 5. Approximate areas surveyed by the major marine mammal survey programs in the eastern Pacific (1958–1976). ———, Smithsonian Institution, Pacific Ocean Biological Survey (1967–1968). National Marine Fisheries Service, Northwest Fisheries Center: ———, Pelagic Fur Seal Investigations (1958–1974); . . . . , Other Cruises (1958–1967). National Marine Fisheries Service, Southwest Fisheries Center: —·—·—, EASTROPAC Tuna Boat Observer Program (1970–1976); – – – – –, Albacore Boat Observer Program (1972–1976). ▓▓▓ , Coastal Marine Laboratory, University of California, Santa Cruz, Bureau of Land Management Survey Program (1975–1976).

mammalogists during infrequent cruises, but many others resulted from eight extensive survey programs of various institutions. Although we have sufficient data from only one of these programs to quantitatively treat sightings with level of effort, the levels of effort for the remaining seven can be subjectively described as follows (the approximate areas surveyed by each of these major programs are shown in Fig. 5).

A. Smithsonian Institution

During 1967 and 1968, the Pacific Ocean Biological survey program conducted at least 45 ship survey cruises in various portions of the North Pacific. Of those cruises, 19 surveyed some portion of the area from the Mexican border to the waters of British Columbia. Surveys covering some portion of the waters from 25°N to 40°N and up to 400 miles seaward were conducted at least once every quarter and almost every month of the year (R.L. Brownell, 1976, personal communication).

In all, observers in this program reported 14 encounters with *L. borealis* in the area from 32°N to 36°N, all during the period from November through May.

B. National Marine Fisheries Service (NMFS)

1. Northwest Fisheries Center (NFC)

In the years from 1958 through 1974 (the latest date for which data were available for examination), the pelagic fur seal program conducted extensive ship surveys in the northeastern Pacific from the latitude of southern Washington (about 49°30′N), north and east to the Canadian and Alaskan coasts, northwest to the western Aleutians, and into the Bering Sea. Observers maintained continuous daylight watches, weather permitting, and logged all marine mammal sightings. During those 16 years, surveys were conducted at least once each month. Despite the hundreds of ship survey days, there were no sightings of *L. borealis* in any of these areas (Hiroshi Kajimura, NMFS, Seattle, 1975, personal communication).

From 1958 through 1967, NMFS also conducted numerous ship surveys within 150 miles of the coast from Washington south to the tip of Baja California, studying seals and cetaceans. Observers maintained continuous daylight watches whenever weather permitted and logged all marine mammal sightings. Cruises were conducted at least once during every quarter, although not during every month of the year. General descriptions of these cruises have been provided by Rice (1963*a*, 1974). Records of marine mammal sightings through 1961, including six observations of *L. borealis*, were reported by Fiscus and Niggol (1965). Records from one additional cruise in 1963 which resulted in numerous sightings of other cetaceans but none of *L. borealis* are reported by Rice (1963b). Fifteen additional records of *L. borealis* from the years 1962 through 1967, one from the Oregon–California border, the remainder from California, are included in this paper.

2. Southwest Fisheries Center (SFC)

From 1970 through 1976, observers placed on board selected tuna boats have logged marine mammal sightings on numerous cruises from San Diego, California, to and from the tropical Pacific yellowfin and skipjack tuna fishing grounds. Despite extensive survey effort, primarily during the first six months of the year, observers reported no sightings of *L. borealis*.

From 1972 through 1975, observers placed on board albacore fishing boats to monitor fishing and to measure oceanographic conditions in productive fishing grounds maintained logs of marine mammal sightings, weather and time permitting, in the waters between about 28°N and about 48°N. The eight vessels for which observer data were available generally began fishing in June or July off northwestern Baja California or southern California and continued to move northward with the fishing through mid- to late October. Twenty-one observations of *L. borealis* resulting from this program, increasing in numbers as one moves north and later in the season, are included in this paper.

C. University of California

During 1975 and 1976, the Coastal Marine Laboratory, University of California, Santa Cruz, California, conducted a series of aerial and shipboard surveys off southern California specifically designed to determine densities and seasonal changes in abundance of cetaceans. Aircraft have been utilized primarily for line-transect censuses, ships for more detailed observations, for capture and tagging, and for environmental sampling (K.S. Norris and T. Dohl, 1976, personal communication). Surveys have been evenly spaced throughout the year to ensure approximately equal coverage. Through February 1976, 18 observations of *L. borealis* were reported, all except one from aircraft and all restricted to the months from November through May.

D. Naval Undersea Center

Ship Surveys. Records of cetacean sightings were available from 20 NUC ship survey cruises from 1964 through 1975, covering the area from Point Conception south to the tip of Baja California and sampling each quarter of the year; and from San Diego to Kodiak, Alaska, in April. These surveys resulted in 16 observations of *L. borealis* from 30°31'N northward to 49°48'N, all in the period from October through May.

Although the results of these programs suggest the normal limits of distribution and some seasonal trends in movements of *L. borealis*, they cannot stand alone because of the absence of effort data.

Aerial Survey. From January 1968 through April 1976, aerial surveys were conducted from Point Conception south to 31°30'N and offshore to 121°W. Within that range, surveys were conducted most extensively in the area shown in Fig. 6. The level of effort for the larger area is summarized by month in Table III. Segments of surveys were not considered for analysis if sea surface or weather conditions prohibited observations.

The methods used during the study were summarized by Leatherwood (1974b). They were basically as follows: flights were made in naval H-3 helicopters and S-2 reconnaissance aircraft, in single and twin engine rental aircraft, and in the Goodyear blimp. Surveys were conducted at altitudes ranging from 150 to 300 m, depending on weather conditions. Detailed observations were often made from much lower altitudes, occasionally to as low as 15 m.

Specific flight paths were selected before each flight to ensure approximately equal coverage each month of inshore, offshore, northern, and southern sectors of the study area. These preplanned flight tracks were often modified in flight due to changes in weather, closure of specific areas due to military operations, and mechanical or communications failures aboard the aircraft.

Each time cetaceans were encountered, attempts were made to determine

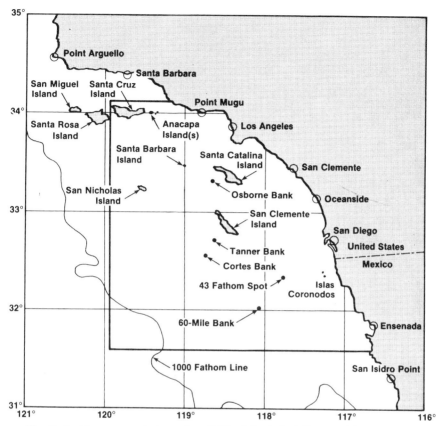

Fig. 6. Aerial survey area in the southern California borderland (January 1968–April 1976).

the species, sketch herd configuration before it was disturbed by the presence of the aircraft, estimate the numbers of individuals, record other marine mammals in association, and note as many other details of the animals' behavior as possible. Photographs were taken of herd configuration and behaviors of particular interest using 35 mm and 70 mm camera systems and lenses of various lengths.

During the eight years covered by the surveys, *L. borealis* were encountered 73 times and an estimated total of 10,834 animals were seen. The earliest observations were in October, the latest in May. Some encounters were recorded for each of the months in between. No sightings were made from June through September.

Because the amount of survey effort varied from month to month, indices of abundance were computed for each month by dividing both the numbers of encounters and the numbers of individuals each month by the area surveyed that

Table III. Aerial Survey Effort by Month for Southern California Study Area January 1968–April 1976

Method/month	July	Aug	Sept	Oct	Nov	Dec	Jan	Feb	Mar	Apr	May	June	Total
Helicopter													
Total nautical mi (L_1)	940	1741	935	1560	265	1835	1365	1530	945	1175	1580	940	14,811
Total sq mi ($L_1 \times 3$)	2820	5233	2805	4680	795	5505	4095	4590	2835	3525	4740	2820	44,433
Fixed wing a/c													
Total nautical mi (L_2)	1180	526	535	235	285	190	375	475	1700	3300	570	430	9,801
Total sq mi ($L_2 \times 3$)	3540	1578	1605	705	855	570	1125	1425	5100	9900	1710	1290	29,403
Goodyear blimp													
Total nautical mi (L_3)	0	0	0	0	0	0	150	0	0	0	0	0	150
Total sq mi ($L_3 \times 3$)	0	0	0	0	0	0	450	0	0	0	0	0	450
Total all methods													
Total nautical mi ($L_1+L_2+L_3$)	2120	2267	1470	1795	550	2025	1890	2005	2645	4475	2150	1390	24,762
Total sq mi [$3 \times (L_1+L_2+L_3)$]	6360	6801	4410	5385	1650	6075	5670	6015	7935	13,435	6450	4170	74,286
% of 20,000-nmi^2 study area covered	32	34	22	27	8	30	28	30	40	67	32	21	

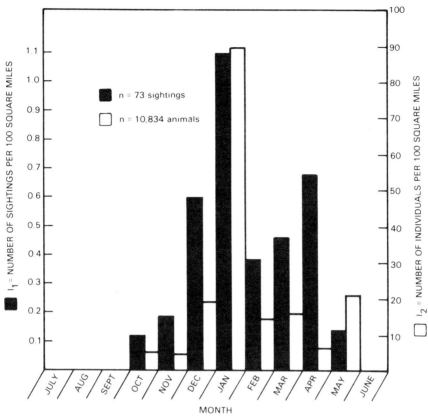

Fig. 7. Indices of abundance of *Lissodelphis borealis* from NUC aerial surveys (January 1968–1976).

month. The area surveyed (see Table III) was estimated by multiplying the total number of linear miles flown (L) by twice the effective perpendicular sighting distance (following Gates, 1969). Because perpendicular distances were not actually measured at the time of the sightings, we used the estimate of 1.5 miles per side as an average for all conditions during which surveys were conducted. Experience with aerial surveys of bottlenosed dolphins, *Tursiops truncatus,* in the northern Gulf of Mexico (Leatherwood and Platter, 1975) and *Delphinus delphis* in both the Black Sea (Zemsky and Yablokov, 1974) and off southern California (Evans, 1975) supported selection of this distance as a conservative average.

The indices of abundance (Fig. 7) show clearly that *L. borealis* are seasonal members of the marine vertebrate fauna of the southern California continental borderland. They appear to arrive there in the fall months, increase in numbers to mid winter, and then decrease towards the late spring and early summer.

Data from all sources combined tend to support this contention. Records from southern California are almost exclusively restricted to the months from October through May. The few sightings south of Point Conception during the summer occurred well off the continental shelf. The one sighting south of 30°N occurred during the periods of peak abundance off southern California.

Although the absence of quantitative data on levels of effort prohibit any detailed analysis, available data suggest an interesting similar trend off central and northern California. Although *L. borealis* have been reported off the continental shelf during almost every month of the year, they have been reported over the continental shelf only during the winter–spring months, the same period when they are found off southern California.

From the available data, *L. borealis* in the eastern North Pacific appear to be normally restricted to the temperate waters between about 30°N and 50°N. Within that range at least portions of the population appear to shift south and inshore during late fall through spring and north and offshore from spring through fall. It is not known how much of the open Pacific they inhabit, although the one published record from 42°47'N, 175°08'E (McGary and Graham, 1960) and the additional sighting at 49°48'N, 154°51'E (this paper) raise the question about continuation of the distribution across the cool currents of the North Pacific.

Although reasons for these apparent seasonal movements in *L. borealis* are not known, several possibilities are worth discussing.

Ekman (1953) suggested that the single most important environmental factor affecting the distribution of marine organisms, in general, is water temperature. Gaskin (1968) reported that the southern right whale dolphin, *Lissodelphis peronii,* had been observed in water ranging from 9°C to 16°C (48.2°F to 60.8°F). Nishiwaki (1966) reported that in the western Pacific, *L. borealis* "seems to frequent the cool temperate waters (around 15°C)." Norris and Prescott (1961) related the occurrence of Dall's porpoise in southern California to the cooling of the waters there to below 64°F and suggested that "the species moves, at its southern limit, in relation to maximum water temperature level." The southern records for *P. dalli* off Baja California were associated with unseasonably cold temperatures there (Leatherwood *et al.,* 1972).

The sightings of *L. borealis* reported in this paper occurred in water ranging in temperature from 46° to 66°F (7.8°C to 18.9°C). Those sightings off southern California for which water temperatures were directly measured at the time of sighting ranged from 54°F to 64°F (12.2°C to 17.8°). Although direct measurements were not obtained at the time of sighting for the records south of 31°N, all three occurred at a time and location when the CALCOFI temperature summaries for the area show extensions of cool waters (60°F; 15.6°C) well south of southern California. These data suggest that *L. borealis,* like *P. dalli,* begins to move into southern California waters in the fall when waters begin to cool below 66°F (18.9°C), venture further south only with intrusions of colder water, and depart when the spring/summer warming trend begins.

Townsend (1935) suggested that the distribution of whales is determined almost solely by movements of food and suitable conditions for reproduction. Tomlin (1960), Norris (1967), and Evans (1975) have all emphasized the close relationships between the distribution of delphinids and their potential food supplies.

The periods of peak abundance of *L. borealis* over the southern California continental borderland correspond very closely with known periods of peak abundance of squid, *Loligo opalescens,* which appears to be an important food item for this species.

Furthermore, during aerial surveys, the species has been seen most frequently around such prominent banks and sea mounts as the 43 Fathom Spot; Sixty Mile, Tanner, Cortes, and Osborne Banks; and San Juan Sea Mount (Fig. 6). These are all rich areas actively used by commercial fishermen and known to be frequented by at least one other delphinid, *Delphinus delphis* (Evans, 1971). During the winter, they support abundant squid populations (C. L. Hubbs, 1975, personal communication). *L. borealis* have also been observed off Pyramid Head, San Clemente Island, and off Catalina Island, perhaps the two most active squid fishing grounds in southern California.

Nishiwaki (1972), in the only published estimates of the size of the entire population of *L. borealis,* suggests that there are over 10,000 individuals. We have observed approximately 3400 individuals in a single survey flight from San Diego to Cortes Banks and back to San Diego by a different route and as many as 2000 individuals in a single aggregation. If one considers the period of peak abundance off southern California, during which there are estimated to be as many as 89 *L. borealis* per 100 square miles, and extrapolates to the total 20,000-square-mile study area (Fig. 6), one obtains an estimate of 17,800 *L. borealis*. Since the surveys were not conducted to produce population estimates and since the effects on survey results of placement of transect legs and aggregations of animals are not clearly known, the estimate cannot be statistically defended and must be regarded as tentative. Nevertheless, during the midwinter months, *L. borealis* appears to be at least the third, if not the second most abundant delphinid, behind *Delphinus delphis* (Evans, 1975) and perhaps *L. obliquidens* (Leatherwood, unpublished data). In the light of these tentative calculations, the estimate of the North Pacific population offered by Nishiwaki (1972) appears overly conservative.

III. HERD SIZES

Nishiwaki (1972) reported that *L. borealis* "generally form schools of 200 individuals or more." Estimates of herd size were recorded for 12 of the 15 published sightings of *L. borealis* in the eastern North Pacific. These estimates

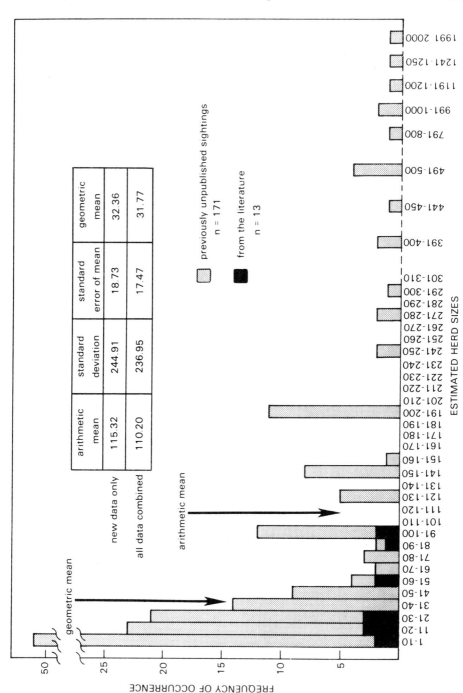

Fig. 8. Frequency of occurrence of herd sizes of *Lissodelphis borealis*.

range from two to 100 individuals about an arithmetic mean of 42.33 ($s_x = 35.59$, $s_{\bar{x}} = 10.27$) and a geometric mean of 27.76.

All observations of *L. borealis* in the eastern North Pacific for which estimates of herd sizes are available, including those 12 previously published records, are summarized by size classes in Fig. 8. These sightings range from one to 2000 animals about an arithmetic mean of 110.2 individuals ($s_x = 236.95$, $s_{\bar{x}} = 17.47$) and a geometric mean of 31.77. Although the data suggests that groups in the eastern North Pacific are smaller than those reported by Nishiwaki (1972) for the species in general, this apparent difference should be tempered by several considerations.

First, it should be noted that each "sighting" was treated as a separate "herd" even if it contained only a single animal. The potential for long-range acoustic contact among cetaceans, as discussed by Payne and Webb (1971), means that a herd could reasonably include animals not visible to the observer at the time of the encounter. Further, any general discussion of delphinid herd sizes should recognize that herds of some species, for example *Delphinus delphis* in the eastern Pacific, have been shown to fragment into small groups at night to feed and then to reassemble into very large herds in the period approaching midday (Evans, 1971). Data available for this study were insufficient to permit determination of the effects on herd size of such variables as time of year, time of day, herd behavior, or bottom topography at the location of the encounter; however, it can be demonstrated that there were no significant differences between herd sizes north of Point Conception and those south of Point Conception or between those off the continental shelf and those over the continental shelf (Mann Whitney U tests all at levels of significance of 0.05).

IV. HERD CONFIGURATION

As used in this chapter, the term "herd configuration" refers to groupings and gross spatial relationships among members of a "herd." This important aspect of the behavior of wild delphinids, variously referred to as formation, geometric formation, group formation, and array, has been discussed for various species by Norris and Prescott (1961), Evans and Bastian (1969), Pilleri and Knuckey (1969), Tayler and Saayman (1972), and Leatherwood (1978). Similarities of these formations in dolphins to those reported for ungulates and some possible functions of the formations are discussed by Wilson (1975).

Configuration is a dynamic feature of a herd of dolphins. During the NUC aerial surveys, the formation of a herd was sketched and described as it appeared when the herd was first encountered. In numerous cases, the aircraft had no noticeable effect and the herd remained in the same pattern throughout the time observed. It was found that when herd configurations could be determined,

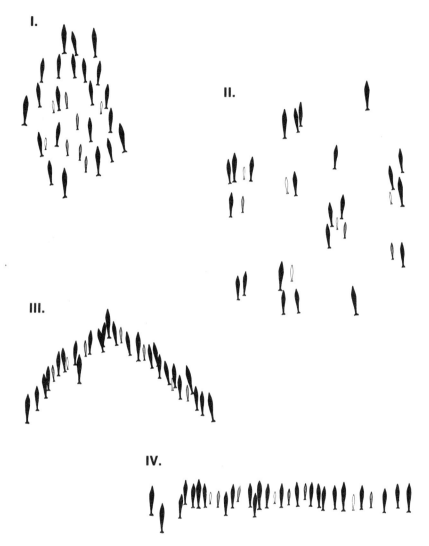

Fig. 9. Basic herd configurations of northern right whale dolphins, *Lissodelphis borealis*.

undisturbed *L. borealis* groups generally occurred in one of four distinct patterns (Fig. 9).

Type I. Tightly packed groups with no clearly identifiable subgroups. Although the outer boundaries of this group appeared dynamic and individuals moved and surfaced independently of each other, all animals appeared to be moving in the same general direction.

Type II. Groups consisting of scattered and distinctly separate subgroups of various sizes. Like type I, all animals appeared to be moving in the same general direction; however, members of each subgroup tended to swim and surface in unison.

Type III. V-shaped formations; lines of animals arranged in a V formation with the vertex of the V in the forward position of the moving group. In V formations, all members of the group appeared to maintain the same positions relative to each other, but seemed to move and surface independently of each other.

Type IV. "Chorus lines." Perhaps only a slight modification of type III, type IV is a configuration in which the animals in a group swam nearly abreast in the same positions relative to each other, surfacing independently.

Of the 70 aerial observations of *L. borealis* during the NUC study, configuration was noted for 37 of the encounters: 14 groups had assumed type I; 18 groups type II; nine groups type III; and seven groups type IV (Fig. 8). Thirty-three of the groups were either not readily classifiable or their configuration was not noted.

Over-all, the data were insufficient to permit any quantitative treatment of the formations by time, position, or number of animals; however, two patterns were clear: (1) *L. borealis* maintained small, tightly packed independent groups when in the company of other delphinid species, with the exception of *Lagenorhynchus obliquidens,* with whom they sometimes formed heterogeneous schools (see p. 112). (2) When *L. borealis* were obviously disturbed by aircraft, they grouped tightly into pattern type I and fled.

V. INTERSPECIFIC ASSOCIATIONS

L. borealis has been reported in the company of eight species of cetaceans (five odontocete and three mysticete) and one species of pinniped (Table IV). Observations included in this paper augment this list of occurrences with two additional odontocete and two additional mysticete species. The frequencies of occurrence of *L. borealis* with other marine mammals in any combination are shown in Fig. 10. Clearly, regional and seasonal variations in abundance of preferred prey species will often draw many different animals to the same relatively small areas and make coincidental observations of several species likely; however, the associations treated in the figure include only those observations in which the species were obviously moving together or were otherwise interacting. Interspecific associations have been reported in widespread locations and among a variety of other marine mammal species (Brown and Norris, 1958; Norris and Prescott, 1961; Fiscus and Niggol, 1965; Kraus and Gihr, 1971; Leatherwood *et al.*, 1972; Perrin, 1975).

Table IV. Published Associations between *Lissodelphis borealis* and Other Species of Marine Mammals

Species	Norris and Prescott (1961)	Fiscus and Niggol (1965)	Brownell (1964)	Leatherwood (1974a)
Globicephala macrorhynchus	X	X		
Grampus griseus		X		
Lagenorhynchus obliquidens	X	X		X
Phocoenoides dalli	X			X
Delphinus delphis				X
Balaenoptera borealis			X	
Megaptera novaengliae			X	
Eschrichtius robustus				X
Zalophus californianus				X

Several interesting behavioral details on *L. borealis* associations have been observed. Leatherwood (1974a) reported *L. borealis* riding pressure waves of *Eschrichtius robustus*. During this study, *L. borealis* were observed riding pressure waves of *Balaenoptera physalus* and clustering about resting *E. robustus*. When *L. borealis* occurred with other small odontocetes, the behavior varied. When they were with any species other than *L. obliquidens*, *L. borealis* maintained small, cohesive subgroups separate from the other species and did not seem to freely intermix with them. When in the company of *L. obliquidens*, they more frequently formed heterogeneous herds and subgroups and tended to bow-ride more readily. It is interesting to note that all four *L. borealis* captured alive for display or research purposes were in the company of *L. obliquidens* at the capture site (Walker, 1975; Morris F. Wintermantel, 1975, personal communication). *L. obliquidens* occur in abundance throughout the range of *L. borealis* in the eastern North Pacific (Leatherwood et al., 1972) and the two species probably have very frequent contact. Other possible reasons for the frequent and close associations between these two species are not known.

VI. SWIMMING BEHAVIOR

Leatherwood et al. (1972) reported that *L. borealis* are generally timid. While many species of small delphinids routinely ride the bow waves of moving vessels, *L. borealis* often remain away from or even actively avoid them. They rode the bow only 17 of the 42 times they were encountered by vessels during this study, although they were often actively pursued. The majority of herds encountered behaved similarly to those described in Norris and Prescott (1961) in that they used a combination of sounding, rapid swimming just below the surface, and fast surface swimming with much jumping to escape the vessel.

During the aerial survey, nine herds were observed actively moving away from vessels in the area.

The species' swimming behavior may be summarized as follows: slow-moving herds, usually exposing just the head and blowhole to breathe, create only slight surface disturbances and are difficult to detect, even under ideal sea conditions. Fast-moving groups, especially those fleeing, appear to use one of two modes of evasive behavior. The first consists of swimming just below the surface, surfacing quickly to breathe, and then resubmerging (Norris and Prescott, 1961; Nishiwaki, 1966). Herds demonstrating this mode of evasive behavior may appear even to a seasoned observer as a breezer of school fish or a herd of sea lions (Peale, 1848; Leatherwood *et al.*, 1972). The second method consists of a series of low-angle leaps described by Brownell (1964) and illustrated by Leatherwood *et al.* (1972). In addition to the low-angle leaps from which they smoothly reenter the water head first, *L. borealis* have been observed to belly

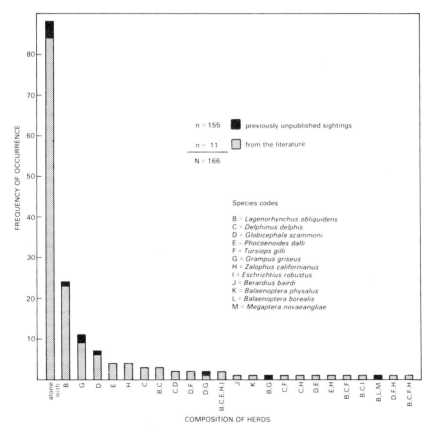

Fig. 10. Associations of *Lissodelphis borealis* with other marine mammals.

flop and side slap, sometimes repeating these jumps 25 times or more over a great distance, and to lobtail upon reentry. Herds traveling in this manner create noticeable surface disturbance (see Plate V in Kasuya, 1971).

If neither mode of evasive behavior is successful, *L. borealis* may sound, surface at some distance, regroup, and then begin running again. Although individuals in slow-moving groups have been observed to breathe at intervals of from 10 to 75 sec, entire herds have been observed to dive for up to 6 min and 15 sec.

Nishiwaki (1966) noted that *L. borealis* using the first escape mode were capable of swimming 20 mph (32 km/hr). Nishiwaki (1972) reported that groups of *L. borealis* move "very slowly," although individuals can swim "very fast," presumably as much as 35 km/hr. Two incidents during this study demonstrate both herd and individual speeds.

A herd of an estimated 150 animals encountered on April 7, 1971, at 33°27'N, 118°40'W was "running" from a large ship. Dropping a "sonobuoy" into the middle of the herd permitted marking a starting point. A second mark, dropped 26 min later, again into the middle of the group, indicated that the herd had traveled over 6.5 miles (10.5 km) at an average speed of about 15 nautical miles per hour.

In April 1971, an observer aboard a naval shore boat traveling at a known speed of 18 knots attempted to overtake a single *L. borealis* for tagging purposes. In the 10 min they were on the same course, the animal clearly outdistanced the vessel.

VII. SOUND PRODUCTION

Fish and Turl (1977) report on recordings made off the coast of southern California in 1975 of a herd of about 200 *L. borealis* in the presence of fin whales, *Balaenoptera physalus*. They reported that in the range 1–40 kHz, the limits of their recording equipment, the herd produced sounds with peak absolute source levels of 170 dB re 1 μPa per 80-Hz band at one meter. They also made the following general observations about the sound production of the species on this occasion: No whistles were recorded. The numerous frequency-modulated "whelps" and "moans" which appeared to the human ear as whistles turned out on later analysis at slower speeds to be a series of high-repetition-rate click trains from a number of animals (J.F. Fish, 1976, personal communication). Only three other recordings were available for even cursory examination.

On January 12, 1961, a research team from Lockheed Corporation recorded signals from a "small" herd of *L. borealis* at 33°33.8'N, 118°26.8'W. Their recording instrumentation consisted of an Ampex 601-2 tape recorder and an M-115 hydrophone, a system with relatively flat frequency response to about 18 kHz. One sample segment from those tapes is shown in Fig. 11A. The sounds of

The Northern Right Whale Dolphin

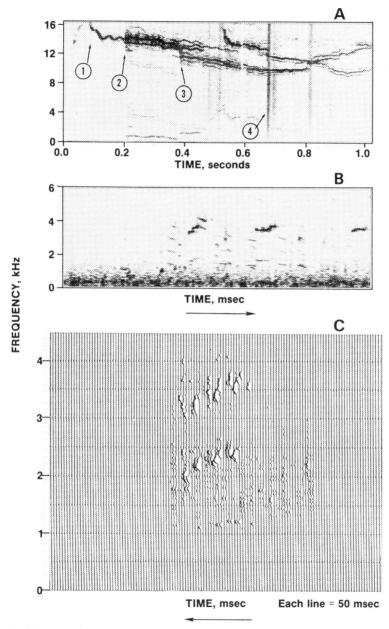

Fig. 11. Visual displays of components of sounds made by *Lissodelphis borealis*: (A) a sonogram of a series of high repetition click trains from a number of animals; (B) a sonogram of a characteristic whistle; and (C) a power spectrum of the same whistle.

L. borealis recorded on that date may be described, with reference to the sonogram, as follows: In general, the sounds were not essentially different from those recorded from ocean/pelagic delphinids of the *Delphinus/Stenella* types, although whistles were far less common in the *L. borealis*. The one whistle illustrated (Fig. 11A1), with energy extending above 16 kHz, blends with a series of rapid, short-duration pulses (Fig. 11A2 and 11A3), of the type described by Fish and Turl (1977). Analysis of these click trains using the method described by Watkins (1967) indicates that the clicks are produced at the rates of 700–1000 pulses/sec. These short-duration, high-repetition-rate pulses are very much like those produced by echolocating common dolphins, *Delphinus delphis,* and beluga whales, *Delphinapterus leucas* (Evans, 1976, personal communication). Pulses of broader spectrum, longer duration, and slower repetition rate were also detected (Fig. 11A4).

On April 7, 1971, and again on April 6, 1972, ANSQS-41 sonobuoys were dropped from naval S-2 reconnaissance aircraft directly in front of moving herds of *L. borealis* and the animals' sounds monitored and recorded. Because limitations of the on-board recording gear prevented any meaningful analysis of the spectral characteristics of the signals, only a few general remarks can be made about the species' sound production on these occasions. No whistles were detected. The only audible sounds from the group consisted of what were identified as echolocation trains, sounds which increased to a crescendo as the herd closed from 800 yards to within as little as a few yards of the hydrophones and then subsided rapidly as they passed. At the time the recordings were made, the clearest signals were noted to have occurred as individual animals were oriented directly towards the hydrophone.

On January 19, 1977, at 32°43′N, 118°45′W, a herd of an estimated 800 *L. borealis* was recorded using a modified ANSSQ41A sonobuoy, a model 42-AFM receiver, and a NAGRA recorder. The total system was sensitive to about 10 kHz. Although some faint echolocation trains were heard, the most prevalent sounds recorded were a series of what may be most accurately described as "clicks," low-frequency sounds consisting of four tones which ranged from 1100 to 4000 Hz, with centers at 1800 and 3000 Hz (Fig. 11B). The relative distribution of energy for those signals is shown in Fig. 11C (D.K. Ljungblad, 1977, personal communication).

VIII. STRANDINGS

Ten beach strandings of *L. borealis* have been reported in the literature (Table I). This paper includes records on 25 additional strandings (Table V). All strandings of this species, including those reported in the literature, were of individual animals. Fraser (1954) reported on a simultaneous stranding of three

Table V. Additional Records of Beach-Stranded Northern Right Whale Dolphins, *Lissodelphis borealis*

Date	Collection no.[a]	Sex	Length (cm)	Location	Source	Disposition of specimen
July 16, 1956	CLH-Lb-01	Male	121.0	2 mi N Scripps Pier, San Diego County, Calif.	Cameron Keppler for C. L. Hubbs	San Diego Museum of Natural History, San Diego, Calif.
May 15, to June 1, 1966		Unknown	Unknown	Pismo Beach, San Luis Obispo County, Calif.	Park Service officials	None preserved
October 7, 1969		Male	132.1	Ano Nuevo State Beach, Santa Cruz, Calif.	B. LeBoeff	Unknown
June 10, 1967	RFG-Lb-01	Unknown	Unknown	Manhattan Beach, Los Angeles County, Calif.	R. F. Green	Ventura College, Ventura, Calif.
March 1, 1970	SLO-02	Male	137.5	6 mi SE Jalama State Beach, San Luis Obispo County, Calif.	Aryan Roest	California Polytechnic State University, San Luis Obispo, Calif.
April 7, 1971	WAW 106	Male	170.0	1 mi N Santa Monica Pier, Los Angeles County, Calif.	W. A. Walker	Natural History Museum of Los Angeles County, Los Angeles, Calif.
August 15, 1971	WAW 131	Male	193.0	1/4 mi S Santa Monica Pier, Los Angeles County, Calif.	W. A. Walker	Natural History Museum of Los Angeles County, Los Angeles, Calif.
Sept. 23, 1971	WAW 139	Male	284.0	Pismo Beach, San Luis Obispo County, Calif.	W. A. Walker	Natural History Museum of Los Angeles County, Los Angeles, Calif.
March 28, 1972	WAW 154	Male	294.6	Bolsa Chica State Beach, Orange County, Calif.	W. A. Walker	Natural History Museum of Los Angeles County, Los Angeles, Calif.
May 22, 1972	C-156	Female	228.7	Palm Park Beach, Santa Barbara County, Calif.	G. V. Morejohn	San Jose State University, San Jose, Calif.

(Continued)

Table V. (*Continued*)

Date	Collection no.[a]	Sex	Length (cm)	Location	Source	Disposition of specimen
August 14, 1972	SLO-Lb-03	Male	220.7	Morro Bay, San Luis Obispo County, Calif.	Aryan Roest	Natural History Museum of Los Angeles County, Los Angeles, Calif.
October 4, 1972	SBMNH-Lb-1	Male	294.6	Goleta Beach, Santa Barbara County, Calif.	Dennis Power	Santa Barbara Museum of Natural History, Santa Barbara, Calif.
Nov. 3, 1972	SLO-Lb-04	Male	307.0	1/4 mi S pier; port of San Luis Obispo, San Luis Obispo County, Calif.	Aryan Roest	California Polytechnic State University, San Luis Obispo, Calif.
Dec. 7, 1972	WAW 176	Male	232.0	1/2 mi S Santa Monica Pier, Los Angeles County, Calif.	W. A. Walker	Natural History Museum Los Angeles County, Los Angeles, Calif.
January 1973	KCB-D14-73	Unknown	Unknown	Boca Soledad, Baja California, Mexico	K. C. Balcomb	Museo Nacional de Mexico Presented on loan to University of California Santa Cruz, Santa Cruz, Calif.
June 15, 1973	WFP 272	Male	267.9	Pacific Beach, San Diego County, Calif.	D. Tomlinson and W.F. Perrin	U.S. National Museum, Washington, D.C.
August 2, 1973	WAW 194	Female	217.0	Will Rogers State Beach, Los Angeles County, Calif.	W. A. Walker	Natural History Museum of Los Angeles County, Los Angeles, Calif.
Sept. 11, 1973	WFP 279	Female	226.4	Scripps Pier, San Diego County, Calif.	W. F. Perrin	San Diego Museum of Natural History, San Diego, Calif.
October 4, 1973	WAW 208	Female	213.0	Will Rogers State Beach, Los Angeles County, Calif.	W. A. Walker	Natural History Museum of Los Angeles County, Los Angeles, Calif.

Date	Specimen	Sex	Length	Location	Collector	Disposition
October 7, 1973	WAW 209	Female	225.5	5 mi N Santa Monica Pier, Los Angeles County, Calif.	W. A. Walker	Natural History Museum of Los Angeles County, Los Angeles, Calif.
October 31, 1973	Cal Acad 2340	Unknown	Unknown	Bolinas Lagoon, Marin County, Calif.	Peter B. Arlen for R. Orr	California Academy of Science, San Francisco, Calif.
July 1974		Female	207.0	Punta Banda, Baja California, Mexico	S. Peterson	Stomach contents only at Natural History Museum Los Angeles County, Los Angeles, Calif.
Sept. 20, 1974	C 179	Female	220.8	Pajaro Dunes, Santa Cruz County, Calif.	G. V. Morejohn	San Jose State University, San Jose, Calif.
September 1974	JSL 241	Unknown	Unknown	75 mi S Magdalena Bay, Baja California, Mexico	J. S. Leatherwood	Specimen not recovered
Nov. 3, 1974	JSL 206	Female	222.2	Coronado Beach, San Diego County, Calif.	J. S. Leatherwood; W. F. Perrin; D. McIntyre	Natural History Museum of Los Angeles County, Los Angeles, Calif.

[a] Key To Specimen Numbers: C = G. V. Morejohn, San Jose State University, San Jose, California; Cal Acad = California Academy of Sciences, Golden Gate Park, San Francisco, California; CLH = C. L. Hubbs, Scripps Institution of Oceanography, La Jolla, California; JSL = J. S. Leatherwood, Naval Undersea Center, San Diego, California; KCB = K. C. Balcomb, Moclips Cetological Society, Moclips, Washington; RFG = R. F. Green, Ventura College, Ventura, California; SBMNH = Santa Barbara Museum of Natural History, Santa Barbara, California; SLO = A. I. Roest, California Polytechnic State University, San Luis Obispo, California; WAW = W. A. Walker, Section of Mammalogy, Natural History Museum of Los Angeles County, Los Angeles, California; WFP = W. F. Perrin, National Marine Fisheries Service, Southwest Fisheries Center, La Jolla, California.

southern right whale dolphins, *Lissodelphis peroni,* in New Zealand. To date, no multiple strandings have been documented for *L. borealis.*

The two southernmost stranding records from Baja California, Mexico (Table V), are most likely of exceptional nature since the species occurs south of about 30° only infrequently during periods of intrusion of unseasonably cool water. Schroeder *et al.* (1973) reported on neuropathologic conditions of two beach-stranded *L. borealis* from southern California. Individual beach strandings are likely to be of primary pathologic nature (Ridgway and Dailey, 1972). We consider it unrealistic to rely on stranding accounts alone when considering normal distribution of marine mammal species.

No correlation between local seasonal abundance and frequency of strandings is apparent. The strandings in the area from San Luis Obispo County to northern Baja California are evenly distributed throughout the year. Reasons for this disparity with records of live animal sightings are not known.

IX. COLORATION

A. Coloration of Adults

Coloration of *L. borealis* characteristically is a black body with vivid white ventral markings extending anteriorly from the caudal peduncle as a narrow band and expanding into a ventral thoracic patch. This thoracic patch extends laterally to the inner aspect of the axillae and converges anteriorly to an apex in the gular region. A small white mark is typically present at the tip of the lower jaw. The flukes are light grey dorsally; ventrally, the distal portions of the flukes are white (Fig. 12).

Norris and Prescott (1961) reported that live right whale dolphins observed from the air appeared brownish-black on the dorsal surface with light brown

Fig. 12. A "normally" colored female *Lissodelphis borealis* (specimen no. WFP-279). Photo by W. F. Perrin.

The Northern Right Whale Dolphin

Fig. 13. A portion of a herd of *L. borealis*, including a calf, off southern California. This plate and the inset illustrate the lighter coloration of calves. Photos by C. A. Hui and D. K. Ljungblad (inset).

flukes. We have also observed that from the air the dorsal surface of the body of *L. borealis* often appears brown to black. This brownish coloration has appeared as a bronze sheen on the black portions of a captive *L. borealis*. We have also observed similar coloration exhibited by captive killer whales, *Orcinus orca,* and Dall's porpoises, *Phocoenoides dalli.* The bronze sheen is probably due to the effects of reflected light. The distinct grey coloration of the flukes is always evident.

B. Coloration of Calves

Peale (1848) described the color of the type specimen, a 4-ft (122-cm) animal, as "black with a white lanceolate spot on the breast." Norris and Prescott (1961) reported that the coloration of young right whale dolphins estimated to be approximately 24–28 in. (61–71 cm) in length, appeared to be identical to that of adults. In all of the observations of calves noted during this study, the calves were much lighter dorsally than adults (Fig. 13), varying under the conditions of observation from almost cream to light grey. Ventral coloration was not observed. The smallest specimen for which photos and a description of coloration were available, a 156.0-cm male (Roest, 1970), was identical in coloration to adults. Color differences between adults and calves, which make calves readily discernible, should be a useful tool in future population surveys.

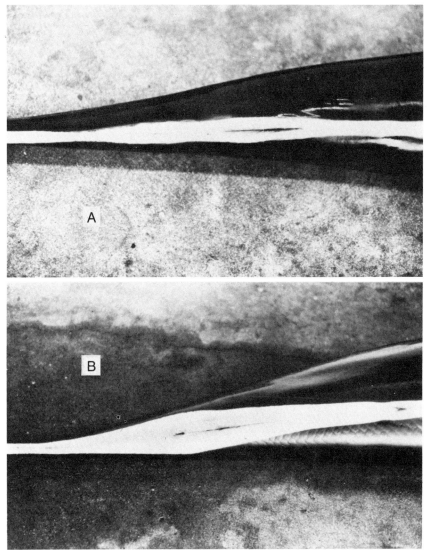

Fig. 14. Close ventral views of the coloration of the genital areas of a male, A (specimen no. WAW-184) and a female, B (specimen no. WAW-208) *Lissodelphis borealis*. Note the differences in the width of the ventral coloration. Photos by W. A. Walker.

C. Color Variation

Variations do occur in the width of the white ventral band and the area encompassed by the thoracic patch. A male collected at Copalis, Washington, demonstrated a white band that converged almost to extinction in the abdominal area (illustrated in Scheffer and Slipp, 1948). The color pattern of males and females differs in that in females the white band shows a pronounced tendency to be considerably wider in the genital area (Fig. 14 A, B).

Published accounts of color pattern variation in this species exist in the literature. Ogawa (1937) reported on two animals (identified as *Tursio peroni*) from Japan. Tobayama *et al.* (1968) published accounts of an additional animal (as *Lissodelphis peroni*) from Ibaragi, Japan. All three specimens exhibited similar color variation in the white ventral markings. Nishiwaki (1972) treated these three animals as a subspecies, *Lissodelphis borealis albiventris*. In December 1972, an additional specimen (WAW 176) demonstrating a color pattern very similar to those previously described from Japan was collected in southern California. On January 19, 1977, four *L. borealis* colored like the four anomalous specimens discussed above were observed and photographed in a herd of an estimated 800 normally colored animals at 32°43′N, 118°45′W (D.K. Ljungblad, 1977, personal communication). The four specimens examined (three from Japan and one from southern California), and the live animals photographed, demonstrate remarkably similar variations from the normal color pattern. Generally, the white ventral band is expanded to extend laterally up the side of the body. The lateral borders of the ventral thoracic patch extended to slightly above the dorsal aspect of the flipper axillae. The convergence at the anterior aspect of the ventral thoracic patch is incomplete, leaving almost the entire lower jaw of a white coloration. A white blaze is present on the lateral aspect of the melon and rostrum. A small portion of the medial aspect of the flippers also show varying amounts of white coloration (Fig. 15 A–D).

Nishiwaki (1972) cited differences in color pattern and dental formulae as the morphologic features separating the two proposed subspecies *Lissodelphis borealis borealis* and *Lissodelphis borealis albiventris*. The dental formulae were 40–43 upper, 42–46 lower, and 37–41, 40–43, respectively. Due to new data presented in this study, we consider the evidence insufficient to differentiate *Lissodelphis borealis* on a subspecific level. The most recently collected animal from California had a dental formula of NA–45 upper, 45–45 lower. The ranges in tooth counts for the 11 newly collected *L. borealis* exhibiting normal coloration in this paper are 36–46, 39–49, 37–46, 37–47 (Table VIII).

Mitchell (1970) considered the Lissodelphinae to demonstrate a primitive, generalized color pattern. Perrin (1972), in his analysis of delphinid coloration, concurred with Mitchell (1970) that the general cape system was the best candidate for the basic, primitive color pattern. The anomalous coloration in the four *Lissodelphis* specimens discussed may be due to genetic failure of the ventral development of this cape system. Another possible explanation is that in

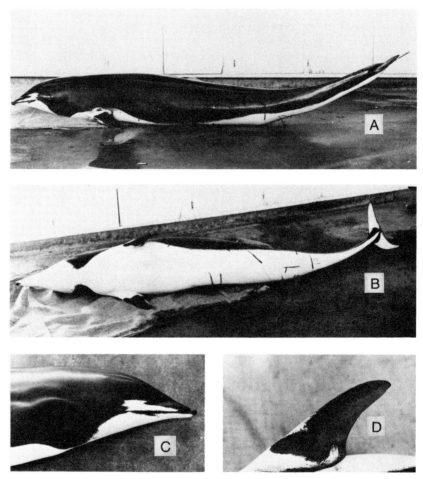

Fig. 15. Views of the lateral surface, A; ventral surface, B; head, C; and dorsal surface of the right flipper, D; of an anomalously colored *Lissodelphis borealis* (specimen no. WAW-176). Ectoparasites are *Penella* sp. Photos by W. A. Walker.

the normal pattern the "dorsal overlay" (terminology of Perrin, 1972) is very dark and extensive, masking the cape; and that the anomalous pattern is the result of genetic deletion of the overlay.

X. REPRODUCTION

Reproductive data from three male *L. borealis* were reported by Harrison *et al.* (1972). This paper includes additional data on ten females and seven males.

Table VI. Body Length and Maturity of Male Northern Right Whale Dolphins, *Lissodelphis borealis*

Collection no.	Date of collection	Length (cm)	Reproductive comments
M-15-66[a]	June 15, 1966	156.0	No spermatogenic activity
WAW 106	April 7, 1971	170.0	No spermatogenic activity
WAW 131	August 15, 1971	193.0	No spermatogenic activity
69-7-1[a]	February 29, 1969	212.0	No spermatogenic activity
WAW 68			
WAW 184	May 18, 1973[b]	219.5	Active spermatogenesis
WAW 176	December 7, 1972	232.0	Active spermatogenesis
WFP 272	June 26, 1973	267.9	Active spermatogenesis
RLB 407[a]	August 11, 1967	281.7	Active spermatogenesis
CLH-Lb-02			
WAW 139	September 23, 1971	284.0	Active spermatogenesis
WAW 154	March 28, 1972	294.6	Active spermatogenesis

[a] Published previously in Harrison *et al.* (1972).
[b] Captive animal; date is date of death.

The reproductive status of available gonad specimens was determined in males by evidence of spermatogenesis during histologic examination of testes and in females by the presence of at least one external ovarian scar on either ovary. Future detailed histologic analysis of available gonadal tissue is planned.

The largest sexually immature male examined was 212.0 cm. The smallest male with indications of testicular activity was 219.5 cm (Table VI). The largest sexually immature female examined was 200.0 cm. The smallest reproductively active female was 201.0 cm (Table VII). One 201.0-cm animal examined was pregnant. One specimen, WFP-279, a 226.4-cm beach-stranding collected in September 1973, was lactating. Perrin (1975) reports male Pacific spotted porpoise, *Stenella attenuata,* and spinner porpoise, *Stenella longirostris,* are

Table VII. Body Length and Maturity of Female Northern Right Whale Dolphins, *Lissodelphis borealis*

Collection no.	Date of collection	Length (cm)	Reproductive comments
RLB 443	March 28, 1968	200.0	No ovarian scars present
MBF 167	January 2, 1969	200.0	No ovarian scars present
DWR 1969-80	March 9, 1969	201.0	Pregnant; 26.0 cm fetus
WAW 208	October 4, 1973	213.0	Ovarian scars present
WAW 194	August 2, 1973	217.0	Ovarian scars present
C 179	September 20, 1974	220.8	Ovarian scars present
JSL 206	November 3, 1974	222.2	Ovarian scars present
WAW 209	October 7, 1973	225.0	Ovarian scars present
WFP 279	September 11, 1973	226.4	Ovarian scars present; lactating
C 156	May 22, 1972	226.7	Ovarian scars present

slightly longer than females at sexual maturity. Available data indicate that this may also occur in *L. borealis,* although due to the small sample size (total of 20 animals with reproductive data), this conclusion should be regarded as tentative.

XI. FOOD HABITS

Scheffer and Slipp (1948) and Norris and Prescott (1961) reported cephalopod beaks from the stomachs of beach-stranded *L. borealis* from

Table VIII. Number and Kinds of Otoliths Found in Stomachs of Four *Lissodelphis borealis* from California

Kinds of fishes	Number of otoliths				Occurrence	
	RLB 407*	WAW 139	1969-80	USNM 39567	Number	% of Total
Bathylagidae						
Leuroglossus stilbius			221		221	23.24
Centrolophidae						
Icichthys lockingtoni	2				2	0.21
Melamphaidae						
Scopelogadus bispinosus	1				1	0.11
Melamphaes lugubris			1		1	0.11
Merlucciidae						
Merluccius productus	2				2	0.21
Myctophidae						
Symbolophorus californiensis		9			9	0.95
Diaphus theta	23	71	22		116	12.20
Lampadena urophaos	7				7	0.74
Lampanyctus cf. *ritteri*	5				5	0.53
Lampanyctus ritteri		22	29		51	5.36
Stenobrachius leucopsarus			6		6	0.63
Tryphoturus mexicanus	5		154		159	16.72
Ceratoscopelus townsendi			122		122	12.83
Tarletonbeania crenularis		131			131	13.77
Unidentifiable myctophids	107				107	11.25
Paralepididae						
Lestidium ringens			4		4	0.42
Scomberesocidae						
Cololabis saira				7	7	0.74
Total	152	233	559	7	951	100.00

Washington and southern California, respectively. Fitch and Brownell (1968) reported on fish remains from the stomach of one southern California beach-stranded specimen.

Stomach contents from 11 additional *L. borealis* have been available for analysis (nine beach-stranded and two collected at sea). Although identification of food remains is still in progress, preliminary data suggest that stranded cetacean stomach contents should be interpreted with a great deal of caution. Fish remains identified from the stomachs of some of the stranded specimens include a pronounced combination of near-shore and offshore fishes. Many of the near-shore fish species are not representative of the normal known distribution of *L. borealis* and were probably ingested just prior to stranding.

Of the 11 stomachs examined, five (45%) contained no fish remains. Evidence of ingested squid was present in all instances.

Fish species identified from otoliths removed from three additional *L. borealis* stomachs are presented in Table VIII. Comparison of these samples with that presented in Fitch and Brownell (1968) reveal striking similarities in that the lanternfish family, Myctophidae, made up 75% of the total identifiable otoliths recovered. In addition, three other mesopelagic fish families, Bathylagidae, Melamphaidae, and Paralepididae, are represented and collectively with the family Myctophidae, comprise 99% of the total sample. In one stomach sample (USNM 39576) the Pacific saury, *Cololabis saira,* was the only fish species represented. This species is epipelagic in habit and had to be consumed at or very near the surface.

XII. PARASITISM

Two genera of parasites have been reported from *L. borealis: Nasitrema* sp., the air sinus trematode (Neiland *et al.*, 1970) and the larval cestode *Phyllobothrium* sp. (Dailey and Brownell, 1972). Neuropathologic conditions in two beach-stranded *L. borealis* caused by *Nasitrema* sp. were reported (Schroeder *et al.*, 1973). This genus of air sinus trematode was implicated in the death of one captive *L. borealis* taken for live display purposes (Walker, 1975).

Extensive examinations have been conducted on seven of the beach-stranded *L. borealis* included in this paper. Those examinations have provided considerable new information on the role of parasites in individual strandings of this species of dolphins. Numerous new host records have also been established (Dailey and Walker, 1978).

XIII. MORPHOMETRICS

Morphometric data on 23 additional animals are presented in this report (Table IX). Daugherty (1965) and Hall and Kelson (1959) reported the maximum

Table IX. External Morphological Data on 24 Specimens of Northern Right Whale Dolphins

Description of measurement	CLH-LB-02 (♀) wt., 113.4 kg		RLB-443 (♂)		MBF 167(♀) wt., 65.5 kg		MBF 168(♂)	
	cm	%	cm	%	cm	%	cm	%
1. Length total (tip of upper jaw to fluke notch)	281.7	100.0	200.0	100.0	200.0	100.0	213.0	100.0
2. Length, tip of upper jaw to center of eye	33.6	11.93	29.5	14.75	32.0	16.0	33.0	15.5
3. Length, tip of upper jaw to apex of melon	4.9	1.74	4.5	2.25	6.5	3.25	4.5	2.11
4. Length of gape (tip of upper jaw to angle of gape)	23.7	8.41	20.5	10.25	22.0	11.0	23.3	11.3
5. Length, tip of upper jaw to external auditory meatus	37.0	13.13	34.0	17.0	37.0	18.5	36.7	17.23
6. Center of eye to external auditory meatus (direct)	4.7	1.67	(R)5.0	2.5	5.0	2.5	4.6	2.16
7. Center of eye to angle of gape (direct)	9.1	3.23	—	—	11.0	5.5	10.3	4.84
8. Center of eye to center of blowhole (direct)	—	—	—	—	17.5	8.75	17.9	8.4
9. Length, tip of upper jaw to blowhole	34.3	12.18	32.0	16.0	33.0	16.50	30.1	14.13
10. Length, tip of upper jaw to anterior insertion of flipper	58.7	20.84	52.0	26.0	57.5	28.75	55.7	26.15
11. Length, tip of upper jaw to umbilicus	149.8	53.13	115.0	57.5	118.5	59.25	120.0	56.34
12. Length, tip of upper jaw to center of genital aperture	176.9	62.80	145.0	72.5	147.5	73.75	143.0	67.14
13. Length, tip of upper jaw to center of anus	206.3	73.23	150.0	75.0	155.5	77.75	157.0	73.71
14. Projection of lower jaw beyond upper	1.3	.46	1.5	0.75	—	—	0.7	0.33
15. Thickness of blubber mid-dorsal	—	—	1.0	0.5	1.5	0.75	1.0	0.47
16. Thickness of blubber mid-lateral at mid-length	1.5	.53	1.3	0.65	2.0	1.00	1.6	0.75
17. Thickness of blubber mid-ventral at mid-length	—	—	1.5	0.75	1.75	0.88	1.8	0.85
18. Girth, intersecting axilla	118.0	41.89	104.0	52.0	86.0	43.00	91.2	42.82
19. Girth, maximum	107.0	37.98	—	—	100.0	50.00	96.4	45.26
20. Girth, on a transverse plane intersecting anus	43.0	15.26	—	—	40.0	20.00	36.6	17.18
21. Dimensions of eye height	—	—	—	—	1.5	0.75	1.0	0.47
Dimensions of eye length	—	—	—	—	2.5	1.25	2.5	1.17
22. Length of mammary slits right	—	—	1.5	0.75	1.5	0.75	—	—
Length of mammary slits left	—	—	—	—	1.5	1.25	—	—

The Northern Right Whale Dolphin

	CACM(♂) wt., 15.0 kg		WAW-68(♂) 57.7		SLO-02LB(♂) 21.0		WAW-106(♂)	
23. Length of genital opening	12.0	4.26	15.0	7.5	12.5	6.25	13.5	6.34
Length of anal opening	3.5	1.24	—	—	—	—	—	—
24. Dimensions of blowhole width	2.9	1.03	—	—	3.0	1.50	3.2	1.5
Dimensions of blowhole length	—	—	—	—	2.0	1.00	1.5	0.70
25. Length, flippers (anterior insertion to tip)	28.1	9.98	28.0	14.0	24.5	12.25	26.0	12.21
26. Length, flipper (axilla to tip)	—	—	20.5	10.25	19.0	9.50	18.0	8.45
27. Width, flipper (maximum)	9.3	3.30	9.0	4.5	8.0	4.00	7.9	3.71
28. Width, flukes (tip to tip)	32.6	11.57	36.0	18.0	36.0	18.00	32.2	15.12
29. Distance from nearest point on anterior border of flukes to notch	10.5	3.73	9.5	4.75	10.5	5.25	10.0	4.69
30. Depth of notch between flukes	1.5	0.53	3.0	1.5	2.5	1.25	2.0	0.94
31. Tooth counts								
Upper left	46		—		—		—	
Upper right	46		—		—		—	
Lower left	49		—		—		—	
Lower right	47		—		—		—	
1. Length total (tip of upper jaw to fluke notch)	132.1	100.0	212.0	100.0	137.5	100.0	170.0	100.0
2. Length, tip of upper jaw to center of eye	—	—	32.0	15.09	20.5	14.91	22.5	15.15
3. Length, tip of upper jaw to apex of melon	—	—	4.5	2.12	—	—	2.0	1.35
4. Length of gape (tip of upper jaw to angle of gape)	12.7	9.61	23.0	10.85	14.0	10.18	15.5	10.44
5. Length, tip of upper jaw to external auditory meatus	—	—	36.0	16.98	—	—	—	—
6. Center of eye to external auditory meatus (direct)	—	—	5.5	2.59	—	—	—	—
7. Center of eye to angle of gape (direct)	—	—	9.5	4.48	—	—	9.0	6.06
8. Center of eye to center of blowhole (direct)	—	—	16.0(L)	7.55	—	—	14.5(L)	9.76
9. Length, tip of upper jaw to blowhole	—	—	33.0	15.57	20.5	14.91	24.0	16.16
10. Length, tip of upper jaw to anterior insertion of flipper	34.3	25.97	55.0	25.94	36.0	26.18	43.0	28.96
11. Length, tip of upper jaw to umbilicus	—	—	120.0	56.60	78.0	56.73	91.0	61.28

(Continued)

Table IX. (Continued)

Description of measurement	Measurement cm	%	Measurement cm	%	Measurement cm	%	Measurement cm	%
12. Length, tip of upper jaw to center of genital aperture	—	—	142.0	66.98	93.0	67.64	108.0	72.73
13. Length, tip of upper jaw to center of genital aperture	—	—	157.5	74.29	100.0	72.73	117.5	79.12
14. Projection of lower jaw beyond upper	0.64	0.48	1.0	0.47	—	—	0.5	0.34
15. Thickness of blubber mid-dorsal	—	—	—	—	—	—	—	—
16. Thickness of blubber mid-lateral at mid-length	—	—	—	—	—	—	—	—
17. Thickness of blubber mid-ventral at mid-length	—	—	—	—	—	—	—	—
18. Girth, intersecting axilla	—	—	95.0	44.81	—	—	73.0	49.16
19. Girth, maximum	—	—	67.0	31.60	—	—	73.0	49.16
20. Girth, on a transverse plane intersecting anus	—	—	40.0	18.87	—	—	32.0	21.55
21. Dimensions of eye height	—	—	0.9	0.42	—	—	—	—
Dimensions of eye length	—	—	2.8	1.32	1.7	1.24	—	—
22. Length of mammary slits right	—	—	—	—	—	—	—	—
Length of mammary slits left	—	—	—	—	—	—	—	—
23. Length of genital opening	—	—	11.5	5.42	—	—	8.0	5.39
Length of anal opening	—	—	3.0	1.42	—	—	—	—
24. Dimensions of blowhole width	—	—	3.0	1.42	2.5	1.83	—	—
Dimensions of blowhole length	—	—	2.0	0.94	—	—	—	—
25. Length, flippers (anterior insertion to tip)	—	—	28.5	13.44	22.0	16.0	23.0	15.49
26. Length, flipper (axilla to tip)	—	—	19.0	8.96	16.0	11.64	17.0	11.45
27. Width, flipper (maximum)	—	—	9.0	4.85	6.5	4.73	7.0	4.71
28. Width, flukes (tip to tip)	23.5	17.79	36.5	17.22	26.0	18.91	25.0	16.84
29. Distance from nearest point on anterior border of flukes to notch	—	—	10.5	4.95	—	—	9.0	6.06
30. Depth of notch between flukes	—	—	2.5	1.18	2.0	1.45	2.0	1.35
31. Tooth counts								
Upper left	—		40		—		42	

The Northern Right Whale Dolphin

	WAW-131(♂)		WAW-139(♂)		WAW-154(♂)		C-156(♀)	
Upper right	—		40		—		41	
Lower left	—		43		—		42	
Lower right	—		42		—		40	
1. Length total (tip of upper jaw to deepest part of notch)	193.0	100.0	284.0	100.0	294.6	100.0	228.7	100.0
2. Length, tip of upper jaw to center of eye	28.0	14.51	33.0	11.62	34.0	11.54	32.5	14.21
3. Length, tip of upper jaw to apex of melon	4.0	2.07	5.0	1.76	6.0	2.04	—	—
4. Length of gape (tip of upper jaw to angle of gape)	18.5	9.59	24.0	8.45	25.0	8.49	22.3	9.75
5. Length, tip of upper jaw to external auditory meatus	30.5	18.3	37.5	13.20	38.5	13.07	—	—
6. Center of eye to external auditory meatus (direct)	4.0	2.07	5.0	1.76	5.0	1.7	—	—
7. Center of eye to angle of gape (direct)	10.0	5.18	10.5	3.70	10.0	3.39	—	—
8. Center of eye to center of blowhole (direct)	15.5(L)	—	17.5(R)	6.16	18.0(R)	6.11	18.4(R)	8.05
			15.0(L)	5.28	16.0(L)	5.43	16.5(L)	7.21
9. Length, tip of upper jaw to blowhole	27.0	13.99	33.5	11.80	33.0	11.2	30.7	13.42
10. Length, tip of upper jaw to anterior insertion of flipper	30.5	28.50	64.5	22.5	64.0	21.72	61.4	26.85
11. Length, tip of upper jaw to umbilicus	109.0	56.48	155.5	54.75	156.0	52.45	131.0	57.20
12. Length, tip of upper jaw to center of genital aperture	131.0	67.88	197.0	69.37	198.0	67.21	161.3	70.53
13. Length, tip of upper jaw to center of anus	143.5	74.35	222.0	78.17	220.0	74.68	170.4	76.08
14. Projection of lower jaw beyond upper	1.0	0.52	1.3	0.46	1.2	0.41	—	—
15. Thickness of blubber mid-dorsal	1.0	0.52	—	—	—	—	1.0	0.44
16. Thickness of blubber mid-lateral at mid-length	—	—	—	—	—	—	0.8	0.35
17. Thickness of blubber mid-ventral at mid-length	—	—	—	—	—	—	—	—
18. Girth, intersecting axilla	86.0	44.56	82.0	28.87	104.0	35.30	96.9	42.37
19. Girth, maximum	86.0	44.56	93.0	32.75	110.0	39.34	—	—
20. Girth, on a transverse plane intersecting anus	36.0	18.65	36.5	12.85	46.0	15.61	—	—
21. Dimensions of eye height	1.0	0.52	1.0	0.35	1.0	0.54	1.0	0.44
Dimensions of eye length	2.0	1.04	2.5	0.88	3.0	1.02	2.4	1.05
22. Length of mammary slits right	—	—	—	—	—	—	2.2	0.96
Length of mammary slits left	—	—	—	—	—	—	2.2	0.96

(Continued)

Table IX. *(Continued)*

Description of measurement	Measurement cm	%	Measurement cm	%	Measurement cm	%	Measurement cm	%
23. Length of genital opening	12.5	6.48	20.0	7.04	20.5	8.49	14.9	6.52
Length of anal opening	1.5	0.78	3.0	1.06	2.5	0.85	1.2	0.52
24. Dimensions of blowhole width	3.0	1.55	3.0	1.06	3.3	1.12	—	—
Dimensions of blowhole length	1.0	0.52	1.5	0.53	1.5	0.51	—	—
25. Length, flippers (anterior insertion to tip)	25.1	13.01	39.0	13.73	28.0	9.5	27.2	11.89
26. Length, flipper (axilla to tip)	19.0	9.84	20.5	8.8	19.5	6.62	19.8	8.66
27. Width, flipper (maximum)	7.3	3.78	8.5	2.99	9.0	3.05	8.3	3.63
28. Width, flukes (tip to tip)	3.5	18.13	34.0	11.97	34.0	11.54	33.5	14.65
29. Distance from nearest point on anterior border of flukes to notch	10.5	5.44	10.5	3.70	10.5	3.56	—	—
30. Depth of notch between flukes	2.0	1.04	1.4	0.49	2.0	0.68	1.9	0.83
31. Tooth counts								
Upper left	36		46		43		—	
Upper right	37		45		44		—	
Lower left	39		45		46		—	
Lower right	37		46		43		—	
	LBV-14(♂)		SLO03Lb(♀)		SBNHM-LB-1(♂)		SLO-041b(♂)	
1. Length total (tip of upper jaw to fluke notch)	220.7	100.0	222.2	100.0	294.6	100.0	307.0	100.0
2. Length, tip of upper jaw to center of eye	33.0	14.95	28.8	12.96	—	—	32.5	10.59
3. Length, tip of upper jaw to apex of melon	4.7	2.13	5.1	2.30	—	—	—	—
4. Length of gape (tip of upper jaw to angle of gape)	23.5	10.65	23.2	10.44	34.3	11.64	21.0	6.84
5. Length, tip of upper jaw to external auditory meatus	—	—	—	—	—	—	—	—
6. Center of eye to external auditory meatus (direct)	—	—	7.9	3.56	—	—	—	—
7. Center of eye to angle of gape (direct)	—	—	—	—	—	—	—	—

The Northern Right Whale Dolphin

	1	2	3	4	5	6	7	8
8. Center of eye to center of blowhole (direct)	—	—	10.5(L)	4.73	—	—	—	—
9. Length, tip of upper jaw to blowhole	30.6	13.86	30.0	13.50	35.6	12.08	37.0	12.05
10. Length, tip of upper jaw to anterior insertion of flipper	56.5	25.6	63.2	28.44	66.0	22.40	68.0	22.15
11. Length, tip of upper jaw to umbilicus	—	—	131.8	59.32	—	—	—	—
12. Length, tip of upper jaw to center of genital aperture	163.0	73.86	163.6	73.63	200.7	68.13	200.0	65.15
13. Length, tip of upper jaw to center of anus	—	—	171.0	76.96	221.0	75.05	224.0	72.96
14. Projection of lower jaw beyond upper	—	—	1.1	0.50	—	—	2.5	0.81
15. Thickness of blubber mid-dorsal	—	—	—	—	—	—	—	—
16. Thickness of blubber mid-lateral at mid-length	—	—	—	—	—	—	—	—
17. Thickness of blubber mid-ventral at mid-length	—	—	—	—	—	—	—	—
18. Girth, intersecting axilla	92.5	41.91	—	—	—	—	—	—
19. Girth, maximum	39.6	17.94	—	—	—	—	—	—
20. Girth, on a transverse plane intersecting anus	—	—	—	—	—	—	—	—
21. Dimensions of eye height	—	—	—	—	3.2	1.09	—	—
Dimensions of eye length	—	—	—	—	3.8	1.29	—	—
22. Length of mammary slits right	—	—	—	—	—	—	—	—
Length of mammary slits left	—	—	—	—	—	—	—	—
23. Length of genital opening	—	—	—	—	19.1	6.48	12.0	3.91
Length of anal opening	—	—	—	—	4.4	1.49	—	—
24. Dimensions of blowhole width	—	—	—	—	—	—	2.5	0.81
Dimensions of blowhole length	—	—	—	—	—	—	—	—
25. Length, flippers (anterior insertion to tip)	30.0	13.59	25.5	11.48	26.7	9.06	27.0	8.79
26. Length, flipper (axilla to tip)	19.0	8.61	21.1	9.50	—	—	16.0	5.21
27. Width, flipper (maximum)	8.5	3.85	7.1	3.20	9.5	3.22	9.0	2.93
28. Width, flukes (tip to tip)	38.0	17.22	33.0	14.85	35.6	12.08	32.0	10.42
29. Distance from nearest point on anterior border of flukes to notch	—	—	10.4	4.68	—	—	—	—
30. Depth of notch between flukes	2.0	0.91	1.7	0.77	2.5	0.85	1.5	0.49
31. Tooth counts								
Upper left	—	—	41		—	—	—	—
Upper right	—	—	42		—	—	—	—
Lower left	—	—	41		—	—	—	—
Lower right	—	—	42		—	—	—	—

(Continued)

Table IX. *(Continued)*

Description of measurement	WAW-176(♂) cm	%	WAW-184(♂) cm	%	WFP-272(♂) cm	%	WAW-194(♀) cm	%
1. Length total (tip of upper jaw to fluke notch)	232.0	100.0	219.5	100.0	267.9	100.0	217.0	100.0
2. Length, tip of upper jaw to center of eye	31.5	13.58	31.0	14.12	32.2	12.03	33.0	15.2
3. Length, tip of upper jaw to apex of melon	6.0	2.59	4.5	2.05	4.0	1.49	5.0	2.3
4. Length of gape (tip of upper jaw to angle of gape)	22.0	9.48	23.0	10.48	22.9	8.55	22.5	10.37
5. Length, tip of upper jaw to external auditory meatus	34.0	14.66	36.5	16.63	—	—	35.0	16.13
6. Center of eye to external auditory meatus (direct)	4.5	1.95	5.0	2.28	—	—	4.5	2.07
7. Center of eye to angle of gape (direct)	10.0	4.31	9.5	4.3	10.0	3.73	10.5	4.84
8. Center of eye to center of blowhole (direct)	17.0(R) / 16.5(L)	7.1 / 8.2	18.0(R) / 16.0(L)	7.29 / 8.2	15.5(L)	5.79	17.0(R) / 15.5(L)	7.83 / 7.14
9. Length, tip of upper jaw to blowhole	33.0	14.22	32.0	14.58	31.8	11.87	30.5	14.06
10. Length, tip of upper jaw to anterior insertion of flipper	53.0	22.84	55.0	25.05	61.0	22.77	59.0	27.19
11. Length, tip of upper jaw to umbilicus	130.0	56.03	125.0	56.95	154.9	57.82	12.0	55.3
12. Length, tip of upper jaw to center of genital aperture	150.0	64.66	148.0	67.42	188.0	70.18	15.3	70.51
13. Length, tip of upper jaw to center of anus	174.0	75.0	164.0	74.7	210.8	78.39	16.2	74.65
14. Projection of lower jaw beyond upper	2.5	1.00	1.5	0.68	1.0	0.37	1.2	0.55
15. Thickness of blubber mid-dorsal	—	—	—	—	—	—	—	—
16. Thickness of blubber mid-lateral at mid-length	—	—	—	—	—	—	—	—
17. Thickness of blubber mid-ventral at mid-length	—	—	—	—	—	—	—	—
18. Girth, intersecting axilla	9.6	41.38	—	—	76.20	28.44	10.4	47.93
19. Girth, maximum	9.8	42.24	—	—	77.5	28.93	13.4	47.93
20. Girth, on a transverse plane intersecting anus	4.0	17.24	—	—	39.4	14.74	3.8	17.51
21. Dimensions of eye height	—	—	—	—	1.0	0.37	—	—
Dimensions of eye length	—	—	—	—	2.5	0.93	—	—

The Northern Right Whale Dolphin

		WFP-279(♀)		WAW-208(♀)		WAW-209(♀)		JSL-206(♀)	
22.	Length of mammary slits right	—	—	—	—	—	—	2.5	1.06
	Length of mammary slits left	—	—	—	—	—	—	2.0	0.92
23.	Length of genital opening	—	—	14.5	6.6	16.5	6.17	16	7.37
	Length of anal opening	—	—	—	—	—	—	2.0	0.92
24.	Dimensions of blowhole width	3.0	1.29	3.0	1.37	—	—	3.0	1.38
	Dimensions of blowhole length	1.5	0.65	1.0	0.46	—	—	1.5	0.69
25.	Length, flippers (anterior insertion to tip)	28.5	12.28	29.0	13.21	29.2	10.9	29.0	13.36
26.	Length, flipper (axilla to tip)	20.0	8.62	19.5	8.89	20.3	7.58	21.0	9.68
27.	Width, flipper (maximum)	9.0	3.88	8.0	3.64	9.0	3.36	8.5	3.92
28.	Width, flukes (tip to tip)	35.0	15.09	36.0	16.40	34.9	13.03	38	17.51
29.	Distance from nearest point on anterior border of flukes to notch	5.0	2.16	10.5	4.78	10.5	3.92	10.5	4.84
30.	Depth of notch between flukes	2.0	.86	2.3	1.05	1.5	.56	3.0	1.38
31.	Tooth counts								
	Upper left	—		46		—		45	
	Upper right	45		45		—		44	
	Lower left	45		45		—		46	
	Lower right	45		44		—		47	

		WFP-279(♀) wt., 81.4 kg		WAW-208(♀)		WAW-209(♀)		JSL-206(♀)	
1.	Length total (tip of upper jaw to fluke notch)	226.4	100.0	213.0	100.0	225.5	100.0	222.2	100.0
2.	Length, tip of upper jaw to center of eye	30.0	13.25	30.0	14.08	24.5	10.86	28.8	12.96
3.	Length, tip of upper jaw to apex of melon	5.1	2.25	6.0	2.82	7.0	3.10	5.1	2.30
4.	Length of gape (tip of upper jaw to angle of gape)	22.0	9.72	21.5	10.09	24.0	10.64	23.2	10.44
5.	Length, tip of upper jaw to external auditory meatus	35.5	15.68	33.5	15.73	39.0	17.29	—	—
6.	Center of eye to external auditory meatus (direct)	4.5	1.99	4.5	2.11	5.0	2.21	—	—
7.	Center of eye to angle of gape (direct)	10.4	4.59	9.5	4.46	11.0	4.88	7.9	3.56
8.	Center of eye to center of blowhole (direct)	16.0	7.07	15.5 (L)	7.28	16.5 (L)	7.32	10.5	4.73
9.	Length, tip of upper jaw to blowhole	30.5	13.47	32.0	15.02	35.0	15.52	30.0	13.50
10.	Length, tip of upper jaw to anterior insertion of flipper	59.5	26.28	53.5	25.12	61.0	27.05	63.2	28.44

(Continued)

Table IX. *(Continued)*

Description of measurement	Measurement		Measurement		Measurement		Measurement	
	cm	%	cm	%	cm	%	cm	%
11. Length, tip of upper jaw to umbilicus	127.5	56.32	116.0	54.46	122.0	54.10	131.8	59.32
12. Length, tip of upper jaw to center of genital aperture	156.5	69.13	150.0	70.42	166.0	73.61	163.6	73.63
13. Length, tip of upper jaw to center of anus	165.5	23.10	158.0	74.18	173.0	76.72	171.0	76.96
14. Projection of lower jaw beyond upper	1.0	0.44	1.3	0.61	1.75	0.78	1.1	0.50
15. Thickness of blubber mid-dorsal	1.1	0.49	—	—	—	—	—	—
16. Thickness of blubber mid-lateral at mid-length	0.8	0.35	—	—	—	—	0.7	0.30
17. Thickness of blubber mid-ventral at mid-length	—	—	—	—	—	—	—	—
18. Girth, intersecting axilla	97.0	42.84	96.0	45.07	104.0	46.12	—	—
19. Girth, maximum	99.0	43.73	102.0	47.89	104.0	46.12	—	—
20. Girth, on a transverse plane intersecting anus	42.0	18.55	39.0	18.31	42.0	18.63	—	—
21. Dimensions of eye height	1.2	0.53	—	—	—	—	—	—
Dimensions of eye length	2.5	1.1	—	—	—	—	—	—
22. Length of mammary slits right	2.2	0.97	3.0	1.41	3.0	1.33	—	—
Length of mammary slits left	2.2	0.97	2.5	1.17	2.5	1.11	—	—
23. Length of genital opening	17.0	7.5	16.0	7.51	20.0	8.87	—	—
Length of anal opening	2.0	0.89	—	—	—	—	—	—
24. Dimensions of blowhole width	2.7	1.19	3.0	1.41	3.0	1.33	5.0	2.25
Dimensions of blowhole length	1.5	0.66	1.5	0.70	1.0	0.44	—	—

The Northern Right Whale Dolphin

25. Length, flippers (anterior insertion to tip)	28.6	12.63	27.0	12.68	29.0	12.86	25.5	11.48
26. Length, flipper (axilla to tip)	20.1	8.88	17.5	8.22	22.0	9.76	21.5	9.68
27. Width, flipper (maximum)	8.6	3.80	8.0	3.76	8.5	3.77	7.5	3.38
28. Width, flukes (tip to tip)	37.7	16.65	35.5	16.67	37.0	16.41	33.0	14.85
29. Distance from nearest point on anterior border of flukes to notch	10.3	4.55	10.0	4.69	11.0	4.88	—	—
30. Depth of notch between flukes	2.5	1.10	3.0	1.41	1.5	.67	—	—
31. Tooth counts								
Upper left	41		42		44		41	
Upper right	42		41		45		42	
Lower left	41		43		45		41	
Lower right	42		43		45		42	

length of *Lissodelphis borealis* as 8 ft (247.8 cm). Nishiwaki (1972) reports *L. borealis* to reach approximately 2.3 m. One 281.7-cm mature male has been reported (Harrison *et al.*, 1972). Six specimens included in this report (all males) were over 260 cm in length (Table VIII). The largest animal collected was a beach-stranded 307.0-cm male. No females over 230 cm have been collected. It appears from the specimens examined to date that males may reach a greater maximum length than females.

ACKNOWLEDGMENTS

Numerous other colleagues contributed data and/or specimens for this report. We thank all the following: R. L. Brownell, Jr., R. L. DeLong, and K. C. Balcomb summarized the Smithsonian surveys; Clifford Fiscus, Dale Rice, and Hiroshi Kajimura summarized programs of the NMFS Northwest Fisheries Center; K. S. Norris, T. D. Dohl, and L. Hobbs the University of California records; and W. F. Perrin records of the NMFS Southwest Fishcries Center. We are also indebted to John E. Fitch for the identification of fish otoliths.

Sources of incidental records are identified in Table II; sources of data on the various specimens in Table V.

NUC aerial surveys were conducted as part of the Independent Research Project entitled Marine Mammal Populations. J. S. Leatherwood and W. W. Evans principal investigators. Flights were generously provided by DEPCOM-OPTEVFORPAC.

Robert F. Green, Ventura College, was a collaborator on early stages of this project. His assistance and contributions are sincerely appreciated.

We are also indebted to Mary Jean Walker for her typing and editorial assistance.

REFERENCES

Brown, D. H., and Norris, K. S., 1956, Observations of captive and wild cetaceans, *J. Mammal.* **37(3):**311–326.

Brownell, R. L., 1964, Observations of odontocetes in central California waters, *Nor. Hvalfangst-Tid.* **3:**60–66.

Dailey, M. D., and Brownell, R. L., 1972, A checklist of marine mammal parasites, in: *Mammals of the Sea; Biology and Medicine* (S. H. Ridgway, ed.), pp. 528–589, Charles C. Thomas, Springfield, Illinois.

Dailey, M. D., and Walker, W. A., 1978, Parasitism as a factor in single strandings of southern California cetaceans, *J. Parasitol.* (in press).

Daugherty, A., 1965, Marine Mammals of California, California Department of Fish & Game, 86 pp.

Ekman, S., 1953, *Zoogeography of the Sea,* Sedgwick and Jackson, London.

Evans, W. E., 1971, Orientation behavior of delphinids; radiotelemetric studies, *Ann. N.Y. Acad. Sci.* **188:**142–160.

Evans, W. E., 1975, Distribution, Differentiation of Populations and Other Aspects of the Natural History of *Delphinus delphis* in the Northeastern Pacific, PhD dissertation, University of California at Los Angeles, 145 pp.

Evans, W. E., and Bastian, J., 1969, Marine mammal communication; social and ecological factors, in: *The Biology of Marine Mammals* (Harald T. Andersen, ed.), pp. 425–475, Academic Press, New York.

Fiscus, C. H., and Niggol, K., 1965, Observations of cetaceans off California, Oregon, and Washington, *U.S. Fish Wild. Serv. Spec. Sci. Rep. Fish.* **498**:iii–27.

Fish, J. F., and Turl, C. W., 1977, Acoustic Source Levels of Four Species of Small Toothed-Whales, Naval Undersea Center Technical Publication No. 547, 13 pp.

Fitch, J. E., and Brownell, R. L., 1968, Fish otoliths in cetacean stomachs and their importance in interpreting feeding habits, *J. Fish. Res. Bd. Can.* **25(12)**:2561–2574.

Fraser, F. C., 1954, The southern right whale dolphin, *Lissodelphis* peroni (Lacepede), *Bull. Br. Mus. (Nat. Hist.) Zool.* **2(11)**:339–346 (and 1 plate).

Gaskin, D. E., 1968, Distribution of Delphinidae (Cetacea) in relation to sea surface temperatures off Eastern and Southern New Zealand, *N. Z. J. Mar. Freshwater Res.* **2–3**:527–534.

Gates, D. E., 1969, Simulation studies of estimators for the line transect sampling method, *Biometrics* **24**:135–145.

Gilmore, R. M., 1951, The whaling industry. Whales, dolphins, and porpoises, in: *Marine Products of Commerce* (D. K. Tressler and J. McW. Lemon, eds.), pp. 680–715, Reinhold, New York.

Guiget, C. J., and Schick, W. J., 1970, First record of a right whale dolphin from British Columbia, *Syesis* **3**:188.

Hall, E. R., and Kelson, K. R., 1959, *The Mammals of North America*, II, Ronald Press, New York.

Harrison, R. J., Brownell, R. L., and Boice, R. C., 1972, Reproduction and gonadal appearances in some odontocetes, in: *Functional Anatomy of Marine Mammals*, Vol. I (R. J. Harrison, ed.), pp. 361–429, Academic Press, New York.

Hawley, R., 1958–1960, *Whales and Whaling in Japan,* Vol. I(1), Kawakita Printing (Miscellanea Japonica), Kyoto, Japan.

Hershkovitz, P., 1966, Catalog of Living Whales, Smithsonian Institute Bulletin No. 246, 259 pp.

Kasuya, T., 1971, Consideration of distribution and migration of toothed whales off the Pacific coast of Japan based upon aerial sighting record, *Sci. Rep. Whales Res. Inst.* **23**:37–60 (and 7 plates).

Klumov, S. K., 1959, Commercial Dolphins of the Far East, U.S. Fish and Wildlife Service, Marine Mammal Biol. Lab., Seattle mimeo, translation by L. V. Sagen, 14 pp.

Kraus, C., and Gihr, M., 1971, On the presence of *Tursiops truncatus* in schools of *Globicephala melaena* off the Faroe Islands, in: *Investigations of Cetacea III (I)* (G. Pilleri, ed.), 181 pp., Institute of Brain Anatomy, Berne, Switzerland.

Leatherwood, J. S., 1974a, A note on gray whale behavioral interactions with other marine mammals, *Mar. Fish. Rev.* **36(4)**:50–51.

Leatherwood, J. S., 1974b, Aerial observations of migrating gray whales, *Eschrichtius robustus,* off southern California, 1969–1972, *Mar. Fish. Rev.* **36(4)**:45–50.

Leatherwood, J. S., 1978, Some preliminary impressions on the numbers and social behavior of free-swimming bottlenosed dolphin calves *Tursiops truncatus* in the northern Gulf of Mexico, in: *Proceedings of the Workshop on the Breeding Biology of the Bottlenosed Dolphin* (S. H. Ridgway and K. W. Benirschke, eds.), U.S. Government Printing Office, National Technical Information Service, Washington, D.C.

Leatherwood, S., and Platter, M. F., 1975, Aerial assessment of bottlenosed dolphins off Alabama, Mississippi, and Louisiana, in: *Tursiops truncatus Assessment Workshop* (D. K. Odell, D. B. Siniff, and G. A. Waring, eds.), 156 pp., University of Miami RSMAS publication, Miami, Florida.

Leatherwood, S., Evans, W. E., and Rice, D. W., 1972, The Whales, Dolphins, and Porpoises of the Eastern North Pacific; a Guide to Their Identification in the Water, Naval Undersea Center

Technical Publication No. 282, 175 pp.

McGary, J. W., and Graham, J. J., 1960, Biological and Oceanographic Observations in the Central North Pacific (July–Sept., 1955), U.S. Fish and Wildlife Service Spec. Sci. Rep.: Fisheries No. 358.

Miller, G. S., Jr., and Kellogg, R., 1955, List of North American recent mammals, *Bull U.S. Nat. Mus.* **205**:954.

Mitchell, E. D., 1970, Pigmentation pattern evolution in delphinid cetaceans; an essay in adaptive coloration, *Can. J. Zool.* **48**:717–740 (and 15 plates).

Mitchell, E. D., 1975, Porpoise, Dolphin, and Small Whale Fisheries of the World, Status and Problems, I.U.C.N. Monograph No. 3, Morges, Switzerland.

Neiland, K. A., Rice, D. W., and Holden, B. L., 1970, Helminths of marine mammals, Part I. The genus Nasitrema, nasal flukes of delphinid cetacea, *J. Parasitol.* **56**:305–316.

Nishiwaki, M., 1966, Distribution and Migration of Marine Mammals in the North Pacific Area, 11th Pacific Scientific Congressional Symposium **4**:1–49. (See also Coll. Rep. Ocean Res. Inst., Univ. Tokyo, 1966, **5**:103–123).

Nishiwaki, M., 1972, General biology, in: *Mammals of the Sea; Biology and Medicine* (S. H. Ridgway, ed.), pp. 3–204, Charles C. Thomas, Springfield, Illinois.

Norris, K. S., 1967, Some observations on the migration and orientation of marine mammals, in: *Animal Orientation and Navigation* (R. H. Storm, ed.), pp. 101–125, Oregon State University Press, Eugene.

Norris, K. S., and Prescott, J. H., 1961, Observations of Pacific cetaceans in Californian and Mexican waters, *Univ. of Calif. Pub. Zool.* **4**:291–402.

Ohsumi, S., 1972, Catch of marine mammals, mainly of small cetaceans by local fisheries along the coast of Japan, *Bull. Fish. Res. Lab. Shimizu* **7**:137–166.

Okada, Y., and Hanaoka, T., 1940, A Study of Japanese Delphinidae IV, Sci. Rep. T.B.D. Sect. B. Volume 4, No. 77.

Ogawa, T., 1937, Studies on the Japanese toothed whales, I–IV, *Bot. Zool.* **4**:1936–1937 (in Japanese).

Payne, R., and Webb, D., 1971, Orientation by long range acoustic signalling in baleen whales, *Ann. N.Y. Acad. Sci.* **188**:110–141.

Peale, T. R., 1848, *U.S. Exploring Expedition 1838, 1839, 1840, 1841, 1842 under the Command of Charles Wilkes, U.S.N.*, Vol. 8, Mammalogy and Ornithology, Asherman and Co., Philadelphia.

Perrin, W. F., 1972, Color patterns of spinner porpoises (*Stenella* cf. *S. longirostris*) of the eastern Pacific and Hawaii, with comments on delphinid pigmentation, *Fish. Bull.* **70(3)**:983–1003.

Perrin, W. F., 1975, *Variation of Spotted and Spinner Porpoise (Genus Stenella) in the Eastern Pacific and Hawaii*, University of California Press, Berkeley.

Pike, G. C., and MacAskie, I. B., 1969, Marine mammals of British Columbia, *Fish. Res. Bd. Can. Bull. No. 171*, 54 pp.

Pilleri, G., and Knuckey, J., 1969, The distribution, navigation, and orientation by the sun of *Delphinus delphis* L. in the western Mediterranean, *Experientia* **24(4)**:394–396.

Rice, D. W., 1963a, Pacific coast whaling and whale research, *Trans. N. Am. Wild. Nat. Resource Conf.* **28**:327–335.

Rice, D. W., 1963b, The whale marking cruise of the *Sioux City* off California and Baja California, *Nor. Hvalfangst-Tid.* **6**:153–160.

Rice, D. W., 1974, Whales and whale research in the eastern North Pacific, in: *The Whale Problem* (W. E. Schevill, ed.), pp. 170–195, Harvard University Press, Cambridge, Massachusetts.

Ridgway, S. H., and Dailey, M. D., 1972, Cerebral and cerebellar involvement of trematode parasites in dolphins and their possible role in stranding, *J. Wildl. Dis.* **8**:33–43.

Roest, A. I., 1970, *Kogia simus* and other cetaceans from San Luis Obispo County, California, *J. Mammal.* **51(2)**:410–417.

Scammon, C. M., 1874, The dolphins, in: *The Marine Mammals of North America*, Chapter 9, J. H. Carmany and Co., San Francisco.

Scheffer, V. B., and Slipp, J. B., 1948, The whales and dolphins of Washington State, *Am. Midl. Nat.* **39(2)**:257–337.

Schroeder, R. J., Delliquadri, C. A., McIntyre, R. W., and Walker, W. A., 1973, Marine mammal disease surveillance program in Los Angeles County, *J.A.V.M.A.* **163(6)**:580–581.

Sleptosov, M., 1952, Kittobrazye dal'ne ostochnykh morei (Whales of the far east), *Izv. Tinro*, Vol. 38.

Tayler, C. K., and Saayman, G. S., 1972, The social organization and behavior of dolphins *(Tursiops aduncus)* and baboons *(Papio ursinus)*: Some comparisons and assessments, *Ann. Cape Prov. Mus. (Nat. Hist.)* **9**:11–49.

Tobayama, T. S., Uchida, S., and Nishiwaki, M., 1969, A white-bellied right whale dolphin caught in the waters off Ibaragi, Japan, *J. Mammol. Soc. Jpn.* **4(4)**:112–120 (and 1 plate).

Tomilin, A. G., 1960, The Migrations, Geographical Races, the Thermoregulation and the Effect of the Temperature of the Environment upon the Distribution of the Cetaceans, *Migratssii Zhivotnyykh (Animal Population), Akad. Nauk, SSJR* **2**:3–26.

Townsend, C. H., 1935, The distribution of certain whales as shown by logbook records of American whaleships, *Zoologica* **19**:3–50.

True, R. W., 1889, Contributions to the natural history of the cetaceans. A review of the family of Delphinidae, *U.S. Natl. Mus. Bull.* **36**:1–191.

Walker, E. P., 1968, *Mammals of the World*, Vol. 2, Johns Hopkins Press, Baltimore.

Walker, W. A., 1975, Review of the live capture fishery for smaller cetaceans taken in southern California waters for public display, 1966–1973, in: *Review of Biology and Fisheries of Smaller Cetaceans* (E. D. Mitchell, ed.), pp. 1197–1211, J. Fish. Res. Bd. Can. 32(7).

Watkins, W. A., 1967, The harmonic interval; fact or artifact in spectral analysis of pulse trains, in: *Marine Bioacoustics*, Vol. VII (W. N. Tavolga, ed.), pp. 15–43, Pergamon Press, New York.

Wick, W. Q., 1969, Right whale dolphin from Cape Kiwanda, Tillamook County, Oregon, *Murrelet* **50(1)**:9.

Wilson, E. O., 1975, *Sociobiology; the New Synthesis*, Belknap Press at Harvard University Press, Cambridge, Massachusetts.

Zemsky, V., and Yablokov, A. V., 1974, Catch Statistics, Short History of Exploitation Present Status of *Delphinus delphis, Tursiops truncatus,* and *Phocoena phocoena* in the Black Sea, FAO, ACMRR-MM II-37, La Jolla, California (December 1974).

Chapter 5

THE NATURAL HISTORY OF THE BOTTLENOSE WHALE, *Hyperoodon ampullatus* (FORSTER)

Terje Benjaminsen and Ivar Christensen

Institute of Marine Research
P.O. Box 2906
5011 Bergen-Nordnes, Norway

I. INTRODUCTION

Two species of bottlenose whales, *Hyperoodon*, have been recognized, namely *H. ampullatus* in the North Atlantic and *H. planifrons* in the southern hemisphere (Rice and Scheffer, 1968).

H. ampullatus has been hunted by Norwegian whalers in the North Atlantic during two separate periods. During the first period, which lasted from 1882 to the late 1920s a total of about 50,000 bottlenose were caught. Modern Norwegian whaling for small whales, mainly minke whales (*Balaenoptera acutorostrata*), commenced around 1930, and bottlenose whales are included in the catches (Jonsgard, 1955, 1968).

The first contributions to the natural history of the bottlenose whale were written by Eschricht (1845*a*, *b*, 1849). Valuable information obtained during the first period of bottlenose whaling has been given by Gray (1882), Kükenthal (1888, 1900), Ohlin (1893), Hjort (1902), and Munsterhjelm (1915).

Since 1938 Norwegian whalers have provided biological and technical information for each whale taken. Jonsgard and Øynes (1952) analyzed biological data from 710 bottlenose whales reported in the period 1938–1950; Benjaminsen (1972) studied biological data from 5043 bottlenose caught by Norwegian whalers from 1938 to 1969, as well as 52 whales examined off Iceland in 1967; and Christensen (1973) examined data from 129 bottlenose whales caught off Labrador in 1971.

The present paper is based on information obtained from field work on bottlenose whales off the northeast coast of Iceland in April and May 1967 and off Labrador in May and June 1971. Data reported by Norwegian whalers for whales caught in the period 1938–1972 have also been examined. This study will add to our meager knowledge of the life history and population dynamics of this interesting and important species.

II. DISTRIBUTION
A. Geographic Distribution

Hyperoodon ampullatus is widely distributed in the North Atlantic. The species has also been recorded in the North Pacific, but according to Slipp and Wilke (1953) and Omura *et al.* (1955) it was probably confused with the beaked whale, *Berardius bairdi*. Nishiwaki and Oguro (1971), however, presume that the whales caught off Hokkaido in the Abashiri area in May through June are *Hyperoodon ampullatus*.

Jonsgard and Øynes (1952) referred to several authors and stated that *H. ampullatus* is a migratory species which moves north in spring and south in early autumn. The distribution in winter is poorly known, but they probably stay in lower latitudes of the North Atlantic. According to Ruud (1937), bottlenose whales have been seen as far south as the Cape Verde Islands, and he wrote that several strandings have also been reported on the southwestern coasts of Europe in winter. Juel (1886) stated that they are found as far south as Florida on the western side of the Atlantic, and True (1910) reported three bottlenose whales stranded between Boston and New York.

Gray (1882) reported that in summer he observed bottlenose whales in the Davis Strait as far north as 70°N, down the west coast of Greenland, around Cape Farewell, all around Iceland, along the East Greenland pack ice northward to 77°N, and also west of Spitsbergen and Bear Island. Gray said that he had never seen bottlenose whales in the Barents Sea, but that they undoubtedly are to be found as far east as Novaya Zemlya.

The approximate localities of bottlenose whales caught by Norwegian whalers during the period 1938–1972 are shown in Fig. 1. The figure shows that they have been caught mainly in the waters north and east of Iceland, west of Svalbard, off Labrador, and off northern and western Norway. Catches have also been made in the Denmark Strait and south of Iceland. With the exception of Labrador, these catch localities are the same as those recorded by Risting (1922) for the Norwegian bottlenose whaling from 1882 to 1920.

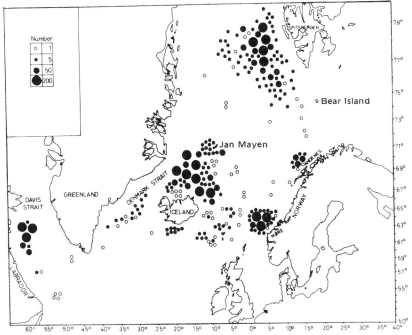

Fig. 1. Localities of bottlenose whales caught by Norwegian whalers in the period 1938–1972. (Modified from Benjaminsen, 1972.)

B. Migration

The following migratory pattern in the northeastern part of the North Atlantic was described by Hjort (1902): The bottlenose is seen as early as March in the waters off the Faroes. The greatest catches are made between Iceland and Jan Mayen at the end of April, in May, and in early June. In these months bottlenose whales are also caught west and northwest of Bear Island. The southern migration starts in the first part of July and whaling ends north of the Faroes in September.

Ohlin (1893) observed bottlenose whales from April 3 to July 10 in a wide area in the Norwegian Sea between 64° and 72° north latitude and between 2°E and 12°W longitude. Munsterhjelm (1915) saw bottlenose whales west of Svalbard from 74° to 78°30'N in May and June.

According to Mitchell (1974), concentrations of bottlenose occur, mainly in early summer, near Sable Island and along the edge of the continental shelf of Newfoundland and Labrador. Norwegian whalers have observed several hundreds of bottlenose off the coast of West Greenland at about 64°N latitude in spring and early summer.

Most of the bottlenose recorded on the coasts of Europe are strandings or seen in late summer and autumn. Eschricht (1849) and Turner (1886) pointed out that the bottlenose has most commonly been captured in the British Isles or on the coasts of France, Belgium, and Holland in September and October. According to Müller (1883), nearly all bottlenose caught on the shores of the Faroes have been taken in August and September, and Collett (1912) mentioned seven strandings of bottlenose on the southern coast of Norway, all of these in autumn. Fraser (1953) reported 46 bottlenose whales stranded on British coasts from 1913 to 1947, 33 of these in the months from July to October. Schultz (1970) summarized 25 strandings of bottlenose whales in the Baltic Sea and on the continental coasts of the North Sea. All but three of these stranded from August to November. Turner (1886) and Fraser (1953) pointed out that the frequent strandings of bottlenose whales on the coasts of northwestern Europe in the autumn occur during their southward migration from the Arctic Ocean.

Benjaminsen (1972) examined the semimonthly catches of male and female bottlenose whales off Svalbard and Iceland as shown in Figs. 2 and 3. He pointed out that seasonal catches are influenced by some of the following whaling regulations: since 1950 no whaling has been permitted between July 1 and 21; in 1952 a 6-month season lasting from March 15 to September 15 was introduced, the 3 weeks protection in July being maintained; and in 1955 further restrictions were enforced, whaling being allowed only from March 15 to June 30 north of 70°N and east of 0°.

In postwar seasons up to the mid-1950s whaling for small whales in the waters west of Svalbard took place also in July and August. However, only two bottlenose whales have been caught in this area after June 30. Benjaminsen (1972) therefore concluded that the majority of bottlenose whales seemed to

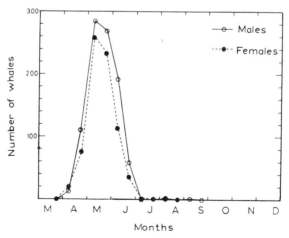

Fig. 2. Semimonthly catches of bottlenose whales off Svalbard in the seasons 1938–1969. (From Benjaminsen, 1972.)

The Natural History of the Bottlenose Whale

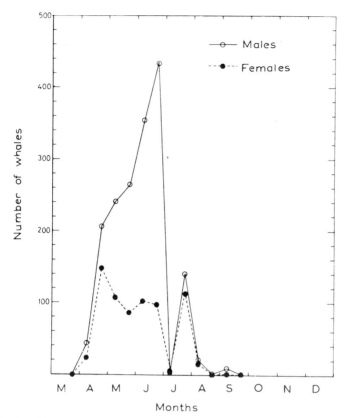

Fig. 3. Semimonthly catches of bottlenose whales off Iceland in the seasons 1938–1969. (From Benjaminsen, 1972.)

leave Svalbard waters before the end of June. He pointed out that only a few bottlenose whales have been taken later than July in Icelandic waters (Fig. 3), even though other species of small whales are caught in this area in August and September, and that most bottlenose whales seem to have left this area by early August. According to Benjaminsen (1972), minor catches have been made off the coast of Norway in September, October, and November, and therefore some bottlenose evidently postpone the migration out of the Norwegian Sea until late autumn.

Occasionally catches of bottlenose whales have been taken off the Faroe Islands since the 16th centruy (Müller, 1883), and some of the Norwegian literature from the Middle Ages mentions the use of the meat and blubber from this species. After the decline of the right whale fishery in the middle of the 19th century, some whalers caught bottlenose whales to fill up their ships. A regular fishery for the bottlenose whale was started by Norwegian whalers in 1882, and this industry expanded until 1896 when 2864 whales were caught by 65 vessels

(Risting, 1922). Thereafter the catches decreased, and in the late 1920s this whaling period ended. In modern Norwegian whaling for small whales, which commenced about 1930, the minke whales, *Balaenoptera acutorostrata,* are the main object, but some of the whalers also specialized in catching bottlenose whales. From 1938 to 1971, a total of 5791 bottlenose were taken in this fishery (Benjaminsen, 1972; Christensen, 1975).

In both periods the main whaling grounds for this species have been east and north off Iceland, west of Spitsbergen, and off the coast of Norway. In 1969 the whaling expanded to Labrador (Christensen, 1975). The catch of the bottlenose whale took place mainly from April to June.

Because of poor prices of meat and blubber from the bottlenose whale in the later years and the increasing prices on minke whale meat, only 20 bottlenose have been caught by Norwegian whalers in the period 1972–1978.

Nothing is known about the status of the stock. Some calculations were done, but more studies on available materials are necessary. Observations done by whalers in 1971 off Spitsbergen tell about herds of bottlenose in this area, and a special sighting cruise on the old bottlenose whaling grounds should be done.

C. Water Depth

Bottlenose whales have been reported to avoid shallow waters (Hjort, 1902; Jonsgard and Øynes, 1952; Benjaminsen, 1972). Data on catch localities support this view. In Fig. 4 the localities of bottlenose whales caught off Labrador are shown. Few whales have been taken over the continental shelf in this area. The greatest number of whales have been caught at depths of more than 1000 m. Benjaminsen (1972) showed that few bottlenose whales were taken at depths of less than 1000 m off the west coast of Norway and that in the other whaling areas as well few bottlenose whales have been caught over the continental shelf. Only one bottlenose whale has been reported caught in the period 1938–1972 in the large shallow waters of the North Sea, and in spite of intensive whaling for small whales in the Barents Sea since the beginning of the 1950s, no bottlenose has ever been reported caught in this shallow area. These data conflict with the statements by several authors that the bottlenose is found as far east in the Barents Sea as Novaya Zemlya.

The fact that bottlenose whales seem to avoid shallow waters during their food migration in the spring and early summer can perhaps be explained by the occurrence and distribution of their prey. Examination of stomach contents has shown that the squid *Gonatus fabricii* seems to be the main diet of this species in the North Atlantic. Unfortunately, very little is known of the biology of this squid. Autumn strandings on the northwest coasts of Europe might be explained by the movement of squid into coastal waters in that season. Some other

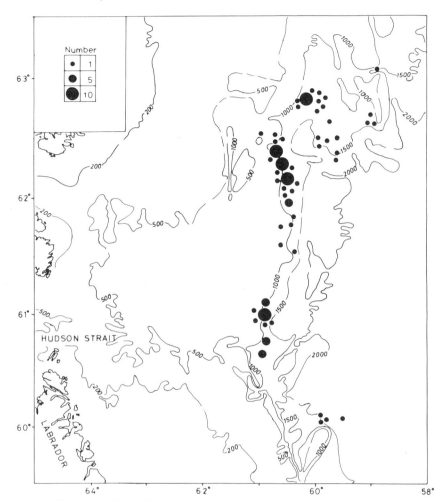

Fig. 4. Localities of bottlenose whales caught off Labrador in the 1971 season.

odontocete whales eating squid also seem to avoid shallow waters. The main catch of sperm whales off South Africa, for instance, is taken outside the continental shelf (Best, 1969b), and Nishiwaki and Oguro (1971) have shown that catches of beaked whales off Japan are taken along the 1000-m contour line.

D. Water Temperature

In the Arctic Ocean, the bottlenose whale stays mainly near the boundaries between the cold polar currents and the warmer Atlantic currents (Kükenthal, 1888; Munsterhjelm, 1915; Nansen, 1924). Malmgren (1864) presumed that the bottlenose does not live in water with temperatures below +3°C. Ohlin (1893), however, saw them most frequently in areas with water temperatures between 0°C and +2.5°C, but also in waters with temperatures between −2°C and −1°C. According to Munsterhjelm (1915), the bottlenose is found mostly in water with temperatures between +2°C and +3°C. It is apparent that they are found at higher temperatures as well. For example, Winn *et al.* (1970) observed 11 bottlenose whales off the Nova Scotia coast, where the surface temperature was +17°C.

Surface temperatures were measured on a cruise to the waters east and northeast of Iceland in April and May 1967. Bottlenose whales were seen in water with surface temperatures between −1.5°C and +0.9°C, most frequently between −1.3°C and −0.9°C.

E. Relation to Ice

According to Gray (1882), the bottlenose is found in the greatest numbers near the edge of the pack ice in the spring and summer, but it is rarely seen among the ice and appears to prefer open bays in the ice. Norwegian whalers, however, have reported that they sometimes have taken bottlenose whales several nautical miles inside the ice edge off Svalbard and Labrador. On one occasion bottlenose whales were caught more than 10 nautical miles into the ice off Labrador. Many of these whales had scratches on their foreheads. These may have been caused by the ice when the whales pushed through for breathing, although other explanations are possible.

F. Segregation

Indications have been found of some geographical segregation between males and females. Gray (1882) caught 203 bottlenose in May and June, with the catch consisting of 96 fully grown males, 56 cows, and 51 younger males. Gray (1941) stated that this catch was taken in bays in the edge of the pack ice south of Jan Mayen. Turner (1886) writes that with very few exceptions the bottlenose whales captured on the coasts of western Europe have been females, frequently accompanied by young calves. According to Collett (1907), about two thirds of the catches in the northern North Atlantic were females and younger males. Munsterhjelm (1915) reported a sex ratio of 5 males to 12 females in catches west of Svalbard in May and June. Of 26 bottlenose whales landed at Scottish whaling stations from April to September, 18 were females (Thompson, 1928). Degerbol (1940) gave a sex ratio of 20 males to 22 females caught by shore-based whalers in the Faroes.

Benjaminsen (1972) showed that in Icelandic waters the number of males is much greater than the number of females in May and June (Fig. 3) and that males prevail in this area during these months. He pointed out that there seemed to be no marked monthly segregation in the Svalbard area (Fig. 2), although males predominate slightly in the catches throughout the hunting season.

Length frequencies of male and female bottlenose whales reported from the major whaling areas from 1938 to 1969 were examined by Benjaminsen (1972) (Fig. 5). The length distributions of females in all localities has its maximum at 24 ft (7.3 m). Males exceeding 28 ft (8.5 m) have more frequently been caught in the waters off Iceland than in other localities.

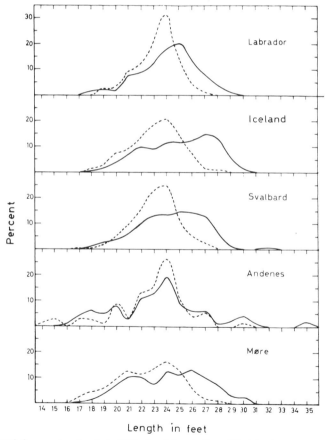

Fig. 5. Length frequencies of male (solid line) and female (dotted line) bottlenose whales caught by Norwegian whalers in different areas in the seasons 1938–1969. (From Benjaminsen, 1973.)

III. BEHAVIOR
A. Social Organization

Information on the social organization of the bottlenose whale has mainly been obtained from observations in the northern North Atlantic in spring and summer. Gray (1882) reported that he saw bottlenose whales mostly in herds of 4–10 animals. In May and June, Kükenthal (1888) most frequently observed them in groups consisting of 3–6 whales. According to Munsterhjelm (1915), they were seen mainly as single animals or in pairs off Svalbard in May, but the number of whales in each group increased through the summer. Nansen (1924) wrote that the bottlenose is a gregarious animal which forms schools composed of several hundred when they are migrating, and in the spring and summer habitat they are most often associated in groups of 2–6 whales. Winn *et al.* (1970) observed bottlenose whales in August off Nova Scotia in three groups consisting of 2, 4, and 5 animals.

Observations on group size of bottlenose whales east and northeast of Iceland in April and May 1967 and off Labrador in May and June 1971 are summarized in Table I. The whales were most frequently seen in paris, but groups of 3–6 and solitary animals also were common. Groups of more than 20 whales were not seen on these cruises.

Table II gives information on the sex, age, and reproductive status of bottlenose whales in groups observed in May and June off Labrador. Solitary animals often were young. In several cases, however, single old bulls were seen without being caught. Solitary old bulls have also been reported by Gray (1882) and Nansen (1924). Groups of two and three whales seem to consist of animals of the same sex and the same or nearly the same age. Associations of four animals seem to be composed mainly of older males. Few observations exist for groups consisting of more than four animals. However, there is some evidence that groups of more than six may consist of maturing whales of both sexes.

Lactating bottlenose whales were most frequently seen alone with their calves, and in some cases two lactating females with their calves formed a group. Ohlin (1893) observed cows alone with their calves.

Table I. Groups of Bottlenose Whales Observed off Iceland in 1967 and Labrador in 1971

	Group size										
	1	2	3	4	5	6	7	8	9	10-15	16-20
Number of groups											
Iceland	6	41	12	5	6	4	1	1	0	7	1
Labrador	10	43	29	15	4	3	2	0	1	1	1

Table II. Sex, Reproductive Status, and Age of Bottlenose Whales Caught off Labrador in Relation to Number of Whales in the Group to which They Belonged

Group size	Reproductive status[a] and age in years of whales caught[b]		
	Males	Females	Males and Females
1	U1, U4, M21	U1, U3, U5	
2	U2, (U5, U6), (M7, M9), (13, M15), M14, M18, M19, M24, M30	(R10, U10), R13, P20, P26	
3	(U3, U3, U3), (U3, U3), U6, M9, M11, M15, (M16, M16), (M16, M24)	U2, U3	
4	U6, M16, M21, M22, M34, M37	P23	
5	M11, M21	(U4, P14)	
6	M15		
7	(M9, M16)		(M12, U7)
9			(U4, U7)

[a] U = sexually immature, M = sexually mature, P = pregnant, R = resting.
[b] Animals caught in the same group are enclosed in parentheses.

B. Care-Giving

Care-giving behavior is well known for the bottlenose whale. It is known that groups of whales will not leave an injured companion until it is dead (Gray, 1882; Southwell, 1884; Ohlin, 1893; Beddard, 1900; Collett, 1907; Munsterhjelm, 1915). Our observations support these reports of attention directed toward individuals in distress. Collett (1907) wrote that if a calf approaches a ship, the mother will swim between the ship and the calf. This particular maternal care behavior has also been observed on several occasions during our cruises to Labrador and Iceland.

C. Diving

According to Scholander (1940), the bottlenose whale is perhaps the deepest and longest diver of all cetaceans. He referred to experienced whalers who have recorded dives lasting from 30 to 45 min, relating that on one occasion a harpooned bottlenose whale remained under water for 2 hr before it emerged for breathing. This exceptionally long diving time has also been observed by Gray (1882). Collett (1907) also referred to a Norwegian whaling captain who with certainty saw a bottlenose stay under for 2 hr. Kükenthal (1888) reported that he saw a wounded bottlenose remain under water for 45 min. Ohlin (1893) stated that animals emerge in the immediate vicinity of the spot where they dive, remaining under water for as long as 1–2 hr. Munsterhjelm (1915) quoted diving times between 10 and 45 min. Gray (1941) wrote that large individuals take out about 500 fathoms (900 m) of line and remain under water for about 1 hr. Winn *et al.* (1970) recorded a maximum dive lasting 15.1 min. Hjort (1902) stated that bottlenose whales usually dive vertically and take out up to 1000 m of line when they are harpooned, and Ohlin (1894) once saw a harpooned bottlenose dive vertically, running out a line of 500 fathoms in 90 sec.

On cruises to Iceland in 1967 and Labrador in 1971 we recorded the following deep diving times for different randomly selected groups: 14, 19, 21, 25, 31, 31, 33, 37, 50, and 70 min.

IV. FEEDING

Squid have been reported to be the main or only food item of the bottlenose whales (J.E. Gray, 1860, D. Gray 1882; Southwell, 1883; Kükenthal, 1888; Ohlin, 1893; Munsterhjelm, 1915). In addition to squid, the following animals have been recorded: sea cucumbers (*Holothuroidea*) (Eschricht, 1846b; Lilljeborg, 1862), sea stars (*Asteroidea*) (Ohlin, 1893), *Thysanopoda* (Collett, 1907; Munsterhjelm, 1915), and herring (*Clupea harengus*) (Ohlin, 1893; Hjort, 1902; Collett, 1907).

The stomach contents of 46 bottlenose whales examined off Iceland in 1967 and of 108 examined off Labrador in 1971 are summarized in Table III. We found that less than 10% of the animals examined off Iceland had eaten fish, while about 50% of the whales caught off Labrador had remains of fish in their stomachs. However, squid seems to be the major food item in both areas.

All the examined squid eaten by bottlenose whales were *Gonatus fabricii*. This species has also been recorded by other authors, e.g., Collett (1912), Munsterhjelm (1915), Gray (1941), and Eschricht (1846b, 1849), who also listed *Loligo* sp. and *Sepia* sp.

The following species of fish were recorded in bottlenose stomachs from Iceland: cusk (*Brosmius brosme*), lumpsucker (*Cyclopterus lumpus*), and redfish (*Sebastes* sp.). Stomach contents from Labrador included: Greenland halibut (*Reinhardtius hippoglossoides*), redfish (*Sebastes* sp.), rabbit-fish (*Chimaera*

Table III. Stomach Contents of Bottlenose Whales Examined off Iceland in 1976 and off Labrador in 1971

	Number of whales with stomach contents				
	Squid only	Squid and fish	Fish only	Various	Empty
Iceland	40	4	0	2	0
Labrador	49	53	1	3	2

monstrosa), piked dogfish (*Squalus acanthias*), ling (*Molva molva*), and skate (*Raja* sp.). One animal from Labrador also had deep-sea prawns (*Pandalus*) in the stomach together with squid and fish, and one whale, a 1-year-old calf, had milk and squid.

Indigestible foreign objects have occasionally been found in bottlenose stomachs. Müller (1883) reported finding a stone as big as a child's hand and a 4-inch piece of wood; Beneden (1888), (from Tomlin, 1967) found stones as big as pears, and Collett (1907) mentions clay and pieces of shell among the stomach contents. Some of these authors were of the opinion that the bottlenose feeds near the bottom and that it sometimes uses the beak as a plough in the mud (Müller, 1883; Ohlin, 1893) so that stones and clay may be swallowed by accident.

Among the bottlenose whales examined by us, two had swallowed pieces of fishing nets; one had 3–4 plastic bags, a sheet of plastic, and a piece of plastic trousers; one had some pieces of plastic sheets; and one had a glove as well as squid and fish in the stomach.

Berzin (1972), who wrote about a variety of man-made objects in sperm whale stomachs, supposed that whales get rid of these objects by belching or regurgitating them.

The bottlenose whale stomach usually contains a mass of horny beaks. One of the whales mentioned by Eschricht (1946*b*) contained beaks from at least 1000 squids. The stomach contents of a female bottlenose which had been caught on

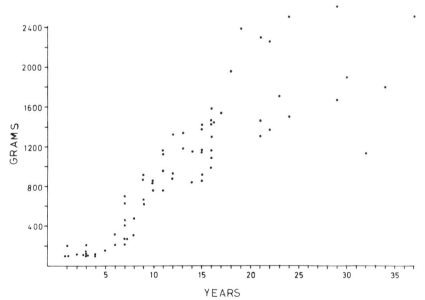

Fig. 6. Relation between the combined testis weight and the age in years of bottlenose whales caught off Labrador.

the coast near Whitstable, Kent, was estimated to have more than half a bushel (18 liters) of beaks, and nothing else (Beardsworth, 1860, quoted from Gray, 1860).

On the cruises to north Iceland in 1967 and Labrador in 1971, stomach-content volumes were estimated for most of the examined animals. Some of these stomachs contained between 20 and 25 liters of beaks and other undigested parts of the squid *Gonatus fabricii* as well as fish.

V. REPRODUCTION

A. Sexual Maturity in Males

From histological examination of testes, Benjaminsen (1972) classified 32 bottlenose whales caught off Iceland as mature or immature and concluded that the length at attainment of sexual maturity seemed to be between 730 and 760 cm. These lengths correspond to ages between 7 and 9 years.

Testes weights were obtained from 74 bottlenose whales caught off Labrador in May and June 1971. The lightest pair weighed 100 g and the heaviest 2620 g. Combined testes weights are plotted against age in Fig. 6. After 5 years the testes increase in weight. The fastest growth of the testes seems to occur

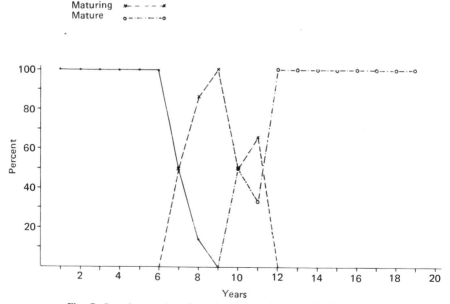

Fig. 7. Sexual maturation of male bottlenose whales caught off Labrador.

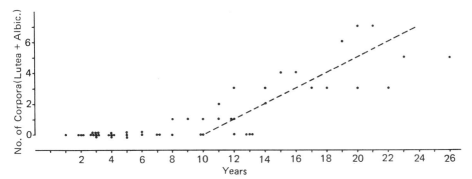

Fig. 8. The accumulation of corpora lutea and albicantia with increasing age of bottlenose whales caught off Labrador. The line represents an accumulation rate of 0.5 per year. (From Christensen, 1973.)

between 8 and 12 years. This indicates that sexual maturity is attained in these age groups.

The age at sexual maturity of whales caught off Labrador was also determined from the histological appearance of the testes. Animals were classified as immature, maturing, or mature, according to the criteria proposed for sperm whales by Aguayo (1963; quoted after Best, 1969a). Twenty seminiferous tubules randomly selected in each testis were determined as mature or immature. The animals were defined as sexually immature if less than 10% of the tubules were mature, maturing if between 10 and 75% of the tubules were mature, and sexually mature if more than 75% of the tubules were mature. Figure 7 shows the relative frequencies of immature, maturing, and mature bottlenose whales in each age group. All males younger than 7 years were immature, maturing whales were between 7 and 11 years, and all animals older than 11 years were determined as mature.

B. Sexual Maturity in Females

Eschricht (1846b) suggests from scanty material that sexual maturity of females is probably attained at a length of about 7 m. Benjaminsen (1972) examined ovaries from 19 bottlenose whales caught off Iceland and presumed that sexual maturity was reached at a length between 22 (6.7) and 23 ft (7.0 m).

Christensen (1973) studied the accumulation of corpora lutea and albicantia in 53 females caught off Labrador. His data (see Fig. 8) show that the age at sexual maturity varies at least between 8 and 13 years. Christensen (1973) found that the corpora accumulate at a rate of 0.5 per year, and that the corpora albicantia seem to persist as bodies which can be recognized in the ovaries of the bottlenose whale throughout its life. Based on these assumptions he traced the corpora back in time to the season when the respective corpora lutea of ovulation were produced and showed that individual females had their first ovulations at ages between 7 and 18 years and that 80% attained sexual maturity at ages from 8 to 12 years.

C. Pregnancy

Benjaminsen (1972) constructed a fetal growth curve based on information on lengths of 251 fetuses given by the whalers, and concluded that the gestation period is about 12 months. This supports the earlier assumptions of a 12-month gestation period in the bottlenose (Ohlin, 1893; Kükenthal, 1900). Benjaminsen (1972) showed that the peak of mating as well as parturition occurs in April. This has also been proposed by Munsterhjelm (1915).

There is only scanty information on the length of the calf at birth. Kükenthal (1900) believed the young to be 11 ft (3.3 m) at birth, and Fraser (1937) and Benjaminsen (1972) presumed the length at birth to be approximately 10 ft (3.0 m).

D. Lactation

It is difficult to determine the average duration of lactation because all the whales examined have been caught in a season of about two months. There is, however, slight evidence that lactation lasts about one year, as a 1-year-old calf had both milk and squid in the stomach.

E. Reproductive Cycle

Only two of the 24 sexually mature females examined were resting. Assuming both pregnancy and lactation to last for about one year, it is likely that most of the females have a 2-year reproductive cycle. This fits well with the accumulation rate of 0.5 corpora per year as shown in Fig. 8.

VI. AGE AND GROWTH

The layered structure in the cementum and dentine of bottlenose whale teeth was demonstrated by Eschricht (1846b). However, tooth layers were not used for age determination until Laws (1953) pointed out that laminated tooth structures might be used for aging. Since then the dentine and cementum growth layers of several species of odontocete whales have been examined.

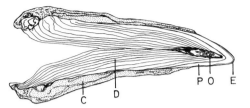

Fig. 9. Outline drawing of a longitudinal tooth section from a bottlenose whale showing the arrangement of the dentine layers. C: cementum, D: dentine, E: enamel, O: osteodentine, P: prenatal dentine. (From Christensen, 1973.)

The teeth of the bottlenose whale consist of layers of enamel, dentine, and cementum. Laminae are formed both in the dentine and the cementum, but the dentine layers are most easily counted. Prenatal dentine is easily distinguished from postnatal dentine because the former has uniform structure and a different yellowish color. Figure 9 shows the arrangement of the dentine layers in bottlenose teeth. The structure of the teeth and the deposition of the dentinal layers were discussed by Christensen (1973), who gave evidence for an annual deposition of the dentine growth layers.

Christensen (1973) examined the age composition of 53 female and 75 male bottlenose sampled off Labrador in 1971. As shown in Fig. 10, the females were from 1 to 27 years old (mean 9.8 years) and the males from 1 to 37 years (mean 13.1 years). The life span of the bottlenose whale therefore is at least 37 years.

Christensen (1973) illustrated the growth of bottlenose whales off Labrador by relating total length and age (Fig. 11). Benjaminsen (1972) presumed the length at birth to be approximately 300 cm, and as shown in Fig. 11 there is a rapid increase in length during the first 2–3 years. The growth rates for males and females are equal until the whales are about 6 years old and have attained a

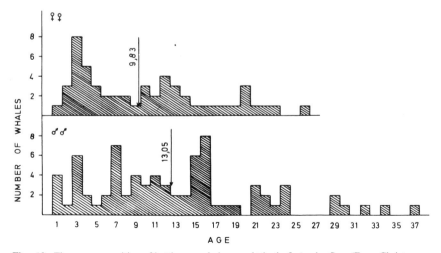

Fig. 10. The age composition of bottlenose whales caught in the Labrador Sea. (From Christensen, 1973.)

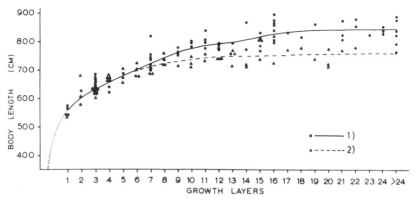

Fig. 11. The growth of bottlenose whales in the Labrador Sea (1: males, 2: females). (From Christensen, 1973.)

length of about 7 m. Growth seems to stop at an age of about 20 years in males and possibly somewhat earlier in females. Christensen (1973) therefore concluded that the length at physical maturity seems to be between 8.5 and 9.0 m for males and about 7.5 m for females. This agrees with the results of Guldberg (1886), who mentioned 30 ft (9.1 m) as maximum length for the bottlenose, and with Munsterhjelm's (1915) estimate of 7.5 m as maximum for the female bottlenose whale.

The difference between male and female growth rates is reflected not only in length, but also in the weight of meat and blubber. Figure 12 shows the weight of muscle and blubber in relation to length in feet. The data were obtained from 473 male and 401 female bottlenose whales caught off the coast of Norway in the period 1938–1969. As shown in the figure, the males yield more meat and grow fatter than the females after a length of about 23–24 ft (7.0–7.3 m). The reason may be that females, which attain sexual maturity at about 23 ft (7.0 m), have to use much of their energy to support the fetus and calf during pregnancy and lactation.

The bottlenose whales off Iceland (Benjaminsen, 1972) and Labrador (Christensen, 1973) seem to follow the same pattern of growth, and the biggest males in both areas were 9 m long.

VII. SUMMARY

This article is based on data collected in different whaling grounds in the North Atlantic in the period 1938–1972. The bottlenose whale is widely distributed in the North Atlantic, but avoids shallow waters such as the continental shelf, the Barents Sea, and North Sea. Feeding migration to the

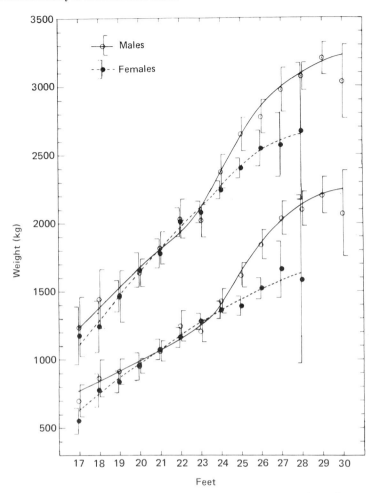

Fig. 12. The relationship between length and the weight of blubber (lower graph), and between length and the weight of meat (upper graph) for male and female bottlenose whales reported caught by Norwegian whalers in the period 1938–1969. The vertical lines indicate 95% confidence intervals.

Arctic waters lasts from the beginning of April to the end of July. Indications have been found of some geographical segregation between males and females. The bottlenose whales were most frequently seen in groups of 2–4 animals. Groups of two and three whales seem to consist of animals of the same sex and same or nearly the same age. Deep diving times of between 14 and 70 min were recorded. The squid *Gonatus fabricii* was the major food item, but several species of deep-water fishes were also found. Evidence was found for an annual deposition of dentine growth layers. Most of the females attain sexual maturity between 8 and 12 years. All males younger than 7 years were immature and all males older than 11 years were determined as mature. Pregnancy lasts for 12 months and the females have a 2-year reproductive cycle.

ACKNOWLEDGMENTS

We are indebted to Mr. Torger Oritsland for help and advice, Dr. Age Jonsgard for valuable comments on the manuscript and for collecting material on a whaling cruise to Iceland, to Mr. Sidney Brown for valuable suggestions on the manuscript, and to Mr. Ingvard Huse who participated in the collection and the processing of material from Labrador. Also thanks to Mrs. Ellen Sophie Lauvas for technical assistance and to Miss Claudia Hamilton for correcting the English text.

REFERENCES

Aguayo, L. A., 1963, Observaciones sobre la madurez sexual del cachalote macho (*Physeter catodon* L.) capturado en aguas chilenas, *Montemar* 3:99–125.
Beardsworth, R. G., 1860, *Illustrated News,* Nov. 18, 1860.
Beddard, F. E., 1900, *A Book of Whales*, Putnam, New York.
Beneden, P. J. van, 1888, Les Ziphoides en mers d'Europe, *Mem. Cour. Acad. Sci. Belg.* 41.
Benjaminsen, T., 1972, On the biology of the bottlenose whale, *Hyperoodon ampullatus* (Forster), *Norw. J. Zool.* **20**:233–241.
Berzin, A.A., 1972, The Sperm Whale, Israel Program for Scientific Translation, Jerusalem. (Translated from Russian: Berzin, 1971).
Best, P. B., 1969a, The sperm whale *(Physeter catodon)* off the west coast of South Africa. 3. Reproduction in the male, *Invest. Rep. Div. Sea Fish. S. Afr.* **72**:1–20.
Best, P. B., 1969b, The sperm whale *(Physeter catodon)* off the west coast of South Africa. 4. Distribution and movements, *Invest. Rep. Div. Sea Fish. S. Afr.* **78**:1–12.
Christensen, I., 1973, Age determination, age distribution and growth of bottlenose whales, *Hyperoodon ampullatus* (Forster), in the Labrador Sea, *Norw. J. Zool.* **21**:331–340.
Christensen, I., 1975, Preliminary report on the Norwegian fishery for small whales: Expansion of Norwegian whaling to arctic and northwest Atlantic waters, and Norwegian investigations of the biology of small whales, *J. Fish. Res. Board Can.* **32**:1083–1094.
Collett, R., 1907, Nogle meddelelser om naebhvalen (*Hyperoodon*) og hvidfisken *(Delphinapterus)*, *Bergens Mus. Arbok* **1906(6)**:1–25.
Collett, R., 1912, *Norges Pattedyr*. H. Aschehoug & Co., W. Nygaard, Kristiania.
Degerbol, M., 1940, Mammalia, in: *Zoology of the Faroes* (A.S. Jensen, W. Lundbeck, and T. Mortensen, eds.), pp. 1–132, Høst, Copenhagen.
Eschricht, D. F., 1846a, Udbytee paa en Reise gjennem det nordvestlige Europa i Sommeren 1846. Sjette Afhandling i Undersøgelser over Hvaldyrene. *Vid. Sel. Nat. Mat. Afh.* **11**:85–138.
Eschricht, D. F., 1846b, Undersøgelser over Hvaldyrene. Fjerde Afhandling. Om naebhvalen. *Vid. Sel. Nat. Mat. Afh.* **11**:321–378.
Eschricht, D. F., 1849, *Zoologisch-anatomisch-physiologische Untersuchungen über die Nordischen Walthiere,* Van Leopold Voss, Leipzig.

Fraser, F. C., 1937, Whales and dolphins, in: *Giant Fishes, Whales and Dolphins* (J.R. Norman and F. C. Fraser, eds.), pp. 203–349. Putnam, London.

Fraser, F. C., 1953, Reports on Cetacea stranded on the British coasts from 1938 to 1947, British Museum (Nat. Hist.), London.

Gray, D. 1882, Notes on the characters and habits of the bottlenose whale (*Hyperoodon rostratus*), *Proc. Zool. Soc. London* 1882:726–731.

Gray, J. E., 1860, On the genus Hyperoodon: The two British kinds and their food, *Proc. Zool. Soc. London.* **28**:422–426.

Gray, R. W., 1941, The bottlenose whale, *Naturalist* **791**:129–132.

Guldberg, G. A., 1886, Naebhvalen, *Naturen* **12**:162–165, 177–179.

Hjort, J., 1902, Fiskeri og hvalfangst i det nordlige Norge, John Griegs Forlag, Bergen.

Jonsgard, A., 1955, Development of the modern Norwegian small whale industry, *Nor. Hvalfangst-Tid.* **44**:697–718.

Jonsgard, A., 1968, A review of Norwegian biological research on whales in the northern North Atlantic Ocean after the Second World War, *Nor. Hvalfangst-Tid.* **57**:164–167.

Jonsgard, A., and Øynes, P., 1952, Om bottlenosen (*Hyperoodon rostratus*) og spekkhoggeren (*Orcinus orca*), *Fauna Oslo* **5(1)**:1–18.

Juel, N., 1886, Nebhvalen, *Nor. Fisk. Tid.* **5**:169–180.

Kükenthal, W., 1888, Bericht über eine Reise in das Eismeer und nach Spitzbergen im Jahre 1886, *Geograph. Blätter* **11(1)**:1–43.

Kükenthal, W., 1900, Die Wale der Arktis, *Fauna Arct.* **1(2)**:179–234.

Laws, R. M., 1953, A new method of age determination in mammals with special reference to the elephant seal (*Mirounga Leonina, Linn.*), *Sci. Rep. Falkld. Isl. Depend. Surv.1* **2**:1–11.

Lilljeborg, W., 1862, Öfversigt af de inom Skandinavien (Sverige och Norrige) anträffade hvalartade däggdjur (Cetacea), *Upsala Univ. Arssk. (Math. Naturvetenskap)* **1862(1)**:1–38.

Malmgren, A. J., 1864, Beobachtungen und Anzeichnungen über die Säugethierfauna Finmarkens und Spitzbergens, *Arch. Naturgesch.* **30**:63–97.

Mitchell, E., 1974, Porpoise, dolphin and small whale fisheries of the world: Definition of the problem, Paper submitted to I.U.C.N. Monogr. series. (mimeo).

Müller, H. C., 1883, Opplysninger om Döglingefangsten paa Faröene, *Vidensk, Medd, Dan. Naturhist. Foren. Khobenhavn* **1883**:48–67.

Munsterhjelm, L., 1915, Anteckningar om Hyperoodon rostratus (Müll.) gjorda under en ishavaresa sommaren 1910, *Tromsø Mus. Arsh.* **37**:1–13.

Nansen, F., 1924, *Blant Sel og Bjørn*, J. Dybwads Forlag, Kristiania.

Nishiwaki, M., and Oguro, N. 1971, Baird's beaked whales caught on the coast of Japan in recent 10 years, *Sci. Rep. Whales Res. Inst., Tokyo* **23**:111–122.

Ohlin, A., 1893, Some remarks on the bottlenose whale (*Hyperoodon*),*Acta Univ. Lund.* **29(8)**:1–14.

Ohlin, A., 1894, Nagra anteckningar om den nutida hvalfangsten i Norra Ishafvet, *Ymer* **1894**:145–164.

Omura, H., Fujino, K., and Kimura, S., 1955, Beaked whale *Berardius bairdi* of Japan with notes on *Zipius cavirostris*, *Sci. Rep. Whales Res. Inst., Tokyo* **10**:89–132.

Rice, D. W., and Scheffer, V. B., 1968, A list of the marine mammals of the world, *U.S. Fish Wildl. Serv. Spec. Sci. Rep. Fish. 431,* 12 pp.

Risting, S., 1922, *Av hvalfangstens historie,* J. Petlitz, Kristiania.

Ruud, J. T., 1937, Bottlenosen, *Nor. Hvalfangst-Tid.* **26**:456–458.

Scholander, P. F., 1940, Experimental investigations on the respiratory function in diving mammals and birds, *Hvalrad. Skr.* **22**:131 pp.

Schultz, W., 1970, Uber das Vorkommen von Walen in der Nord- und Ostsee (ordn. Cetacea), *Zool. Anz.* **185**:172–264.

Slipp, J. W., and Wilke, F., 1953, The beaked whale *Berardius* on the Washington coast, *J. Mammal.* **34(1)**:105–113.

Southwell, T., 1883, On the beaked or bottle-nose whale *(Hyperoodon rostratus)*, *Trans. Norfolk Norwich Nat. Soc.* **3**:476–481.

Southwell, T., 1884, The bottle-nose whale fishery in the North Atlantic Ocean, *U.S. Comm. Fish and Fisheries* **10**:221–227.

Thompson, D'A. 1928, On whales landed at the Scottish whaling stations during the years 1908–1914 and 1920–1927, *Sci. Invest. Fish. Bd Scott.* **3**:3–39.

Tomilin, A. G., 1967, Mammals of USSR and Adjacent Countries, Vol. 9. Cetacea, Israel Program for Scientific Translations, Jerusalem (Translated from Russian: Tomilin, 1957).

True, F. W., 1910, An account of the beaked whales of the family *Ziphiidae*, *Bull. U.S. Nat. Mus.* **73**:1–89.

Turner, W., 1886, On the occurrence of the bottle-nosed whale or beaked whale *(Hyperoodon rostratus)* in the Scottish seas, with observations on its external characters, *Proc. R. Phys. Soc. Edinb.* **9**:25–47.

Winn, H. E., Perkins, P. J., and Winn, L., 1970, Sounds and Behaviour of the Northern Bottle-Nosed Whale, 7th Annual Conference on Biological Sonar and Diving Mammals, Stanford Research Institute, Menlo Park, California (1970) pp. 53–59 (mimeo).

Chapter 6

THE SOCIOECOLOGY OF HUMPBACK DOLPHINS (*Sousa* sp.)

Graham S. Saayman

Psychology Department, University of Cape Town
Rondebosch, C.P., South Africa

and

Colin K. Tayler

Museum, Snake Park and Oceanarium
Port Elizabeth, South Africa

I. INTRODUCTION

Quantitative studies of free-swimming dolphins are of significance for a number of reasons. Whereas information concerning the Delphinidae, the family of smaller toothed whales, has accumulated rapidly during the last two decades, much of our knowledge of the behavior of dolphins has been derived primarily from qualitative observations of bottlenose dolphins (*Tursiops* sp.) under captive conditions (Townsend, 1914; McBride and Hebb, 1948; McBride and Kritzler, 1951; Lawrence and Schevill, 1954; Tavolga and Essapian, 1957; Essapian, 1963; Tavolga, 1966). At present, however, very little is known concerning the basic requirements of dolphins in captivity (Dudok van Heel, 1972; Tayler and Saayman, 1973) and consequently much of what has been written concerning their social behavior and organization is speculative, if not erroneous.

Studies of captive dolphins have been made possible largely as the result of the establishment of public oceanaria where the primary emphasis is upon commercial display of trained animals. Results derived from such studies may be distorted by a variety of factors. Dolphins unresponsive to training procedures are generally rejected, and the colony therefore does not contain representative samples of animals. Furthermore, the age/sex ratios of normal populations of dolphins are not known and therefore cannot be duplicated in captivity. In many institutions captive conditions are grossly inadequate and the death rate is high (Hussain, 1973); thus the possibility of long-term studies on stable populations is often excluded.

Among the most important prerequisites for the maintenance of a healthy breeding colony of dolphins is an adequately spacious pool, the acoustical properties of which should cater to the acute auditory perception of dolphins (Tayler and Saayman, 1973b). Inadequate spatial conditions may lead to abnormally severe aggression. The widely varying composition of free-ranging groups of dolphins (Tayler and Saayman, 1972; Saayman and Tayler, 1973a,b) indicates that provision should be made in captivity for the animals to associate or disperse at will. Ideally, an offending dolphin should be able to retreat from both the sight and sound of a more dominant animal. Furthermore, dolphins rely primarily upon acoustical mechanisms for navigational and discriminatory purposes (Kellogg, 1961; Norris, 1966), but in captivity they are often maintained in small and shallow circular tanks with concrete walls and glass windows. These holding facilities represent, in effect, acoustical reverberation chambers which may grossly disturb an animal with a highly developed auditory perceptual system. The clinically sterile conditions of many oceanaria, although presenting favorable viewing conditions for the public, deprive the dolphins of all contact with marine flora and fauna, the latter representing their prey. In the case of inshore dolphins, which usually inhabit murky seas, crystal clear water in captivity may further inhibit their normal acoustical repertoire. The absence of the above prerequisites may, indeed, lower the physical condition of the animals, a factor in itself likely to distort the results of behavioral studies. In summary, while it is relatively practicable to provide many terrestrial mammals with favorable seminaturalistic conditions in game reserves, it is difficult, if not impossible, for the majority of institutions to reproduce in captivity the necessary prerequisites to cater for the unique socioecological adaptations which the dolphin has made over millions of years.

The above considerations indicate that the interpretation of behavior observed in captivity must be approached with great caution. Moreover, captive bottlenose dolphins display a marked propensity to learn complex behavior sequences by imitation (Tayler and Saayman, 1973a) and thus studies of their behavioral repertoire are fraught with further possible pitfalls of misinterpretation. Problems are further compounded by the difficulties encountered in isolating and defining specific units of behavior and in assessing their function in varying social contexts. Caldwell and Caldwell (1968, pp. 61–62) have written as follows concerning this problem:

> "A complicating factor in studying sexual behavior in this species is the separation of 'play', 'sex' and 'aggression.' These areas are much more clearly defined in invertebrates and the lower vertebrates than they are in bottlenosed dolphins. The lines between these little behavioral boxes created by the human mind become very fuzzy in advanced mammals. One area fades into another both in word definition and in the

actual temporal sequence of the animal's activity. Both dolphins and pilot whales . . . frequently engage in what appear to be fights between adult male and female, in which they give every indication of trying to kill each other. They swim at each other at full speed and bang heads so violently that the sound shakes the tanks and corridors: after this, however, copulation may occur.''

The functions of many of the behavioral sequences seen at close range and in detail in captivity can only be determined by naturalistic studies of dolphins, where behavior can be observed under the appropriate socioecological conditions for which it has been adapted. This applies not only to the interpretation of behavioral interactions but also to vocal communication, an area which has received considerable attention in the Delphinidae. The comments of Evans (1967, p. 164) are pertinent:

"Communication by means of auditory signaling has also been postulated as existing at levels ranging from complex systems bordering on a 'language' or syntactic code to expression of emotional state, territorial calling, identification calling, etc. Most of the information to date has been speculative and based on indirect and in most cases insufficient evidence. . . . Our current knowledge of the social structure within natural populations of cetacea is limited and with some species nonexistent. Without this knowledge, interpretation of results obtained from captive populations is difficult if not impossible to interpret in any meaningful way. Therefore, in reviewing what information is available from captive observations we must keep in mind the little we do know about the ecology and behavior of natural populations of marine mammals and hold any conclusions in abeyance until adequate field data are available.''

Published accounts of comprehensive studies on the social behavior and organization of free-swimming dolphins are virtually nonexistent and the available information must be extracted from diverse and fragmentary observations scattered in the literature. As Evans and Bastian (1969) have shown, much of the information is of an anecdotal and qualitative nature. In the present studies an attempt was made to collect quantitative data on a population of free-swimming humpback dolphins (*Sousa* sp.) over a period of three years in order investigate systematically some of the most elementary questions concerning their social behavior and social organization. Preliminary findings have been published elsewhere (Saayman and Tayler, 1973a,b; Saayman *et al.*, 1972).

II. THE STUDY AREA

Quantitative data presented in this report were derived from the systematic study of humpback dolphins observed from the Robbe Berg peninsula at Plettenberg Bay (34°S 23°E), South Africa. The Robbe Berg is a steep-cliffed peninsula (Fig. 1), approximately 450 m broad, which projects 3.6 km from the mainland in an easterly direction into the sea and forms the southern boundary of Plettenberg Bay. An observer situated at the point—the extremity of the Robbe Berg—had a panoramic view of Plettenberg Bay and of the open sea and could

Fig. 1. The Robbe Berg Peninsula at Plettenberg Bay (from Saayman et al., 1973).

observe the dolphins as they passed directly below him in seas with up to 24 m of water clarity. Two distinct ecological zones were recognized (Figs. 2, 3, and 4):

A. Zone I

The Robbe Berg peninsula forms a barrier to the prevailing southwesterly wind and a boundary between the relatively tranquil waters of the bay and the open sea. The sandy bottom directly below the peninsula deepens abruptly from the occasional rocky outcrops and reefs formed largely by crumbling rock. The almost unbroken sandy beach of the bay sweeps northwards from the Robbe Berg in a gradual curve.

B. Zone II

The southern side of the Robbe Berg forms an undulating coastline with rocky outcrops, deep gullies, and outlying reefs upon which the open sea breaks in strong swells. The Robbe Berg extends in deep submerged reefs in an easterly direction rising to Whale Rock which is also submerged but at times awash, 0.5 km from Robbe Berg Point.

Fig. 2. Zone I at Robbe Berg Point. Dolphins were observed directly beneath the cliffs in the relatively calm and sheltered waters of the bay (from Saayman et al., 1972).

Fig. 3. Zone II at Robbe Berg Point. The open sea breaks upon the rocky coastline. Heavy swells break upon Whale Rock, which is submerged but awash in the right-hand corner of the photograph (from Saayman et al., 1972).

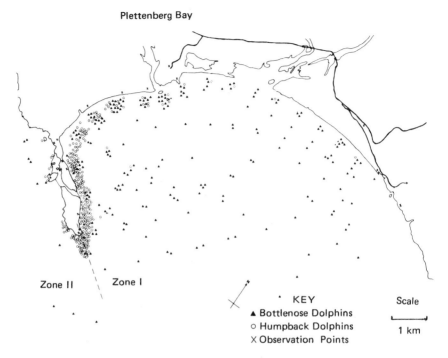

Fig. 4. The distribution of humpback and bottlenose dolphins in Plettenberg Bay. Humpback dolphins remained close inshore whereas bottlenose dolphins entered all sectors (from Saayman et al., 1972).

III. METHODS

A systematic watch was maintained for dolphins at Plettenberg Bay during the first two weeks of spring, summer, autumn, and winter seasons between November 1970 and July 1973. Since the conditions for observing humpback dolphins were most favorable at Robbe Berg Point, and since experience showed that these dolphins tended to frequent the vicinity of the point, systematic vigils were conducted most intensively in this area. In addition, dolphins could be followed by vehicle for a distance of 14.5 km and watched from several vantage points as they progressed along the coastline of the bay. A number of qualitative observations of dolphins, made in Algoa Bay (34°S 25°E) are also included in this report.

Binoculars (20 × 60 and 8 × 35) were used during all observation sessions. The duration of behavioral events was recorded on a tape recorder and the data were then transcribed into field notebooks and transferred to scoring sheets and punched cards. Dolphin progressions were plotted against topographical aerial photographs of the coast. Daily recordings of sea temperatures were made in the surf zone. Wind velocities were measured with a Dwyer wind meter. Photographs were taken with a 35-mm Pentax camera using 55- and 250-mm lenses. Behavioral sequences were filmed with a Minolta 8D 10 ciné-camera with a continuously variable 7- to 70-mm telephoto zoom lens.

IV. STATISTICAL ANALYSIS OF RESULTS

Records were kept of the duration of vigils for dolphins as well as of the duration of sightings of dolphins per day. The duration in minutes of behavioral events was calculated for each daily observation session. A number of statistical comparisons were made, using rates derived from these data. Differences were tested for significance using a single-factor analysis of variance with harmonic means for unequal sample sizes (Winer, 1962).

V. THE DOLPHINS

The humpback dolphin (Fig. 5) is observed frequently along the southeastern cape coast, but its distribution and behavior have not been previously studied in

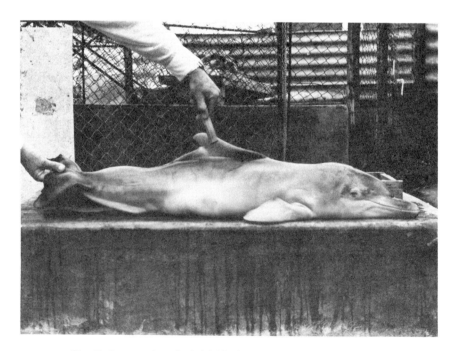

Fig. 5. A neonate humpback dolphin. Photograph by Graham J.B. Ross.

this area. Its taxonomic status is by no means settled; it is described as *Sotalia lentiginosa* by Tietz (1963) and *Sousa plumbea* or *lentiginosa* by P.J.H. van Bree (personal communication).

The humpback dolphin is readily distinguished by the notably small, sickle-shaped dorsal fin which is situated upon the elevated mid-dorsal section. The "hump" is the most characteristic morphological feature, from which the common name "humpback" is derived. The species has a long, slender rostrum which appears first as the dolphin surfaces for air. As the animal sounds after rising steeply from a depth during feeding, the back is strongly arched, displaying the prominent hump.

There have been no systematic age-determination studies of *Sousa*. In the absence of precisely determined standards, it was necessary to rely upon field criteria to categorize the age classes of the animals. At present there are no known morphological characteristics to discriminate reliably between the sexes at sea; whereas bulls with penile erections were seen occasionally, it rarely proved possible to identify the individuals when they surfaced for air.

A. Calf I

Young calves were off-white in color and were approximately one third the length of adults. In general, these calves were closely accompanied by adults and remained alongside and slightly behind the dorsal fin of the adult in the characteristic manner of delphinid young. Whereas they were able to keep pace and to maintain high speeds while in this position, very young calves appeared to be unable to judge the surface when swimming slowly since they rose steeply for air, appeared to overshoot, and cleared the water completely reentering belly first with a splash. Calf Is frequently adopted the suckling position beneath the mother. Occasionally they positioned themselves crosswise with the belly on top of the rostrum of an adult, presumably the mother, which then agitated the belly of the calf with sideways movements of the head. Similar sequences have been observed and filmed in captive bottlenose dolphins in the Port Elizabeth Oceanarium, and were perhaps performed for the relief of gastric discomfort.

Two stranded neonates measured 97 and 108 cm and weighed 7.9 and 15.5 kg, respectively; the former specimen however, was dehydrated and emaciated (G.J.B. Ross, personal communication).

B. Calf II

The coloration of the older calves, approximately half the length of adults (Fig. 6) had commenced to deepen to gray. Calf IIs frequently swam independently of adults but remained in their vicinity. At this stage, swimming movements were well coordinated, and the calves displayed the full pattern of adult social behavior, including airborne somersaults and inverted postures

Fig. 6. Two adult humpback dolphins and a calf II (from Saayman et al., 1972).

adopted by adults during mating (see pp. 194–196). Suckling was seen less frequently and the calves occasionally carried fish in their mouths.

C. Juvenile

Dolphins approximately two thirds of adult length were classified as juveniles. The light color characteristic of calves had been replaced by a uniform gray at the juvenile stage. These animals were clearly less dependent upon adults than were either the calf I or II groups. The impression was gained that juveniles tended to form subgroups which generally remained in the vicinity of adults. For example, a number of juveniles remained beneath the observation point while the adults moved slowly from the observer's view. Shortly after the last of the adult dolphins had disappeared beyond a promontory, the juveniles displayed a marked "startle" reaction and the water boiled as they formed up into a tightly knit formation and proceeded at high speed in the direction taken by adults.

D. Grayback

These dolphins were not yet of maximum length or girth. The color pattern was a uniform slate gray on the dorsal and lateral surfaces with a characteristic unblemished and unscarred appearance, sometimes displaying a purplish sheen. A stranded specimen measured 2.71 m, but testis size indicated that it was probably a sexually subadult animal (G.J.B. Ross, personal communication). Graybacks displayed full independence of movement and association and were behaviorally indistinguishable from adult dolphins.

E. Whitefin

Adult humpback dolphins were characterized by a remarkable girth and the whitening of the dorsal fin and adjacent areas. In some animals, this whitening extended to the tip of the rostrum and flukes and appeared to be an index of increasing age. A striking feature of adult humpback dolphins was the occurrence of prominent deep scars on the flanks and dorsal surface, particularly posterior to the dorsal fin. Some animals were individually identifiable by virtue of such scars (Fig. 7). Misaligned mandibles were seen in two cases, one of which was a captive animal in the Port Elizabeth Oceanarium; the other specimen was seen frequently in Plettenberg Bay. An adult bull which stranded at Durban measured 2.79 m and weighed 287.1 kg. It bore scars of a deep sharkbite on the flank.

Fig. 7. The deep scar posterior to the dorsal fin permitted the individual identification of the whitefin adult dolphin (TF) in the foreground.

VI. RESULTS

A. Range

The appearance of dolphins close inshore initially seemed to be an irregular and sporadic phenomenon, and there was some doubt as to the feasibility of studying these marine predators systematically from coastal vantage points suitable for close-range observations. At the outset it was therefore necessary to determine whether the dolphins were essentially nomadic, ranging more or less at random over a virtually unbounded marine habitat.

Humpback dolphins were seen regularly close inshore from several observation points in Plettenberg Bay, the majority of which provided only long-range observations with a limited field of view owing to the low angle of inclination. Experience, however, revealed that these dolphins frequented the vicinity of Robbe Berg Point, where they were observed for a total of 181.6 hr or 27.7% of the 655.3 hr spent watching for them in this area between November 1970 and July 1973. Daily vigils at the point averaged 367.4 ± 11.08 ($\bar{X} \pm$ SE) minutes per day over the entire period of study; this rendered an average return of 101.8 ± 10.3 minutes of observation of the dolphins per day. An observer situated at the point for approximately 6 hr/day could, therefore, expect to observe humpback dolphins from an eminently suitable vantage point for 1.7 hr, thus providing the opportunity to surmount some of the limitations imposed upon the accumulation of quantitative data on the naturalistic behavior of these marine mammals by the nature of their extensive three-dimensional aquatic habitat.

Moreover, the restriction of group movements of humpback dolphins to within 1 km of the shore was a striking feature of their behavior. This distribution is illustrated for Plettenberg Bay in Fig. 4; the study area was divided into grid squares of 0.8 km and dolphin progressions were plotted as the animals entered each individual grid. Humpback dolphins, in the majority of cases, remained close inshore within 250 m of Robbe Berg and just seawards of the breaking waves. In other areas of Plettenberg Bay, as well as in Algoa Bay, where the seabed was sandy with outcrops of isolated reefs, humpback dolphins moved systematically from one outcrop to the next; different groups followed similar routes and lingered over the same reefs where they remained submerged for comparatively long periods. This distribution contrasted markedly with that of bottlenose dolphins, which moved along similar routes close inshore but were also observed in deep water with almost equal frequency (Fig. 4). Whereas bottlenose dolphins are considered to be inshore delphinids, they have been identified 34 km offshore in water depths of more than 1000 fathoms (Ross, 1973). The tendency of humpback dolphins to frequent specific areas such as Robbe Berg Point, where they could be studied from elevated observation points, and the relative restriction of their range close inshore made this species a particularly suitable subject for behavioral studies of a delphinid.

Dolphins at sea present very few distinctive morphological features, as they surface to breathe, that would permit reliable distinction between individuals, and this is true even when they are studied at close range from optimum observation points. It was not possible, therefore, to determine the extent of the area normally traversed by schools of humpback dolphins during the course of their routine day-to-day hunting and maintenance activities. Moreover, the restriction of suitable close-range observations by the topographical features of the coastline to the single locality of the Robbe Berg exacerbated this problem. Under such circumstances, the boundaries of the home range of schools of dolphins can perhaps only be established with complete accuracy by the development of radio-tracking and telemetry devices to trace the movements of specific individuals (see Evans, 1971).

However, in this study it was possible to identify several humpback dolphins by virtue of prominent scars (Fig. 7), permitting limited speculation concerning the range of the animals. Data on the occurrence of three of the most conspicuously scarred individuals observed from the Robbe Berg on consecutive days of study during each successive field expedition are shown in Fig. 8. These data indicate that at least two of the known individuals frequented the vicinity of the Robbe Berg during all three years of study, irrespective of season. All three dolphins were seen in the same groups on three separate occasions, suggesting that they belonged to the same school. Rough seas, cloudy skies, rain, reflected sunlight, the speed of group progressions, and the protracted periods of time dolphins spent beneath the surface—particularly when feeding—frequently imposed difficult conditions of observation and precluded the possibility of individual identification. Moreover, vigils were not systematically conducted from the Robbe Berg throughout the hours of daylight during all study periods, and we have no information concerning the movements of humpback dolphins at night. Notwithstanding the cumulative effect of these limiting factors, the three dolphins were positively identified on 40.2% of the days of study. This evidence strongly suggested that these individuals were part of a school which habitually ranged over the Plettenberg Bay and adjacent areas throughout the year. It was not possible to determine the extent of their range, but the fact that the identifiable dolphins were on occasions sighted on successive days (e.g., autumn 1972, Fig. 8) suggested that the school exploited a limited and familiar terrain.

B. Seasonal Occurrence

The results presented in the preceding section indicate that the Plettenberg Bay area formed part of a range regularly utilized by at least one school of

The Socioecology of Humpback Dolphins

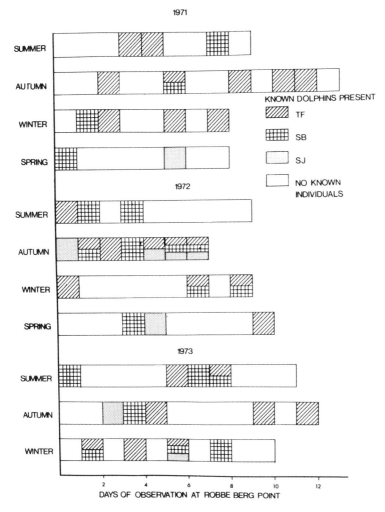

Fig. 8. The occurrence of three individually identifiable humpback dolphins on successive days of observation from the Robbe Berg. The three dolphins were positively identified on 40.2% of the days of study and were present in the area for three consecutive years irrespective of season.

humpback dolphins. Seasonal changes in environment are associated with migratory behavior in many cetaceans, and it initially seemed probable that the home-range utilization of the humpback dolphin might show similar seasonal variations; whereas very little is known concerning the taxonomy and geographical distribution of *Sousa*, it is generally accepted that this genus is limited to tropical waters (Gaskin, 1972; Nishiwaki, 1972). It was therefore possible that Plettenberg Bay, falling within the subtropical region, verges on the limits of its distribution, and thus it was necessary to determine whether humpback dolphins frequented the area in the same numbers throughout the year, despite seasonal fluctuations in sea temperatures, or whether the dolphins extended their range and migrated to warmer waters during the winter.

Two calculations were made in order to answer these questions: (1) The number of dolphins observed per sighting was divided by the number of sightings obtained per season (Table I). The mean ($\pm$ SE) number of dolphins observed per sighting at Robbe Berg Point ranged between 3.91 ± 1.07 in the autumn of 1973 and 13.6 ± 2.11 in the winter of 1972. In general, more dolphins were observed per sighting during the winter months, although this effect was most marked in the 1971–1972 study period. The data were grouped in relation to season for all three years of study (Fig. 9, upper part). Groups of dolphins tended to be larger in winter than during the spring ($F = 3.56$; $df\ 3,152$; $P < 0.05$). (2) A sighting rate was calculated by dividing the number of sightings per day by the number of hours of daily vigils for dolphins (Table II). The mean ($\pm$ SE) sighting rates ranged between 0.14 ± 0.05 during the summer of 1971 and 0.43 ± 0.11 during the autumn of 1972. The data were grouped in relation to season for all three years of study (Fig. 9; lower part); the differences observed were not significant ($F = (F = 1.81$; $df\ 3,103$; $P < 0.25$).

These results confirmed the fact that the dolphins remained within the study area throughout the periods of investigation and corroborated the evidence obtained from observations of identifiable individuals (Fig. 8). However, groups of dolphins tended to be larger in the winter than in the spring. Furthermore, a strong subjective impression, derived from efforts to obtain quantitative data and cinematographic records of social behavior patterns, suggested that the dolphins frequented the vicinity of the Robbe Berg for much shorter periods of time in spring than in winter.

In order to validate this impression, the duration of daily sightings of dolphins was adjusted for the length of vigils per day, which varied in duration from season to season, owing to climatic factors such as rainfall and differences in the number of daylight hours as well as to the fact that concomitant studies of bottlenose dolphins were carried out in areas of Plettenberg Bay other than Robbe

Table I. Number of Humpback Dolphins per Sighting at Robbe Berg Point in Relation to Season (Means $\pm$ SE)

	Number of dolphins per sighting			
	Spring	Summer	Autumn	Winter
1970–1971	6.40 ± 2.38	6.00 ± 1.40	5.20 ± 1.18	7.00 ± 1.46
1971–1972	4.20 ± 1.02	5.70 ± 0.87	9.82 ± 1.37	13.67 ± 2.11
1972–1973	4.17 ± 0.93	4.80 ± 0.72	3.91 ± 1.07	6.52 ± 1.40

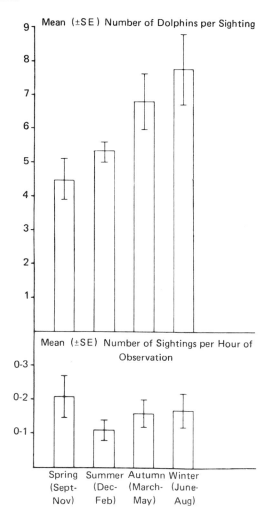

Fig. 9. The occurrence of humpback dolphins in Plettenberg Bay in relation to season. The dolphins were present throughout the year, but the size of groups tended to increase during the winter months.

Berg Point during the initial stages of the investigation (Saayman, et al., 1973). The means (± SE) of this index of sighting duration were consistently higher for the winter expeditions than for those obtained in spring (Table III). The data were grouped in relation to season for all three years of study (Fig. 10, upper part) and confirmed the subjective impression that the dolphins remained in the Robbe Berg Point area for significantly longer periods of time in winter than in spring ($F = 4.97$; $df\ 3,103$; $P < 0.01$). The mean daily sea surface temperatures increased significantly in the summer months ($F = 48.2$; $df\ 3,131$; $P < 0.001$), but the variations between the spring and winter temperatures were not marked (Fig. 10, lower part).

Thus, humpback dolphins did not appear to undertake any large-scale

Table II. Number of Sightings of Humpback Dolphins per Hour of Observation at Robbe Berg Point in Relation to Season (Means ± SE)

	Number of sightings per hour of observation			
	Spring	Summer	Autumn	Winter
1970–1971	0.27 ± 0.10	0.14 ± 0.05	0.21 ± 0.06	0.31 ± 0.09
1971–1972	0.23 ± 0.09	0.19 ± 0.05	0.43 ± 0.11	0.16 ± 0.02
1972–1973	0.39 ± 0.11	0.27 ± 0.06	0.19 ± 0.05	0.39 ± 0.10

migrations despite significant seasonal fluctuations in sea temperatures. However, groups of dolphins tended to be larger in winter. Whereas their geographical range appeared to be relatively limited, the movements of the dolphins appeared to be less restricted to the region of the Robbe Berg Peninsula during the spring than during the winter. Since spring and winter sea temperatures did not differ markedly, this differential utilization of the habitat did not seem to be directly related to variations in seasonal water temperatures.

C. Social Structure

Estimates of the numbers and densities of populations of humpback dolphins in the Plettenberg Bay area were severely hampered by the irregular movements of small groups of dolphins up and down the coast, by the difficulties encountered in recognizing individuals (and thus distinguishing between schools), and by the marked instability and temporary nature of groups which frequently altered in composition from one sighting to the next. Furthermore, the lack of any distinctive morphological features for discriminating between the sexes in the field made it impossible to determine the sex ratios of the groups observed. The problems are illustrated in an example extracted from field notes:

> 3rd May, 1972. 1015 hr. Robbe Berg Point. 7 humpback dolphins swim NNE from the open sea into the Bay. Composition: 1 whitefin, 2 grayback adults, 2 juveniles, 1 calf I, 1 calf II.

Table III. Duration of Sightings of Humpback Dolphins at Robbe Berg Point in Relation to Season (Means ± SE)

	Duration of sightings per hour of observation			
	Spring	Summer	Autumn	Winter
1970–1971	0.17 ± 0.07	0.27 ± 0.09	0.28 ± 0.05	0.33 ± 0.06
1971–1972	0.06 ± 0.02	0.16 ± 0.05	0.62 ± 0.09	0.38 ± 0.08
1972–1973	0.14 ± 0.06	0.22 ± 0.04	0.11 ± 0.04	0.36 ± 0.09

The Socioecology of Humpback Dolphins

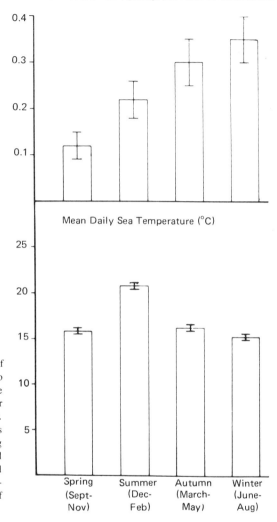

Fig. 10. The duration of sightings of humpback dolphins in relation to season. The dolphins frequented the vicinity of Robbe Berg Point for longer periods during the winter. Since sea surface termperatures increased significantly only during the summer months, seasonal variations in water temperatures did not appear to be directly responsible for the differential utilization of the habitat.

1021. A single whitefin swims NNE into the bay.

1052. 8 humpbacks appear from the direction taken by the previous 8 dolphins. They swim SSE towards the point.

1111. 2 grayback adults split off from this group, change direction and return NNE along the Robbe Berg.

1120. The remaining 6 dolphins swim rapidly SSE, round the point into the open sea, remaining submerged for long periods so that no reliable assessment of composition is obtained. It is not possible to determine whether this group contains the same animals seen at 1015.

1155. 5 humpbacks round the point from the open sea and swim NNE into the Bay. Composition: 1 whitefin, 1 juvenile, 1 calf I, 1 calf II, 1 unidentified.

1201. A further 8 humpbacks round the point and follow the other 5 NNE into the Bay. Composition: 1 whitefin, 4 grayback adults, 3 unidentified.

1215. A large whitefin follows the groups NNE into the bay. The animals swim directly away and disappear along the Robbe Berg.

1350. A whitefin and two grayback adults swim past at high speed and disappear SSE around the point into the open sea.

1415. 11 humpbacks approach the point, swimming SSE. Composition: 2 whitefins, 4 grayback adults, 1 juvenile, 1 calf I, 2 calf IIs. 1 unidentified. The group is engaged in social interactions and courtship-like behavior, circling directly below the observation point.

1510. A single grayback adult rounds the point from the open sea. It swims towards a whitefin, turning its head from side to side as if echolocating. It turns on its side and the whitefin rises from beneath it, belly up and pushes the grayback clear of the water. Three other dolphins join this pair, one of them performing a somersault while airborne. The dolphins chase and rub against each other and are joined by the calf I, which somersaults through the air as it approaches.

1524–1609. The dolphins circle beneath the observation point in two clear subgroups, separated by approximately half a kilometer. The three calves form a play-group with a single grayback adult.

1615. The group containing most of the adults rounds the point into the open sea. The calves and the grayback follow slowly, still engaged in play activity.

It is likely that at least some of the dolphins observed in the individual sightings obtained during this 6-hr period were the same animals, and the presence of two young calves in the majority of groups seen supports this impression. This example illustrates the typical alternation of group movements between the relatively sheltered waters of the bay and the rough seas breaking on the seaward side of the peninsula. Dolphins approaching the point from either the bay or the open sea often joined groups which were already present and then engaged in social interactions, as in the incident cited at 1510 hr. Dolphins sometimes left their groups, either by swimming rapidly ahead and leaving a slowly progressing group behind or by changing direction completely and departing on an independent course as did the two grayback adults at 1111 hr. Frequent increases or decreases in the total number of dolphins counted in the area (for example between 1155 and 1215 and at 1415 hr) suggested that groups observed at widely separated intervals of time formed part of a larger "school" which regularly dispersed into smaller subunits or "groups" characterized by their temporary nature and fluctuating membership.

Confirmation of this general impression was derived from sightings of the prominently scarred and identifiable dolphins. These animals characteristically traveled or interacted with a variety of companions in groups of unstable and variable composition. The composition of groups accompanying two of the most readily identifiable individuals (TF and SB) is shown in Table IV. The size and composition of groups varied considerably for both dolphins even on successive days of observation during the same season (e.g., May 1972). Dolphins TF and SB occasionally traveled alone but, in the majority of cases, were accompanied by both mature and immature individuals in mixed groups ranging apparently at

Table IV. Composition of Groups Accompanying Two Individually Recognizable Humpback Dolphins (Frequencies Include the Known Individual)

Date	Adults		Juveniles	Calves		Unidentified	Total
	Whitefins	Graybacks		I	II		
Known individual: TF							
1970 Nov. 2	2	3	0	0	1	0	6
1971 Feb. 8	3	4	0	1	1	0	7
Feb. 9	2	5	0	2	0	0	9
May 4	2	0	3	1	0	0	6
May 7	3	5	3	1	1	3	16
May 10	2	0	0	0	0	0	2
May 12	3	0	1	0	0	0	4
Aug. 6	3	10	3	2	1	1	20
Aug. 8	4	3	3	0	3	0	13
1972 Feb. 8	2	1	0	0	0	0	3
May 3	3	0	0	1	2	2	8
May 4	3	3	1	0	2	0	9
May 7	6	6	3	1	1	0	17
May 8	3	8	3	1	1	0	16
May 9	6	11	0	2	1	0	20
July 5	1	2	0	2	0	14	19
July 12	1	3	0	1	1	14	20
Nov. 12	1	1	1	0	0	0	3
1973 May 7	1	0	0	0	0	0	1
May 10	5	3	2	2	0	0	12
May 12	1	0	0	0	0	0	1
July 7	3	1	0	0	1	0	5
July 12	1	1	1	0	0	0	3
Known individual: SB							
1970 Nov. 6	1	1	0	0	0	0	2
1971 Feb. 11	3	3	2	1	1	0	10
May 7	3	5	3	1	1	3	16
Aug. 4	1	1	2	0	1	0	5
1972 Feb. 11	1	5	0	2	1	0	9
May 3	3	0	0	1	2	2	8
May 5	1	1	0	0	0	6	8
May 8	3	8	3	1	1	0	16
May 9	1	2	0	0	0	0	3
July 12	1	3	0	1	1	14	20
July 14	2	1	1	0	1	5	10
July 16	3	5	4	3	0	6	21
1973 May 6	4	1	0	0	2	13	20
July 10	1	1	0	0	0	0	2

random between 2 and 21 members. On the average, groups associated with the dolphins TF and SB, respectively, numbered 9.6 ± 1.40 and 9.9 ± 1.10 ($\bar{X} \pm SE$) individuals.

These group sizes were somewhat larger than the average size of groups calculated for the population as a whole from 211 reliable counts of humpback dolphins sighted in Plettenberg Bay during the course of the study; the mean ($\pm SE$) of the frequency distribution of groups of humpback dolphins of varying sizes, shown in Fig. 11, was 6.5 ± 0.38. There were a number of further differences between the grouping tendencies of the two identifiable dolphins and the population as a whole. Inspection of Fig. 11 reveals that humpback dolphins showed a marked propensity to travel either singly or in pairs, such sightings accounting for 27.5% of the total number, whereas dolphins TF and SB were seen alone or with a single companion on only five occasions or 13.5% of the total number of times they were identified in reliably counted groups. Furthermore, groups comprising less than 10 dolphins accounted for 76.8% of the total number of sightings shown in Fig. 11 whereas the corresponding figure for dolphins TF and SB was 56.8% These differences suggested that individual dolphins might differ significantly in their affiliative behavior, some classes of animals tending to maintain contact with larger, mixed groups. However, the relatively infrequent number of sightings of the two specific individuals may represent a biased sample and such conjectures remain purely speculative at this time.

A number of further insights into the school structure of humpback dolphins may be derived from an analysis of the composition of 145 groups of varying

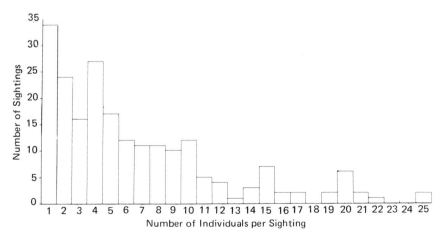

Fig. 11. The distribution of the number of sightings of humpback dolphins in relation to the size of groups observed (211 sightings).

sizes in which the age classes of all the animals present were determined (Fig. 12). This distribution, with a mean ($\pm$ SE) of 5.3 ($\pm$ 0.43) dolphins per sighting, closely resembled the distribution determined for the population as a whole (Fig. 11). It was clear that adult animals were predominantly responsible for the characteristic of independent movement, since dolphins traveling either singly or in pairs were, in the majority of cases (82.6%), adult animals. Immature dolphins rarely traveled on their own or in the company of single adults, suggesting that cows and their immature offspring tended to associate with groups containing more than one adult. Groups containing adult and immature dolphins became more frequent as group sizes increased, and the relatively infrequently observed groups numbering 10 or more dolphins were predominantly mixed groups containing all age classes. Whereas a single group consisting entirely of eight adult dolphins was seen on one occasion, exclusively adult groups containing more than four dolphins were relatively rare. These findings suggested that dependent young were present in the population throughout the year and that the available number of adults dispersed over the habitat in groups of relatively restricted size, accompanied by a proportion of the immature animals in the school.

The largest groups of humpback dolphins seen interacting as a single unit in Plettenberg Bay numbered 25 individuals (Fig. 11). Granting the assumption that these animals formed part of a single school, this group size appeared to approach the upper limits of the social unit in the area. The majority of individually recognizable dolphins were usually identified when groups numbering 20 or more members were seen, supporting the supposition that the relatively small groups wandering independently over the habitat formed part of the larger social unit which only infrequently coalesced to form the complete school. Accurate estimations of the composition of the largest groups observed were unfortunately not obtained, but in two independent counts of groups containing 20 and 21 dolphins, respectively, all of the individuals present were determined. The age class composition of these groups, counted nine months apart, was remarkably similar (Table V). Adult dolphins outnumbered immature individuals by more than two to one, although, if the whitening of the dorsal fin is indeed an age-linked characteristic, only a small proportion of the groups were dolphins of advanced maturity.

An attempt was made to further elucidate the maximum number of dolphins per age class in order to achieve some approximation of the social structure of the school. Thus, the largest number of individuals per age class observed traveling together in a single group was extracted from independent counts made for each year from November 1970 to July 1973 and a hypothetical school structure was established for each year of study (Table VI). There was, in general, good agreement between these estimates, which, on the average, approached the maximum number of 25 dolphins actually observed together. Furthermore, the estimates of school numbers remained reasonably constant from year to year, as

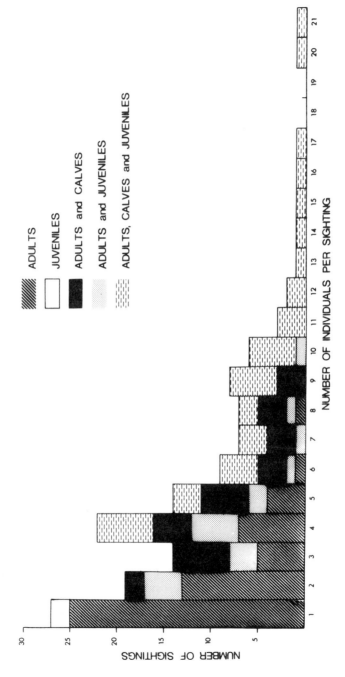

Fig. 12. The composition of groups of humpback dolphins of varying sizes, in which the age classes of all of the animals present were determined (145 sightings).

Table V. Size and Age Class Composition of Two Groups of Humpback Dolphins, Based on Two Independent Counts in which All Group Members Were Identified

Date	Adults			Immatures				Group total
	Whitefins	Graybacks	Total	Calf I	Calf II	Juveniles	Total	
6.8.71	3	10	13	2	1	4	7	20
8.5.72	4	12	16	1	1	3	5	21
Mean ± SE	3.5 ± 0.5	11.0 ± 1.0	14.5 ± 1.5	1.5 ± 0.5	1.0 ± 0.0	3.5 ± 0.5	6.0 ± 1.0	20.5 ± 0.5
Percent	17.0	53.7	70.7	7.3	4.9	17.0	29.3	

Table VI. Estimated Size and Age Class Composition of a School of Humpback Dolphins, Based on Largest Number of Dolphins Identified per Class in Independent Counts over Four Successive Years

Date	Adults			Immatures				Group total
	Whitefins	Graybacks	Total	Calf I	Calf II	Juveniles	Total	
1970	3	7	10	1	2	7	10	20
1971	4	10	14	3	3	6	12	26
1972	6	12	18	3	2	4	9	27
1973	5	5	10	3	2	4	9	19
Mean ± SE	4.5 ± 0.6	8.5 ± 1.6	13.0 ± 1.9	2.5 ± 0.5	2.3 ± 0.3	5.3 ± 0.8	10.0 ± 0.7	23.0 ± 2.0

might be expected in a stable population if the transfer of immature dolphins to the adult classes proceeded at a normal rate.

An estimation of the age class composition of the school of dolphins, derived from the data in Table VI, is shown in Fig. 13. Grayback adults constituted the largest age class, and, together with the whitefin adults, the school contained proportionately more mature dolphins than juveniles and calves. The ratio of mature to immature dolphins, based upon this estimate, is in the same direction as that obtained by counting all of the members in the two relatively large groups presented in Table V.

D. Birth Periodicity

A total of 75 calves, estimated by their size, color, and behavior to be less than six months of age, were sighted during the study in Plettenberg Bay. It was

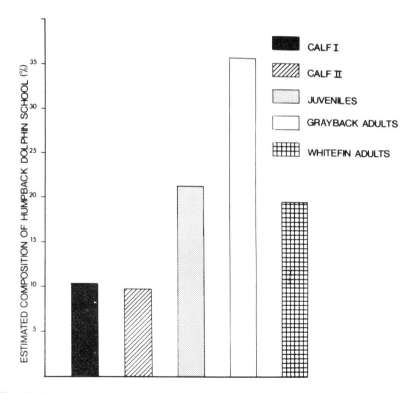

Fig. 13. The estimated social structure of a hypothetical school of humpback dolphins. The age class composition, derived from data presented in Table VI, indicated that there were proportionately more adult than immature dolphins. Newly born calves accounted for approximately 10% of the population.

probable that the same calves were counted on several successive occasions, since it was generally not possible to discriminate reliably either between the young animals or between the adults accompanying them. During the successive years of study, the maximum number of calf Is observed together in a single group did not exceed three (Table VI) and only two calf Is were observed in the large group of 20 individuals positively identified in August 1971 (Table V). Thus, young calves accounted for approximately 10% of the population of humpback dolphins in Plettenberg Bay (Fig. 13).

Calves estimated to be less than one month of age were seen only during the summer months of January and February. In the Plettenberg Bay area, three individual groups containing such individuals were sighted in each successive summer season, whereas two dolphins thought to be newborn were seen together in a single group in February 1971. Furthermore, three calves less than one month old were observed in a single group of humpback dolphins in Algoa Bay in February 1973. This suggested that a birth season for this species might be confined to the summer months; therefore, birth dates were calculated for the 75 calf Is observed by working backwards from the date of the sighting and allowing one month for possible error of age estimation on either side of the birth date so derived. For example, if a calf thought to be three months old was observed in May, the day of birth was plotted at the corresponding date in February and allowance was made for an error factor of one month on either side of the estimated birth date. The distribution of estimated birth dates is shown in relation to season in Fig. 14. The data indicate that births occurred throughout the year and that there was no discrete birth season. However, a birth peak was apparent, since a high proportion (65.3%) of births was concentrated within the summer period between the months of December and February. The data therefore indicate a long calving period from spring through the autumn months, and, furthermore, two calves were estimated to have been born during the winter months in two separate seasons.

Stranded neonates of *Sousa* have been recorded relatively infrequently. Two calves were found on the beach in the Cape Receife area within three days of each other (October 14 and 17, 1970), providing definite evidence that calving occurs in spring (G.J.B. Ross, personal communication). Furthermore, one of two cow humpback dolphins maintained with a bull in the Port Elizabeth Oceanarium was shown on autopsy to contain a fetus conceived in January; given a 12-month gestation period, this calf would have been born in summer.

Social behavior including precopulatory behavior and mating activity, was seen at all times of the year and did not vary quantitatively in relation to season. There was no significant difference between rates calculated for the duration of social interactions per daily observation period when the data were grouped in relation to the time of year over all three years of study ($F = 1.53$; $df\ 3, 103$; $P < 0.25$).

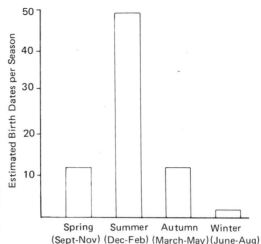

Fig. 14. The estimated times of births of humpback dolphin calves. There appeared to be no circumscribed birth season, although a peak of births was apparent in the summer months.

E. Maintenance Activities and Social Behavior

Humpback dolphins generally moved along the surf zone in small groups just beyond the breaking waves. They very rarely entered the surf and only exceptionally were seen to use wave energy (surfing). During group progressions the dolphins were usually deployed in a tightly knit formation and gave the impression of slow and orderly movement. Traveling and feeding groups engaged in little surface activity and often remained submerged for long periods; after sounding the dolphins moved considerable distances before rising to breathe again, and this sometimes made counting of individuals and tracking of movements difficult. In general, groups of humpback dolphins tended to act as a unit, making it possible for us to categorize the behavior of the group as a whole for quantitative purposes. Descriptions of the four main classes of activity are given below.

1. Group Progressions

Group movements were scored when the dolphins were clearly in transit and swam rapidly in the same direction from one area to another without evidence of feeding or social activities (Fig. 15). Such group movements were often sustained for long periods, and the dolphins disappeared from view along the coastline without changing direction. A progression rate, based upon 58 observations, was determined by plotting the movements of dolphins along the coast against topographical aerial photographs: the average speed of humpback dolphin progressions was calculated as 80.6 ± 5.5 m/min ($\overline{X} \pm$ SE).

Fig. 15. A traveling group of humpback dolphins moving close to the Robbe Berg in Zone I.

2. Feeding Behavior

Groups of humpback dolphins generally dispersed widely to feed, and the animals appeared to capture fish on an individual basis. When feeding over reefs, the dolphins might remain submerged for periods up to 3 min. The energy of strong swells or waves breaking close to rocky outcrops was sometimes used by dolphins to gain momentum for the dive. On surfacing, a dolphin might take half a dozen breaths while swimming slowly, rising intermittently from a shallow depth before commencing the next dive. Dolphins rising steeply from a depth often emerged carrying fish in their mouths.

Occasionally humpback dolphins captured fish close to the surface. Hunting at a shallow depth was usually seen after a marked drop in the water temperature associated with strong southeasterly winds, and the consequent large-scale movement of warm-water-dwelling pelagic fish which were forced into the bay by the encroachment of cold water from the open sea. These conditions were highly favorable for local rock fishermen angling for game fish such as yellowtail (*Seriola pappei*) and leerfish (*Hypacanthus amia*). The predatory behavior of dolphins at such times was characterized by long-jumping and high-speed chasing, the dolphins swimming inverted, belly uppermost. Large shoals of fish were seen to run ahead of the dolphins, which pursued them in open water or followed them close to the rocks where the chase was continued within a few centimeters of the jagged rocky ledges along the length of the Robbe Berg. The dolphins seized their prey behind the neck with a sideways movement of the head and then manipulated the fish with the tongue so as to swallow them head first. A young juvenile was once seen to capture seven fish, approximately 15 cm long, in the space of 13 min.

3. Social Behavior

The social behavior of humpback dolphins was observed for a limited period of time in captive conditions in addition to the extensive observations made under normal conditions during the present field studies. The most striking features of intragroup social interactions as described below

Social interactions between individuals included several forms of bodily contact with associated display swimming, leaping, and chasing (Fig. 16). Bodily contact varied both in form and in intensity. Gentle caressing and massaging of the partner with flippers and flukes was frequently seen when two dolphins swam slowly together, one animal inverted directly beneath the other.

Fig. 16. Social activity in a group of humpback dolphins. In the pair on the left, one animal swims inverted, belly uppermost, as it chases its partner. Note the flexibility of the torso of the animal in the foreground as it turns toward its partner.

More violent bodily contact occurred during vigorous collisions (both underwater and while airborne), when one dolphin rubbed the length of its body past another while swimming at speed, or when one dolphin leaped clear of the water and landed upon its partner.

Social interactions were often initiated when a dolphin lay motionless at the surface on its back and then struck the water forcibly several times with the dorsal surface of its flukes. This behavior was varied when the dolphin hung vertically in the water, head downward, protruding the tail and part of the peduncle, and then brought the ventral surface of the flukes down sharply several times in rapid succession, striking the water with an audible report (Fig. 17).

A spectacular form of display occurred when individuals performed inverted backward somersaults: the dolphin swam inverted, reared its tail out of the water and, becoming completely airborne, performed a full backward somersault and reentered the sea flukes first, belly down, and facing the direction from which the approach was made (Fig. 18). The point of reentry was often in the vicinity of another dolphin, and a chase frequently ensued. During high-speed chases, dolphins frequently leaped clear of the water and, twisting sideways, fell heavily to strike the water with the lateral surface.

The general pattern of precopulatory activity in *Sousa* resembled that of In-

Fig. 17. Fluke beating during social activities. A dolphin protrudes its flukes and part of the peduncle prior to striking the surface in the vicinity of the animal to its left.

Fig. 18. A group of dolphins engaged in high-speed chasing. The animal on the left has just completed an airborne somersault and is slowly sinking, tail downward, flippers extended.

dian Ocean bottlenose dolphins *(Tursiops aduncus)*, described in Saayman *et al.* (1973). Copulation in the humpback dolphin, however, differed from that in the bottlenose dolphin in a number of respects: intromission by bottlenose dolphins lasted only a few seconds and was accompanied by vigorous pelvic thrusting. Furthermore, sexual approaches by bull bottlenose dolphins were made from a variety of directions, whereas the pattern of copulatory behavior in the humpback dolphin was more stereotyped. In the majority of cases observed in captivity, the bull humpback dolphin displayed an erection as it swam inverted beneath a slowly swimming cow. The bull then approached the cow and, rising beneath her, flippers touching, effected intromission. The pair then glided together without any thrusting or bodily movements. The animals swam thus for approximately 20–30 sec, whereupon they separated and rose simultaneously for air and again engaged as before. Copulatory episodes were seen at any time of the day and might be repeated at intervals for several hours. The positions of the partners were occasionally transposed and the cow swam inverted beneath the bull during mating sequences. In the wild, a cow was seen to lie motionless, belly upwards just beneath the surface, the partner approaching to make belly contact, whereupon the pair swam slowly together in the manner observed in captivity. Although this behavior was identified as copulation between adult animals in captivity, similar behavior sequences were observed between all age classes in the wild, including young calves and whitefin adults, and it was noticeable that the partners frequently changed position after surfacing to breathe. The interchangeability of position therefore precluded the possibility of determining the sex of the mating partners on the basis of their behavior alone.

Unusually lively and protracted social interactions were seen when two groups of humpback dolphins coalesced or when single newcomers joined groups already present. The majority of the dolphins present engaged in behavior sequences involving extensive bodily contact, inverted swimming, somersaulting, leaping, and chasing. A conspicuous feature of such social interactions was the tendency of participating animals to form a compact group which circled slowly in a clockwise direction. At such times the vigor of social interactions was attenuated as the animals gradually sounded out of sight in a spiral formation. This more sedate behavior alternated with high-speed pursuits involving three or more dolphins, in which the tempo of interactions was again accelerated.

Social interactions sometimes involved group activities centered on animate or inanimate objects. For example, a juvenile was seen to carry and throw with a flicking movement of the head an object which resembled a sea shell. The object was repeatedly thrown and retrieved while a group consisting of a whitefin, a grayback, and two other juveniles closely accompanied the dolphin which retained possession of the object in an encounter which lasted 23 min. On several occasions the juvenile lay motionless on its back, holding the object in its mouth, while the other dolphins milled around and under it, sometimes pushing it clear of the water from beneath. During throwing sequences, the whitefin was seen to bite at the juvenile in the vicinity of the neck and mouth and much bodily contact was observed which closely resembled the behavior seen during courtship interactions. When the object was thrown, all dolphins dived steeply together, but the same animal always emerged at the surface carrying the object in its mouth. Very similar interactions have been observed and filmed in the colony of bottlenose dolphins in the Port Elizabeth Oceanarium (Tayler et al., 1970).

4. Resting Behavior

Resting behavior was scored when the dolphins remained in one area without evidence of feeding or social interactions. The dolphins moved slowly in a compact group with a drifting or gliding motion, rising slowly to breathe while circling over the same area.

5. Distribution of Activities

The distribution of the total amount of time dolphins were observed at the point during the course of the study is shown in relation to the four categories of behavior—group progressions, feeding, social behavior, and resting episodes—in Fig. 19. The average duration of the respective activities observed

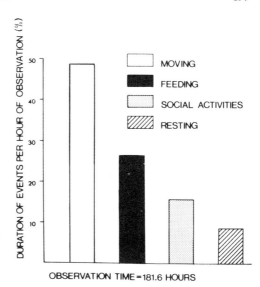

Fig. 19. The distribution of the social and maintenance activities of humpback dolphins during 181.6 hr of observation of the animals at Robbe Berg Point.

at the Point per day is shown in Fig. 20. The data indicate that the dolphins were generally occupied in some energetic activity; only 9% of the observation time was spent in resting episodes. Dolphins were primarily occupied in moving from one place to another and almost 50% of the time they were observed was spent in group progressions. Feeding activity was observed for approximately half an hour per day, and accounted for 26.5% of the total observation time.

An observer interested in collecting data on social behavior in these cetaceans received a relatively low return on daily vigil time, since social interactions were prevalent during only 15.7% of the time dolphins were present at the point, or for about 20 min per daily observation period. These findings suggested that social interactions in this species formed only a small proportion of diurnal activities, particularly as the Robbe Berg Point area was chosen as the major study site partially because social behavior was observed so infrequently at other localities along the coast. However, no information is available on the nocturnal behavior of humpback dolphins under normal conditions, although limited qualitative observations on their behavior in captivity suggested that, in contrast to *Tursiops aduncus,* social interactions might occur quite frequently at night, particularly during the period of the full moon.

F. Utilization of Habitat

Observations of humpback dolphins in different sectors of Plettenberg Bay indicated that the marine habitat was not homogeneous, since the dolphins responded differentially to some ecological features of the environment. Where the seabed was sandy, groups of dolphins characteristically progressed in com-

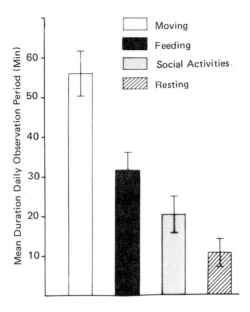

Fig. 20. The average duration of social and maintenance activities of humpback dolphins observed per day at Robbe Berg Point.

pact traveling formation just beyond the surf zone. Dolphins moving parallel to the coastline along the long stretches of open beach which swept northward from the Robbe Berg peninsula seldom changed direction and the progressions were sustained at a steady pace. The dolphins occasionally lingered in the vicinity of gullies scoured out by the effect of localized rip currents (reported to be rich fishing areas) but soon proceeded in the original direction. These groups appeared to be in transit from one focal sector of their range to another, and experience indicated that such group movements terminated in areas where the sandy seabed gave way to rocky outcrops and reefs where the dolphins dispersed to hunt.

The Point of the Robbe Berg peninsula provided an excellent testing case for the hypothesis that humpback dolphins distributed their activities differentially in response to the utility value of specific areas. It was possible to observe the behavior of the animals as they passed directly from zone I, the sandy-bottomed, sheltered bay, into zone II, characterized by submerged reefs on the seaward side of the peninsula. Differential responses of the dolphins to the geographical features of the Robbe Berg are illustrated by an example extracted from field notes:

> 8th February, 1971. 1150 hr. Robbe Berg Point. Wind south easterly, 16 mph. 8 humpback dolphins swim SSE towards the point. Composition: 3 whitefins, 4 juveniles, 1 calf I. Known individual TF is present. Two whitefins lead the group. TF swims inverted, belly up, beneath a juvenile.
> 1230. The group approaches the rough swells breaking round the Point. TF and partner are engaged in courtship-like activity, chasing at high speed and vigorously rubbing past each other.

1237. The group closes up in compact formation with the calf in the center as the dolphins round the Point into the open sea.

1243–1255. The group remains compact, some dolphins swimming inverted as they chase fish close to the surface.

1257. The group rounds the Point, moving slowly NNE into the quiet waters of the bay. Two adults swim together on the periphery of the group: one swims inverted beneath the other and both surface simultaneously to breathe. One dolphin then immediately inverts and swims directly beneath the partner with flippers clasping its flanks.

1321. The group disappears, swimming slowly NNE along the Robbe Berg.

1407. 8 humpback dolphins swim SSE towards the Point. TF and a small calf I are present, but there is also a calf II, indicating that a change in group composition has taken place. The calf I suckles and then returns to its habitual position behind the dorsal fin of the mother.

1430. The dolphins round the Point and disperse widely as they enter zone II, several swimming inverted at speed. A shoal of fish runs ahead of the dolphins which form up in a line behind them. The dolphins feed intensively, some of them leaping clear of the water in the chase. The group acts as a unit, two dolphins occasionally leaping together. The young calf remains in position near the dorsal fin of the mother, its movements closely attuned to hers. A 2-m hammerhead shark is active in the center of the group of dolphins; it appears to be feeding no more than 10 m from the nearest dolphin which is a large whitefin.

1507. A dolphin somersaults, its flukes flashing clear of the water and, as its head appears at the surface, it snaps its jaws shut on a fish. A dolphin lies belly up on the surface and beats the water with three successive strokes of the flukes. A juvenile surfaces holding a fish.

1514. It is now high tide. Feeding at the surface gradually declines, and the dolphins commence to dive deeply and remain submerged for long periods of time.

1519. The calf I swims at speed, raking over the dorsal fin of an adult, which responds by knocking the calf almost clear of the water with a heavy side-swipe of the flukes.

1525. The group returns to zone I. Dolphin TF remains in zone II near Whale Rock.

1539. TF rounds the point and swims NNE after the other animals.

1545. A grayback and a calf II follow TF into the bay; the calf II carries a small fish which hangs from the side of its mouth.

1547. A large shoal of game fish, probably yellowtail, swim SSE from the bay into the open sea.

1550. Two large game fish, probably leerfish, swim SSE into the open sea.

1559. 12 humpback dolphins are now counted in zone I. There are two distinct groups, and courtship-like activity involving chasing, rubbing, leaping and inverted swimming occurs in both of them. A dolphin protrudes its flukes above the water and slaps them down four times in succession. Another grayback adult lies on its side on the surface and beats the water with its flukes. A third adult swims directly towards the grayback and they sound together.

1611. The dolphins round the Point into zone II in a compact group. They disperse and immediately commence to dive deeply. A single dive lasts approximately 2 min, the animal surfaces, takes five to six breaths at approximately 30-sec intervals and sounds once more. Occasionally a dolphin surfaces with a fish in its mouth.

1621. The dolphins return to zone I in a compact group. The animals circle slowly, spiraling clockwise, rising slowly to breathe, barely moving their flukes, and then submerge without strong forward propulsion. Their movements are not characterized by a definite directionality.

1645. A pair of dolphins swim together, one inverted beneath the partner. The other animals continue to circle slowly in two separate groups.
1703. A dolphin lies on its side at the surface and beats the water with three successive strokes of the flukes. It then swims directly to another dolphin, inverts beneath it, and there is a great flurry of water as they are joined by a third animal and a chase ensues.
1724. TF moves towards zone II and 11 other dolphins slowly follow.
1730–1740. The whole group is dispersed over zone II, diving deeply and remaining submerged for long periods.
1830. The observer leaves as the light fades. The dolphins are still feeding intensively in zone II.

It was apparent that the dolphins exploited the geographical features of the Robbe Berg and utilized its shelter for social interactions and resting episodes which alternated with periods of feeding activity over the reefs in the open sea in the vicinity of the peninsula. The duration and frequency of group progressions, feeding behavior, social behavior and resting episodes are shown in relation to the two ecological zones in Fig. 21 and Table VII.

1. Group Progressions

Movements of groups of dolphins occurred more frequently in zone I (Table VII) and therefore the time spent in transit was proportionately greater in this sector (Fig. 21). The average duration of group progressions was significantly longer in zone I than in zone II ($P < 0.01$), and this probably reflected the tendency of groups to wander slowly up and down the Robbe Berg in the sheltered waters of the bay in the interim between feeding and social activities. Group progressions in zone II were typically more direct and rapid, and the dolphins passed out of sight along the seaward side of the peninsula without the frequent changes of direction which often characterized progressions in zone I. This suggested that group progressions in zone II occurred when the dolphins were on a set course, either entering the bay or en route to other hunting grounds further along the seaward stretch of the coast. Whereas similar direct and uninterrupted progressions also occurred in either direction in zone I, it was clear that, in many cases, the dolphins moved slowly up and down the Robbe Berg lingering in the shelter of zone I in the vicinity of a rich feeding ground.

2. Feeding Behavior

The dolphins spent proportionately more time feeding over the reefs of zone II (Fig. 21). Feeding episodes occurred more frequently in zone II than in zone I with a significantly longer average duration ($P < 0.01$) and a maximum uninterrupted period of 139 min (Table VII). The feeding pattern typically consisted of protracted diving over the reefs in zone II. Feeding close to the surface, as described in the above field example, was rarely seen in zone II and was usually associated with the movement of pelagic fish into the bay as the result of strong southeasterly winds and a dramatic decrease in water temperature. The dolphins fed relatively infrequently in zone I and, in this sector,

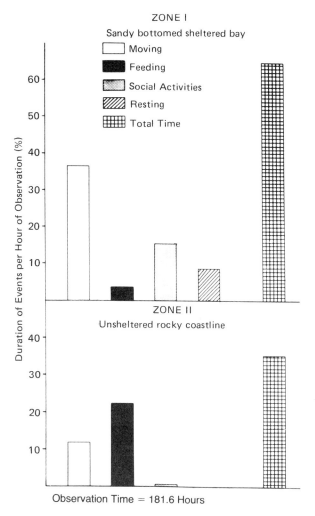

Fig. 21. The duration of maintenance activities and social behavior of humpback dolphins in relation to zones I and II. The dolphins spent proportionately more time in zone I, occupied in group progressions, social interactions, and resting episodes, which alternated with periods of feeding over the reefs in zone II.

hunting was in general characterized by the pursuit of fish close to the rocks of the peninsula a few meters below the surface.

3. Social Behavior

The dolphins spent proportionately more time engaged in social activities in the sheltered zone I (Fig. 21). Episodes of social behavior occurred more

Table VII. Duration and Frequency of Episodes of Maintenance Activities, Social Behavior, and Group Progressions of Humpback Dolphins in Relation to Two Ecological Zones at Robbe Berg Point (Means ± SE)

	Zone I (sandy bottomed sheltered bay)	Zone II (unsheltered rocky coastline)	F ratio	df	P
Mean duration of group progressions (min)	15.1 ± 0.78	11.8 ± 0.90	7.56	1371	< 0.01
Range	2–60	2–50			
Number of episodes	261	112			
Mean duration of feeding behavior (min)	13.4 ± 1.70	27.2 ± 2.76	8.53	1128	< 0.01
Range (min)	4–45	2–139			
Number of episodes	33	97			
Mean duration of social behavior (min)	22.2 ± 2.07	10.4 ± 2.27		Not applicable	
Range (min)	2–262	5–15			
Number of episodes	79	5			
Mean duration of resting episodes	18.0 ± 2.26			Not applicable	
Range (min)	2–93	Nil			
Number of episodes	53				

frequently in zone I than in zone II with a longer average duration and maximum uninterrupted period of 262 min (Table VII).

4. Resting Episodes

All resting occurred in the sheltered zone I (Table VII). Resting periods averaged 18.0 ± 2.26 min ($\bar{X} \pm$ SE) with a maximum uninterrupted duration of 93.0 min. In all, proportionately more of the 181.6 hr of direct observation of humpback dolphins at the Point was spent by the dolphins in zone I (Fig. 21).

G. Diurnal Activity Cycle

It was not possible to observe the nocturnal activities of free-swimming humpback dolphins. During daylight hours the time of arrival of humpback dolphins in the study area tended to be quite variable, and initially there was no easily discernible pattern of activity in relation to the time of day.

In order to test this impression quantitatively, the duration of sightings of humpback dolphins per day was adjusted for the number of hours of vigil at four "day-periods" between first light and sunset. Secondly, daily rates for four patterns of activity (group progressions, feeding behavior, social behavior, and resting episodes) were calculated by dividing the duration of each activity by the number of hours the dolphins were present at each day-period. The results are presented in Table VIII. The data show that the dolphins frequented the study area at the same rate throughout the day. Moreover, the distribution of social behavior and maintenance activities showed no very close relationship to the time of day. However, feeding periods tended to be longer in the early part of the day than at midday and in the early afternoon ($P < 0.05$).

H. Influence of the Tidal Cycle upon Behavior

There was a marked influence of the tidal cycle upon the activities of humpback dolphins in the vicinity of Robbe Berg Point. Protracted observations suggested that the entire schedule of the dolphins was determined largely by a primary effect of the tides upon feeding activity. This effect is illustrated in fig. 22. The total amount of time dolphins were present at the point was divided into four behavioral categories (group progressions, feeding behavior, social behavior, and resting episodes), and the distribution of social behavior and maintenance activities was considered in relation to the tidal rhythm. Daily rates for the four patterns of activity were calculated and subjected to statistical analysis (Table IX).

Table VIII. Duration per Hour of Observation of Episodes of Maintenance Activities, Social Behavior, and Group Progressions of Humpback Dolphins in Relation to Time of Day (Means ± SE)

	Time of day (hr)						
	−0859	−1159	−1459	−1800−	F ratio	df	P
Duration of sightings/hour	0.23 ± 0.05	0.29 ± 0.04	0.23 ± 0.03	0.26 ± 0.03	0.82	3303	>0.25
Group progressions	0.58 ± 0.09	0.62 ± 0.05	0.64 ± 0.05	0.53 ± 0.06	0.96	3303	>0.25
Feeding behavior	0.41 ± 0.09	0.23 ± 0.04	0.21 ± 0.04	0.30 ± 0.05	2.71	3303	<0.05
Social behavior	0.40 ± 0.03	0.10 ± 0.03	0.10 ± 0.03	0.13 ± 0.04	0.79	3303	>0.25
Resting episodes	0.01 ± 0.03	0.08 ± 0.03	0.09 ± 0.03	0.03 ± 0.02	1.70	3303	<0.25

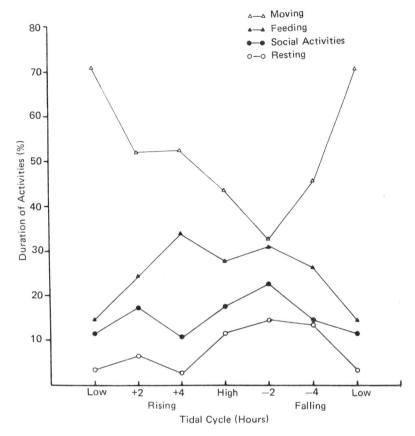

Fig. 22. Rhythmic variations in the duration of maintenance activities and social interactions in relation to the tidal cycle.

The rising tide was associated with a significant decrease in group progressions ($P < 0.01$), whereas feeding activity and resting episodes showed a reciprocal increase ($P < 0.05$); social behavior varied together with both feeding activity and resting episodes, but the peak at 2 hr after high tide was not significantly higher than the low values recorded at low tide ($P > 0.25$).

The interpretation of these results must be considered in relation to the differential exploitation by the dolphins of the two ecological zones at the Point (see pp. 197–203). At low tide, group progressions predominated and the dolphins moved in compact traveling formation in transit from one area to another. As the tide commenced to rise, however, feeding activity increased and, two hours before high tide, the proportion of time spent feeding was more than double the value recorded at low tide. The dolphins fed intensively for a period in zone II, and then entered zone I and engaged in either social activities or resting episodes.

Table IX. Duration per Hour of Observation of Episodes of Maintenance Activities, Social Behavior, and Group Progressions of Humpback Dolphins in Relation to Tidal Cycle (Means ± SE)

	Tidal cycle (hr)									Significant difference[a]
	Rising		High	Falling		Low				
	+2	+4	+6	−2	−4	−6				
Column No.	(1)	(2)	(3)	(4)	(5)	(6)	F ratio	df	P	
Group progressions	0.64 ± 0.05	0.62 ± 0.05	0.58 ± 0.05	0.44 ± 0.06	0.58 ± 0.05	0.74 ± 0.05	3.1	5740	< 0.01	4<u>352</u>16
Feeding behavior	0.18 ± 0.04	0.32 ± 0.05	0.25 ± 0.04	0.27 ± 0.06	0.23 ± 0.05	0.14 ± 0.04	2.36	5740	< 0.05	6<u>15342</u>
Social behavior	0.12 ± 0.03	0.08 ± 0.03	0.12 ± 0.03	0.15 ± 0.04	0.08 ± 0.03	0.10 ± 0.03	0.57	5740	> 0.25	None
Resting episodes	0.07 ± 0.02	0.02 ± 0.02	0.08 ± 0.03	0.14 ± 0.04	0.11 ± 0.04	0.04 ± 0.02	2.25	5740	< 0.05	2<u>6135</u>4

[a] Columns underlined in common have values that are not significantly different from each other.

Thus, group progressions declined systematically as the tide commenced to rise and reached the lowest levels 2 hr after high tide. At this time, resting episodes increased significantly and the high levels of feeding activity continued to be sustained. The occurrence and duration of social activities appeared to be less dependent upon the tidal cycle, but showed a close correlation with resting episodes, reaching maximum values 2 hr after high tide, or approximately twice the levels recorded at low tide.

I. Interactions with Other Animals

1. Fish

There is very little reliable information concerning the species of fish preyed upon by humpback dolphins, but their tendency to hunt in the vicinity of reefs indicated that they fed mainly upon reef-dwelling organisms. In captivity in Port Elizabeth, however, humpback dolphins accepted any commercially available species, as did bottlenose dolphins (see Tayler and Saayman, 1972). Captive humpback dolphins spent much time poised vertically, head downwards, above the reefs in the tank and investigated crevices and crannies with their long rostrums, snapping with a sideways motion of the head at any rock-dwelling fish which emerged. This behavior possibly gave some indication as to their mode of hunting at a depth under normal conditions.

Humpback dolphins frequently swam past and ignored large shoals of mullet (*Mugilidae*) which were often seen moving slowly just beneath the surface close to the rocks along the Robbe Berg. However, the stomach of a large bull humpback dolphin, which stranded in tropical waters near Durban, was found to contain eight mullet, and the dolphin carried a partially decomposed mullet in its mouth. The dolphins usually ignored large pelagic predators such as leerfish (*Hypacanthus amia*) and yellowtail (*Seriola pappei*) which were greater than 1 m in length. However, humpback dolphins occasionally hunted at the surface at times when pelagic fish were plentiful (see pp. 192–193), but it was not possible to identify the prey.

Mr. A. Crawford-Brunt, a well-known fisherman at Plettenberg Bay, once observed two adult humpback dolphins apparently playing with a fish approximately 1 m long, thought to be a pampano (*Trachinotus blochi*). One of the dolphins held the fish in its mouth, clasping it midway across the body, and carried it some 90 m from the Robbe Berg, where it was released. The fish, apparently injured or exhausted, skittered along on its side towards the rocky ledges of the peninsula, but, as it neared the white foam close to the rocks, the dolphins accelerated, recaptured the fish, and carried it again further out into the bay. This procedure was repeated several times as the dolphins progressed slowly out of sight along the peninsula.

2. Birds

Two species of cormorants *(Phalacrocorax capensis)* and *(Phalacrocorax carbo)* roosted in large colonies on the rocks above sea level near Robbe Berg Point. Dolphins and cormorants frequently hunted in close proximity and on one occasion approximately 20 cormorants were scattered over an area in the vicinity of a group of 12 feeding humpback dolphins. However, an adult dolphin and a calf II once chased a cormorant from the water. In another incident, an adult dolphin avoided the direct approach of a cormorant, which pursued the dolphin for some distance underwater.

When humpback dolphins were feeding near the surface, black-backed gulls *(Larus dominicanus)* were often in evidence, scavenging fragments of fish. The gulls dropped into the water or sat on the surface in the vicinity of the dolphins and sometimes took flight when the dolphins surfaced beneath them. A gull was once seen to snatch two small fish from a juvenile humpback dolphin which was hunting near the surface.

Jackass penguins *(Spheniscus demursus)* were common in the study area and were seen at intermittent intervals near feeding groups of dolphins. A calf II on one occasion chased a penguin for 40 sec, snapping at its tail feathers. The bird was in no way harmed, and the dolphin ultimately disengaged. This incident closely resembled similar encounters often observed between captive bottlenose dolphins and penguins.

3. Southern Right Whales

Southern right whales *(Eubalaena australis)* migrate to the southeastern Cape coast in spring and early summer to calve and mate in sheltered bays (Best, 1970; Saayman and Tayler, 1973*b*). Once two whales were observed lying side by side, close to the Robbe Berg, drifting with the surface current. Two adult humpback dolphins swam past within a few meters of the whales, without any visible reaction. On another occasion in Algoa Bay, a single whitefin humpback dolphin accompanied a southern right whale for about half an hour as the whale swam slowly close to the breaking surf. The dolphin remained close to the head of the whale, without making physical contact. The pair of animals changed direction several times, their movements closely attuned.

4. Bottlenose Dolphins

Bottlenose *(Tursiops* sp.) and humpback dolphins sometimes followed similar routes close to the shore, and a number of encounters between the two were observed. On three occasions single humpback dolphins progressed in the center of groups of bottlenose dolphins of which they appeared to be fully integrated members. On two further occasions, large groups of bottlenose dolphins passed at high speed through the area where the humpback dolphins were feeding. In one instance, the humpback dolphins sounded deeply and remained submerged until the bottlenose dolphins had passed, whereas they displayed no noticeable avoidance reaction on the second occasion.

Groups of bottlenose dolphins appeared aggressive to lone humpback dolphins in their vicinity in two separate encounters. Several bottlenose dolphins would change direction abruptly and chase an individual at high speed for a short distance. A playlike encounter, between approximately 13 juvenile and two adult bottlenose dolphins and a group of seven adult humpback dolphins, was observed in Algoa Bay. The entire ensemble swam at high speed in a tightly knit group in the surf zone. The behavior displayed, with much underwater action and many long-jumps and side-flops, was characteristic of bottlenose rather than of humpback dolphins. The latter, in this case were behaviorally indistinguishable from the bottlenose dolphins. The energetic and high-speed interaction was sustained for 75 min and was terminated when the bottlenose dolphins withdrew and joined their main group which had remained some 2 km from the coast.

5. Cape Fur Seals

Groups of Cape fur seals *(Arctocephalus pusillus)*, as well as lone individuals, frequented Plettenberg Bay all year round and were often seen in the vicinity of humpback dolphins, approaching within 8 m without apparent reaction from either genus. However, two seals, one an adult bull, once swam through a group of feeding dolphins which immediately dispersed and regrouped only after the seals had passed. A film was obtained of an unusual form of group activity, in which individual humpback dolphins took turns swimming at high speed towards the rock face of the Robbe Berg, veering off, always to the left, barely avoiding contact with the rock, while the other dolphins remained passive nearby. At this time, an immature Cape fur seal appeared close to the rocks and was immediately put to flight and pursued for a considerable distance by a single dolphin.

6. Sharks

Humpback dolphins always ignored small hammerhead sharks (*Sphyrna zygaena*) which were frequently seen in their vicinity in the summer. Indeed, hammerhead sharks, approximately the same size as adult dolphins, occasionally moved unmolested through groups of dolphins or remained in their vicinity for protracted periods even when young calves were present.

The reaction of humpback dolphins to sharks, however, was not always placid as illustrated by the following incident, witnessed and filmed from the Robbe Berg on May 9, 1972 at 0940 hr.

A group of 10 humpback dolphins progressed into the bay from the open sea in a compact traveling formation. The group contained two known individuals

(TF and SB), three whitefins, a calf, and several graybacks. As the dolphins passed from view, two whitefins and a calf returned, swimming SSE at high speed 50 m from the peninsula. A 4- to 5-m shark, swimming in the same direction, parallel to the Robbe Berg, was seen moving slightly ahead of the dolphins, which increased their speed, approached the shark, and forced it into a sand-bottomed cove directly beneath the observers. Swimming at high speed in a triangular formation, the dolphins approached within 1 m of the shark, which swirled around and lunged towards them, thus making its escape, whereupon it was again pursued in the same direction and cornered in another cove. No physical contact was seen, and the shark again made its escape and continued at high speed towards the open sea, with the dolphins in close pursuit. The dolphins then disengaged and returned in a leisurely manner in the direction taken by their original group. The shark was thought by the observers to be a blue pointer *(Carcharodon carcharias)*, many of which have been captured by anglers in this area.

On another occasion, a group of eight humpback dolphins, again including TF and a calf, appeared to avoid a 4- to 5-m hammerhead shark when it turned towards them and followed them for a few meters. When the shark resumed its course in the opposite direction, two of the adult dolphins and the calf left the group and followed the shark for about 100 m, whereupon they rejoined the group without incident.

7. Killer Whales

Killer whales *(Orcinus orca)*, known to prey upon other Cetacea (Martinez and Klinghammer, 1970), were occasionally seen in the study areas. Three killer whales were observed off Shark Rock, which abuts the mainland in Algoa Bay, an area frequented by humpback dolphins. Two of the killer whales displayed large, triangular dorsal fins, one of which curled to one side at the tip. The third animal, being about half the length of the adults, was apparently a calf. When first seen the killer whales were engaged in courtship-like activity, rubbing against one another, with a considerable amount of fluke beating by the largest individual. This activity abated as the killer whales approached Shark Rock where two groups of humpback dolphins, comprising a total of about 20 individuals, were gathered. The killer whales approached within a few meters of Shark Rock, but remained just beyond the breaking waves. The dolphins circled quietly, almost in contact with Shark Rock and always above the submerged reefs in its vicinity. As the killer whales withdrew and progressed slowly out to sea, the dolphins slipped away in small groups which swam away, all in the same direction, very close to the coast.

The inshore distribution of humpback dolphins and their ability to utilize wave action to cross over barely submerged reefs probably ensures the survival of this species when threatened by large killer whales or sharks. Evidence of this ability was filmed off the Robbe Berg when an adult humpback dolphin waited until a swell increased the water depth above an underwater promontory; the dolphin utilized the wave energy to swim over the reef into a gully and then proceeded on its course, having considerably reduced the distance required by a detour around the promontory. In contrast, the prolific pelagic common dolphin (*Delphinus delphis*) displayed precipitate flight in the open sea when pursued by three killer whales in Plettenberg Bay (Fig. 23).

VII. DISCUSSION

Recent studies of animal behavior have focused increasingly upon group processes and on the transactions between the natural group or social unit and the environment to which it has adapted. The emphasis of this area of research, termed "social ethology," is not so much upon " . . . the traditional ethographic study of behaviour patterns shown between conspecific individuals usually studied in dyadic interaction, but upon the relations between individuals and the natural

Fig. 23. Part of a school of approximately 1000 common dolphins (*Delphinus delphis*), pursued by three killer whales (*Orcinus orca*), swim past Robbe Berg Point at high speed on May 2, 1973. The killer whales easily kept pace with the dolphins, remaining some 200 m to the rear of many stragglers which trailed behind the main group of dolphins. The pursuit continued for several miles when the dolphins rounded the Point into the open sea.

group considered as the social environment within which they live and to which they are adapted" (Crook, 1970, p. 197). Moreover, "... ethology attempts to answer questions concerning the functions of a particular behaviour in a given social and ecological environment" (Crook and Goss-Custard, 1972, p. 278). The multidisciplinary approach which characterizes this new development is perhaps nowhere more clearly illustrated than in the rapidly expanding field of primatology (Altmann, 1967). Field studies of nonhuman primates have flourished for a variety of reasons, not least among which is the relevance they are thought to have for theories concerning the evolution of man. Consequently much of the speculation concerning the factors which condition social structure and behavior is derived from socioecological studies of the relatively accessible terrestrial primates. However, as increasing numbers of studies of other higher mammals become available, it has become apparent that a complex level of social organization is not necessarily unique to the higher primates. This finding indicates that fundamental assumptions concerning the adaptiveness of social systems should be subjected to comparative assessment.

The dolphin has excited considerable scientific interest because of its large, complex and highly convoluted cerebral cortex and its reputedly high intelligence. Captive bottlenose dolphins, for example, display a remarkable propensity to learn complex behavior sequences and sounds, including the elementary use of tools, and to practice and elaborate these motor patterns in the absence of the animal "model" originally stimulating the imitative behavior, without any apparent reinforcement apart from the performance of the activity itself (Tayler and Saayman, 1973a). This high level of learning capacity suggests that dolphins might be expected to have evolved a complex form of social relationships. In order to obtain some measure of insight into the nature of dolphin social systems, we have elsewhere compared knowledge of dolphins, based upon qualitative observations, to that existing on the extensively studied terrestrial baboons (Tayler and Saayman, 1972). We speculated at that time that "... it is possible that the social organization of bottlenose dolphins will ultimately be shown to have more in common with that described for the more highly evolved chimpanzee." The present systematic and quantitative studies of humpback dolphins represent the culmination of a series of investigations on the social behavior and socioecology of cetaceans, and an attempt will be made here to assess the extent to which the above speculation is justified.

Humpback dolphins ranged within 1 km of the coast, and thus the economy of their social system appeared to be very largely conditioned by the restriction of their primary environment to waters close inshore. In particular, the dolphins generally hunted in the vicinity of reefs, suggesting that their preferred prey consisted largely of reef-dwelling fish. Although neither the feeding rate nor the species of prey of humpback dolphins is well known, the average daily food consumption of captive bottlenose dolphins (*Tursiops aduncus*) varied between 11.4 kg for an adult cow and 13.6 kg for an adult bull (Tayler and Saayman, 1972); the feeding rate of two *Tursiops truncatus* cows was 4.2% of body weight per day, or 7.3 and 13.6 kg, respectively (Sargent, 1969). Assuming that the food requirements of humpback dolphins were of a similar order, the exploitation of the relatively limited inshore food resources might be expected to play a pivotal role in the regulation of their social structure, social organization, and routine maintenance activities. Indeed, a relationship between school size and feeding strategy was indicated by the marked tendency for the size of groups of delphinids in the Plettenberg Bay area to increase in relation to their inshore–offshore distribution (Saayman et al., 1972). Humpback dolphins, forming the smallest groups, were located only very close to the coast. Bottlenose dolphins occurred in groups of intermediate size, averaging about 140 individuals, but schools numbering up to 500 animals were frequently seen deployed in four or five groups over several square kilometers. *Tursiops* frequented deep water far out at sea but also penetrated the surf zone to a depth of about 1 m. In contrast, pelagic dolphins (*Delphinus delphis* and *Stenella coeruleoalba*) formed the largest schools, averaging approximately 700 individuals per sighting, and were generally seen only far out at sea. These findings suggested that the increase in the school size of *Tursiops* was related to the ability to exploit the pelagic food supply in addition to utilizing inshore food resources. Humpback dolphins, on the other hand, appeared to be almost entirely dependent upon inshore resources, whereas only the prolific pelagic food sources were capable of sustaining the vast schools of pelagic dolphins. Thus, the restriction of the prey of humpback dolphins to reef-dwelling species may have limited the size of their schools; this suggestion is further reinforced by the tendency of the dolphins to disperse in small groups along the coast. This feeding strategy may have offered the advantage of permitting individuals to hunt on an individual basis over a wide area undisturbed by competition from other school members, particularly if the prey itself were widely scattered over the home range and largely restricted in exploitable numbers to isolated reefs and rocky outcrops.

Granting the apparent preference of humpback dolphins for reef-dwelling prey, necessitating hunting at considerable depths in a habitat affording the fish refuge in crevices and crannies, predatory activity in animals which probably eat up to 5% of body weight per day must require a considerable expenditure of energy. Under these conditions, any ecological factors influencing the accessibility of prey species might be expected to exert a profound influence upon the maintenance activities of the dolphins. Indeed, three potential factors were identified in this study: the influence of strong southeasterly winds and the diurnal and tidal cycles. The influx of warm-water-dwelling pelagic fish into the bay in association with marked decreases in water temperatures as the result of strong southeasterly winds influenced the technique of predatory activity, since in contrast to the normal deep-diving pattern of feeding, the dolphins took fish close to the surface. These observations indicated that predation in these marine predators might be opportunistic, as indeed it is in some terrestrial carnivores, the animals taking whatever suitable prey is most readily available.

However, the influence of winds exerted only an intermittent influence upon predatory activity, whereas the diurnal cycle and the tidal cycle, in particular, appeared to exert a systematic effect upon the entire daily schedule of the animals. The behavior of bottlenose dolphins in Plettenberg Bay has been shown to be markedly influenced by the diurnal rhythm (Saayman et al., 1973): the dolphins were sighted most frequently in the bay in the early morning and in the late afternoon, during which times high feeding rates were observed. The frequency of social interactions reached maximum levels after the morning peak in feeding activity and declined in the late afternoon when feeding again became prevalent. Similar variations in social interactions were observed in the captive colony of bottlenose dolphins in the Port Elizabeth Oceanarium, and Powell (1966) found that the light–dark cycle influenced the incidence of vocal activity in captive *Tursiops truncatus*. Layne (1958) noted that freshwater dolphins in the Amazon made use of lakes most frequently in the early morning and in the late afternoon, at which times the dolphins appeared to be most actively feeding. It was not clear whether the diurnal variations in the behavior of free-ranging bottlenose dolphins in Plettenberg Bay were related to the accessbility of prey governed by the diurnal activity cycles of food-fish. The predatory behavior of humpback dolphins, however, differed from that of bottlenose dolphins in that the humpback dolphins generally hunted at a depth in the vicinity of reefs close to the shore, apparently on an individual basis, whereas bottlenose dolphins hunted in groups in which the coordinated herding of the food-fish, either close to the shore or in deeper waters, was sometimes a prominent feature (Tayler and Saayman, 1972).

Moreover, the technique of diving deeply for fish was not observed in bottlenose dolphins in Plettenberg Bay, although *Tursiops truncatus* were reported by Russian observers (Tomilin, 1967) to submerge for up to 7 min, whereupon they surfaced to breathe several times at short intervals, thus suggesting some versatility in hunting techniques in this species in relation to ecological conditions. Although the prey of humpback dolphins is not known, some observations by Penrith (1972) on the behavior of reef-dwelling fishes may suggest some explanations for the fluctuations in the feeding activities of the dolphins in relation to the diurnal and tidal variables. Dageraad fishes *(Chrysoblephus cristiceps)*, for example, show some behavior variations in relation to time of day. Whereas dageraad fishes are found on banks of flat rock throughout the day, they are caught by anglers only in the morning when they are found moving, all in the same direction, slowly across the bank, swimming head downwards, searching for food. Later in the day the fish swim " . . . 1–3 metres off the bottom, in a horizontal position, and the orderly progression of earlier has changed to an apparently aimless milling around." It is not possible to conclude whether or not such changes in the behavior of the food-fish of humpback dolphins affects the ease with which they might be captured, but direct observation of both the prey and its behavior as well as the method of capture by dolphins at a depth may well reveal that factors of this nature are operative. Penrith (personal communication) has suggested the following explanation for rhythmic variations in the feeding activity of dolphins over a submerged reef in relation to the tidal cycle: At low tide, the swell is more forceful, at least down to a depth of 5 m, and more edible organisms tend to be washed out of the reef so that fish are actively feeding close to the rocks. At this time they are fast moving and may thus be more difficult to catch by dolphins. At high tide, however, fish such as steentjies (*Spondyliosoma emarginatum*) tend to shoal in compact groups over the reefs and might thus be more readily taken by dolphins. At present these suggestions are highly speculative, but the tendency of humpback dolphins to spend more time hunting early in the morning and in association with the rising tides rather than at slack water, may have been influenced by fluctuations in the relative accessibility of prey in relation to the diurnal and tidal cycles. Moreover, there is some evidence that the movements and feeding activities of *Tursiops truncatus* are influenced by tidal factors. McBride and Hebb (1948) noted that bottlenose dolphins frequently entered bays and creeks in association with the rising tide and left as the tide commenced to fall. It was suggested that tidal factors might exert a greater influence upon behavior than did the diurnal cycle and that the movements of the dolphins were " . . . presumably a response to the movements of the fishes upon which they feed." Irvine and Wells (1972), who studied the movements of a group of bottlenose dolphins which had been tagged for individual recognition, wrote: "Movements were predictable, usually associated with tidal action. The herd usually entered from the Gulf of Mexico as the tide began to rise. Once well into the Intracoastal Waterway they usually split into twos and threes, apparently feeding in the shallows, before regrouping to return to the Gulf as the tide started to ebb. Other groups of dolphins in other parts of the study area exhibited similar patterns of movement and feeding behaviors associated with the tides." Similarly, Caldwell and Caldwell (1972*a*) reported that the movements of dolphins near St. Augustine, Florida, were influenced by the tides, but noted that there is at present no explanation concerning the causation of this relationship. Some interesting observations by Hoese (1971) indicate that *Tursiops truncatus* utilized the tidal action to hunt fish in tidal creeks at *low* tide: the dolphins chased a group of fish ahead of them and used the resulting bow wave to strand the fish upon a mud bank. The dolphins themselves then slid out of the water onto the bank to seize the stranded prey. The comments of the author indicate that dolphins are indeed capable of exploiting the influence of both seasonal and tidal factors upon the accessibility of their prey: "In winter there are few fish in the salt marsh because of cold weather, and *Tursiops* rarely ascends the tidal creeks. In the marsh in summer *Tursiops* is common, but fish also are numerous then. It may be that in summer, when several dolphins commonly enter the marsh at every low tide, fish are so common that they are easy to capture without elaborate rounding-up procedures . . . . Although *T. truncatus* enters the marsh at other times, the behavior described here necessarily must be limited to within approximately 30 minutes before or after low tide. Otherwise, the mud bank would be inadequately exposed and the fish less concentrated." The evidence presented in the present report on the cyclical variations in the predatory activity of humpback dolphins indicates that systematic studies of the diet of the dolphins, as well as of the determinants of the behavior of the food-fish, are essential to a more complete understanding of the socioecology of these marine predators.

The difficulties encountered in distinguishing bulls from cows and in identifying adequate numbers of individuals (and thus discriminating between independent schools) place restrictions on the conclusions to be drawn from this study on the social structure and organization of humpback dolphins. However, a number of features characteristic of the social system emerged in some detail. It seemed clear that at least one school of humpback dolphins ranged over a relatively restricted area in the vicinity of Plettenberg Bay and that the total membership of this school was approximately 25 individuals (Fig. 11, Tables V and VI). The composition of groups was highly flexible, since associations between known individuals were not stable (Table IV) and the composition of groups directly observed over a period of time fluctuated as dolphins joined or left the group in an apparently random and unpredictable manner. The independence of movement of individuals within the group was further emphasized by the marked tendency of adult dolphins to travel either alone or in pairs (Figs. 11 and 12). However, it gradually became clear that the loosely organized associations between individuals in small groups occurred within the context of a larger social unit or "school." This form of social organization contrasts markedly with that found in a cohesive social system, in which individual members remain closely associated throughout the day and night: social interactions, group movements, and ranging behavior, as well as routine maintenance activities, occur within a restricted spatial area and, in general, the group functions as a tightly knit, unified whole. The highly cohesive social system of some troops of chacma baboons (*Papio ursinus*) provides examples of this form of social organization (Stoltz and Saayman, 1970; Tayler and Saayman, 1972). In contrast, the comparatively loosely organized social groups of chimpanzees (*Pan troglodytes schweinfurthii*), recently studied by a number of primatologists, apparently have a number of features in common with the schooling behavior of humpback dolphins, as the following quotations illustrate: "Chimpanzee groups in the Budongo Forest were not closed social groups. Groups were constantly changing membership, splitting apart, meeting others and joining them, congregating or dispersing" (Reynolds and Reynolds, 1965, p. 396). Similarly Goodall (1965) noted that there were no stable groups other than mother and infant associations in the Gombe Stream population of chimpanzees; she characterized a group as " . . . no more than a temporary association of individuals that may be constant for a few hours or days." Furthermore, Goodall (1965) noted that mature male and female chimpanzees frequently traveled alone and that groups larger than five individuals were seen relatively infrequently; the largest group numbered 23 individuals and was seen only once. Nishida (1968) presented similar data on chimpanzees in the

Mahali Mountains and emphasized the random nature of the composition of groups and the tendency of all individuals, with the exception of very young animals, to travel either singly or in pairs. Groups of chimpanzees observed by Nishida ranged from one to 28 individuals, but groups of between one and five individuals accounted for more than 50 of the total number of groups observed. Thus it appeared that a large group, composed of several adult males and females with their offspring, formed the basic social unit which he characterized as "extensible" in nature: "Indefiniteness of group membership is decisive in both large and small subgroups, although small groups have a greater statistical probability of repeated occurrence than do large ones. . . . While a gorilla group is cohesive, a group of chimpanzees repeats joining and parting. If we think that a cohesive group is a general form in a society of primates, and that a unit-group of chimpanzees often becomes compact under favourable food conditions, this characteristic of the unit-group of chimpanzees can be regarded as being based on the ability by which members of a group can disperse and extend in a wide range if they wish. From this viewpoint, the author wants to term the joining and parting ability of subgroups within a unit-group 'extensibility'" (Nishida, 1968, p. 214). A similar conceptualization of the chimpanzee social system was expressed by Reynolds and Reynolds (1965, p. 423): "It is possible that the 'looseness' and 'instability' found in chimpanzee groups have been exaggerated, because of the obvious dissimilarity from the *spatially compact* type of group organization found in baboons. In fact, one may entertain the hypothesis that chimpanzees possess a social organization so highly developed that it can persist in the absence of immediate visual confirmation normally true for baboons."

It is noteworthy then that the social organization of chimpanzees in many respects closely resembles that observed in humpback dolphins, and, in addition, access to scattered sources of food is considered to be an important ecological correlate of the social system of these frugivorous primates: " . . . fruit is often localised, and moreover, the location where fruit ripens in abundance frequently changes seasonally or annually. . . . To overcome such uncertainty in obtaining fruit, it will be very efficient for members of a group to disperse over a wide range, and if some of them fortunately find a concentration of fruit, to inform other members of it. Thus, extensibility of groups can be thought to have developed in connection with frugivorous habits" (Nishada, 1968, p. 216). "The seasonal ripening, flowering or failure of these crops were observed to have a direct effect upon the movement patterns and sometimes also the behavior of the chimpanzees. Thus when a desirable fruit ripened in a fairly restricted area, this served to aggregate chimpanzees in that area so that temporary associations of thirty or so were seen. Conversely, when a food was ready for eating in most parts of the reserve the chimpanzees were more often encountered in small groups" (van Lawick-Goodall, 1968, p. 172). "The figures support the observational data that the large mixed groups found feeding on fruit trees tend to be temporary congregations of smaller bands that have come to places where food is concentrated. When they leave the tree and travel about the forest, chimpanzees split up into small bands of less than seven individuals . . . " (Reynolds and Reynolds, 1965, p. 398).

Recent field research has shown that marked *interspecific* differences may occur in social organization and behavior in relation to varying habitats (Rowell, 1967), and conclusions drawn from comparisons between orders must therefore remain speculative. Nevertheless, the evidence reviewed above indicates that some striking similarities in the social systems of humpback dolphins and chimpanzees may be related to the scattered and possibly fluctuating nature of the food resources the animals exploit. Moreover, Schaller and Lowther (1969) have drawn attention to some similarities in the social systems of some terrestrial carnivores, such as the lion, hyena, wolf, and wild dog, which in some respects resemble the flexible social organization of chimpanzees and hunting–gathering peoples such as the bushman, as described by Tobias (1964).

Two factors emerge from such comparisons: Firstly, access to food appears to be a primary determinant of the social system as a function of cooperative hunting methods. In the case of humpback dolphins and chimpanzees, a feeding strategy based upon the dispersal of small groups over the habitat might be expected to facilitate the exploitation of food resources. Reynolds and Reynolds (1965) suggested that vocalization and drumming by chimpanzees on the discovery of food may serve to signal the location of the food source to other group members, thus ensuring that the food resource becomes available to the whole group. Whether or not a similar mechanism operates in humpback dolphins is uncertain due to the great practical difficulties encountered in attempting to record phonation at sea in relation to specific activities. However, since auditory communication is a highly developed feature in delphinids, such a signaling system must at present be considered a possibility, particularly as cooperative behavior in dolphins occurs in a number of contexts, ranging from care-giving ("epimeletic") behavior when a group member is injured (Caldwell and Caldwell, 1966) to active cooperation in the herding of food-fish (Tayler and Saayman, 1972). Indeed, Russian observers report that specific sounds are emitted by free-ranging dolphins at particular stages of feeding during fish-herding procedures (Morozov, 1970).

Secondly, comparative studies reveal that the bisexual social unit, in which adult males associate with adult females and their young in cohesive groups persisting throughout the year, is not necessarily unique to the order of primates, as has long been assumed. Recent studies have shown that this comparatively complex form of social structure occurs in other mammals such as the plains zebra and mountain zebra (Klingel, 1972) and social carnivores such as the lion, hyena, and wild dog (Schaller and Lowther, 1969; Schaller, 1972). Since dolphins are relatively long-lived mammals, with a slow maturation process and strong, extended mother–calf ties and, further, since they are relatively large-brained and display complex levels of play and exploratory and imitative behavior in captivity (Tayler and Saayman, 1972, 1973a), it is pertinent to inquire whether their social structure and organization reflect a level of complexity comparable to that characteristic of the higher mammals referred to above. A necessary first step, therefore, is to determine whether the population is structured into relatively cohesive social groups throughout the year or whether bulls form only temporary associations with cows and their offspring during circumscribed seasons when physiological conditions necessary for copulation and conception are prevalent. The evidence, however, is fragmentary, indirect, and at best highly speculative.

It is not possible to distinguish bulls from cows in the field, but the data on the periodicity of births (Fig. 14) indicated that calves were born at all times of the year, although there was a marked birth peak in the summer months. This suggests that bulls remained in the vicinity of cows throughout the year and provides indirect evidence for the coherence of bisexual social groups in humpback dolphins—assuming that the gestation period is 12 months, as it is in bottlenose dolphins (Tavolga and Essapian, 1957; Tayler and Saayman, 1972). Moreover, the constantly fluctuating composition of groups of dolphins suggested that the segregation of the sexes into bachelor or nursery groups, which implies a certain continuity of membership over time, was not a prominent feature of the social organization of humpback dolphins.

Evidence on the social structure of other delphinid species is limited, but a review of the literature by Evans and Bastian (1969) indicates that wide variations in the composition of individual groups may occur both within and between species. Caldwell and Caldwell (1972a) suggested that the small scattered groups of bottlenose dolphins, which together comprise a larger herd, may consist of several adult females and their young, with an adult bull and several subadults of either sex located nearby; this "family group" contrasted with separate groups of juveniles in which sexual segregation apparently occurred. The pilot whales *(Globicephala melaena)* studied by Sergeant (1962) appear to be fairly typical of the delphinid social systems, in that group size varied greatly in relation to circumstances: schools, averaging about 20 individuals, traveled in tightly knit formation but dispersed in scattered groups to feed. Variability in group size and structure was confirmed by examination of whole groups driven ashore, in which adult females generally outnumbered adult males. It is noteworthy that physiological evidence obtained by Sergeant suggested that births might occur at any time of the year, although a birth peak was clearly evident. There was some suggestion that adult bulls might segregate on rare occasions and for limited periods during late summer. The tendency for large schools to disperse into smaller groups, particularly when feeding, appears to be a characteristic feature of the spatial organization of many delphinid species. Schools of *Lagenorhynchus obliquidens,* numbering about 1000 individuals, scattered over approximately four square miles in small groups but became more compact when not feeding; similar observations were made on common dolphins

(*Delphinus* sp.) and pilot whales (*Globicephala scammoni*); the latter species dispersed in scattered groups to feed during the day and then coalesced to travel together at dusk (Brown and Norris, 1956). Similarly, Evans (1971) observed small groups of 5–30 *Delphinus delphis* unite to form one large, tightly knit school which later scattered again into smaller groups. Pilleri and Knuckey (1969) found great variability in the size of schools of *Delphinus delphis* and *Stenella styx* in the western Mediterranean: mixed schools, containing adult males, females, and young, occurred in groups of 30–40 dolphins or in large schools numbering approximately 1000; similarly, *Tursiops truncatus* were seen in groups ranging between eight and 100 individuals.

Thus the concept developed in this report on fluctuating group sizes within the context of a larger school of humpback dolphins in some respects resembles a pattern of grouping tendencies apparent in many delphinids. The findings of the present study are also in agreement with limited qualitative observations on *Sousa* in other geographical ranges: Bruyns (1960) reported that "ridge-backed dolphins" off the coast of India and in the Persian Gulf occurred in groups ranging from four to 20 individuals and, during an expedition to the Persian Gulf and the Indus delta, Pilleri (1973) observed *Sousa* traveling alone, in small groups of five to six individuals, or in mixed schools of up to 16 adult and immature dolphins. The data reviewed above suggest that the temporary fragmentation of schools into smaller groups is a common feature of many delphinid social systems. It is likely, therefore, that small groups of dolphins seen ranging apparently independently form part of a larger school, which itself represents a basic social unit of the total population and, as Perrin (1970, p. 47) has suggested, has " . . . integrity through time, with a high degree of isolation from neighboring schools." Whether or not individuals move freely between groups within the school regardless of age, sex, or time of year, however, remains a highly intriguing question for future research.

If births and, consequently, conceptions were indeed distributed throughout the year, showing only a birth peak and not a clearly circumscribed season of births, and if this implies that adult bulls associated with adult cows throughout the year, then a number of questions occur concerning the nature of the sexual cycle in the humpback dolphin and the possible function of sexual behavior in the maintenance of the social system, apart from its primary relation to the reproduction of the species. However, direct physiological data are completely lacking concerning the reproductive system of the humpback dolphin and, indeed, very little is established concerning the reproductive physiology of other delphinid species, although there is some information on the distribution of births and mating activity. Whereas the majority of births in captive *Tursiops truncatus* in the Marineland of Florida suggested a restricted birth season, oceanaria at other sites did not report the same birth pattern (Caldwell and Caldwell, 1972*a*). Births in free-ranging bottlenose dolphins in the same area, however, occurred in all seasons of the year, although there was a birth peak in spring and early summer (Caldwell and Caldwell, 1972*b*). Similarly, Kasuya (1972) has shown that births in *Stenella coeruleoalba* also occurred during all seasons of the year, although there were apparently two birth peaks per year. Whereas these data indicate that in these species cows were sexually receptive throughout the year, a number of workers have indicated that circumscribed mating seasons occur in *Tursiops truncatus* (McBride and Hebb, 1948; Tavolga and Essapian, 1957; Essapian, 1963; Tavolga, 1966), and Ridgway (1972) has speculated that " . . . the female *Tursiops truncatus* is a seasonally polyestrus animal that reaches the peak of estrus in spring or autumn (or both)" (p. 633). In contrast to the above workers, however, Evans and Bastian (1969) have commented that " . . . observations of captive populations indicate high levels of sexuality throughout the year . . . captive delphinids (*Tursiops and Lagenorhynchus*) regardless of age, sex or time of year, do indeed display sexual activities that are quantitatively and qualitatively quite impressive" (p. 454). Similarly, Caldwell and Caldwell (1967) noted that sexual activity was limited on only two occasions, during and after birth, in nearly two years of observations conducted regularly on the dolphin community at the Marineland of the Pacific.

Thus, the behavioral evidence concerning the occurrence of circumscribed mating seasons is contradictory in nature and is not in accord with the data on the distribution of births. Furthermore, the contention that circumscribed mating seasons occur in *Tursiops truncatus* is largely unsupported by quantitative evidence and contrasts with observations on captive and free-swimming *Tursiops aduncus* which displayed sexual receptivity throughout the year and copulated when pregnant or lactating (Tayler and Saayman, 1972). Moreover, quantitative records of the behavior patterns associated with the courtship and mating activity of an adult bull, an adult cow and their 2-year-old female calf, obtained between March 1970 and February 1971 in the Port Elizabeth Oceanarium, showed that the total number of courtship behavior patterns per observation session did not vary significantly in relation to season ($P < 0.25$) and, furthermore, there were no significant differences ($P < 0.7$) in the frequencies with which courtship-like behavior occurred in relation to season in 119 groups of free-swimming bottlenose dolphins observed for more than 100 hr between October 1970 and July 1973 in Algoa and Plettenberg Bays (Saayman, unpublished observations). These findings supported the subjective impression that copulatory activity in *Tursiops aduncus* was not restricted to a circumscribed period of the year, and they are in accord with the results of the present study on *Sousa*, which indicated that fertile copulation was possible at all times of the year (although more probable in the summer months) and that the frequency and duration of courtship-like behavior was independent of season. On the basis of the above evidence, it seems that the regulation of sexual behavior in these delphinids differs from those mammals characterized by classical "estrous cycles" with clearly defined and restricted periods of receptive behavior associated with ovulation and a particular condition of the genital tract (see Eckstein and Zuckerman, 1956).

Further long-term research is required not only on the physiological and hormonal conditions necessary for the stimulation of sexual receptivity in the delphinidae, but also on the functions of behavior patterns generally categorized as "courtship" or "sexual" in nature. A number of considerations suggested that behaviors characteristically associated with courtship and copulation functioned not only in the context of primary sexual activity, but also in other social contexts which were derived directly from the nature of the social system. For example, in captive bottlenose dolphins, the behavior roles and postures characteristic of courtship were interchangeable between the sexes, so that the bull or the cow of a consorting pair alternately adopted the active or the passive role and either partner might display similar initiating postures in mating contexts. Behavior closely resembling primary sexual activity often occurred in large groups of participating animals, in which more than one bull attempted insertion with a single partner (Saayman and Tayler, 1973b; Saayman et al., 1973). In the present study, the inverted postures characteristic of copulation in *Sousa* were also found to be interchangeable, so that even young calves swam belly-to-belly either above, or inverted below, adult whitefin partners, the inverted roles changing from time to time as the partners surfaced to breathe. Furthermore, social interactions, involving extensive bodily contact and closely resembling courtship activity, frequently occurred when groups of humpback dolphins combined or even when single individuals joined a group. It has been proposed elsewhere (Saayman and Tayler, 1973a,b) that these courtship-like activities may very well function in a "greetings" context, in which the extensive cutaneous contact between all age and sex classes may well serve to reinforce social bonds in a highly social species which disperses in small subgroups for varying periods to feed over an extensive coastal range. It is noteworthy that similar reversals of behavior roles in sexual activities have been reported to occur in several delphinid genera in captivity (Brown et al., 1966). In addition, numbers of captive *Tursiops truncatus* were reported to crowd around copulating partners, young males apparently competing for the cow (Tavolga and Essapian, 1957; Caldwell et al., 1968). Similar observations have been reported by Pilleri (1972) for the free-ranging Indus dolphin (*Platanista indi*). It is therefore possible that in group social interactions of this kind, units of behavior closely resembling those occuring during primary sexual encounters function as an integral part of a "greetings ceremony" within the wider context of an open social system, which may be characteristic of several delphinid species in addition to *Sousa*. In any event this possibility should be considered when attempts are made to categorize and define individual units of behavior as either "sexual," "play," or "aggressive": the contexts within which the behavior occurs represent important determinants of its definition, and these contexts, in many instances, can only be ascertained by the observation of behavior functioning under the appropriate socioecological circumstances to which it has adapted. In addition, the function of courtship-like activity as a cohesive factor in the social organization of these dolphins would partially account for its prevalence throughout the year and would offer some explanation for the apparent emancipation of sexual receptivity and copulatory behavior from a strict dependence upon hormonal and physiological conditions characteristic of "estrus."

In conclusion, the dispersal of small groups of humpback dolphins over a large area close to the coast, in addition to providing an optimum feeding strategy, would have the added advantage of protecting the population from voracious predators such as killer whales. Humpback dolphins would represent relatively inconspicuous targets for predation in contrast to the prolific schools of pelagic dolphins and, moreover, would provide a relatively low return on the energy expenditure of the potential predator. It is not clear whether the spatial dispersion of groups of humpback dolphins would be advantageous in the case of attempted predation by sharks. Although antagonistic encounters with sharks were seen only infrequently, Wood et al. (1970) showed that inshore dolphins (*Tursiops truncatus*) frequently bear severe scars, many of which are attributable to sharks. In the present studies, several humpback dolphins displayed scars of varying severity, some of which were prominent enough to permit individual recognition of the animals. It was not possible to determine whether the wounds were due to shark bites or to possible collisions with jagged reefs or rocky outcrops during vigorous social or hunting activities. However, the encounters observed between dolphins and potential predators indicated that the speed, agility, and group coordination of the dolphins enabled them to cope with sharks while their exploitation of the inshore environment, enabling them to maneuver close to and landwards of outcrops of reefs along the coast, provided them with adequate protection against the larger killer whales. The differential utilization of the ecological zones at Robbe Berg Point (Fig. 21, Table VII) further indicated that the dolphins exploited these features of their home range to full advantage. It is likely that the rich feeding area over the reefs in zone II at the point, together with the relatively calm and sheltered waters of the bay close at hand in zone I, provided a highly suitable habitat for *Sousa* and that this factor accounted very largely for the protracted periods of observation of dolphins which were possible at this site.

ACKNOWLEDGMENTS

This work is supported by a research grant from the South African Council for Scientific and Industrial Research which is gratefully acknowledged. We wish to thank Mr. C.T. Connellan and Mr. V.L. Connett for field and technical assistance.

REFERENCES

Altmann, S. A., 1967, Preface, in: *Social Communication among Primates* (S.A. Altmann, ed.), pp. ix–xii, University of Chicago Press, Chicago.

Best, P.B., 1970, Exploitation and recovery of right whales *Eubalaena australis* off the Cape Province, *Invest. Rep. Div. Sea Fish. S. Afr.* **80**:1–20.

Brown, D.H., and Norris, K.S., 1956, Observations of captive and wild cetaceans, *J. Mammal.* **37**:311–326.

Brown, D.H., Caldwell, D.K., and Caldwell, M.C., 1966, Observations on the behavior of false killer whales, with notes on associated behavior of other genera of captive delphinids, *Los Angeles County Mus. Contri. Sci.* **95**:1–32.

Bruyns, W.F.J., 1960, The ridge-backed dolphin of the Indian Ocean, *Malay. Nat. J.* **14**:159–165.

Caldwell, M.C., and Caldwell, D.K., 1966, Epimeletic (care-giving) behavior in cetacea, in: *Whales, Dolphins and Porpoises* (K.S. Norris, ed.), pp. 755–789, University of California Press, Berkeley.

Caldwell, D.K., and Caldwell, M.C., 1967, Dolphins, porpoises and behavior, *Underwater Nat.* **4**:14–19.

Caldwell, D. K., and Caldwell, M.C., 1968, The dolphin observed, *Nat. Hist., N.Y.* **77**:58–65.

Caldwell, D. K., and Caldwell, M.C., 1972a, *The World of the Bottlenosed Dolphin*, J.B. Lippincott Company, Philadelphia.

Caldwell, M.C., and Caldwell, D.K., 1972b, Behavior of marine mammals, in: *Mammals of the Sea, Biology and Medicine* (S.H. Ridgway, ed.), pp. 419–465, Charles C. Thomas, Springfield, Illinois.

Caldwell, M.C., Caldwell, D.K., and Townsend, B.C., 1968, Social behavior as a husbandry factor in captive odontocete cetaceans, Proceedings of the Second Symposium on Diseases and Husbandry of Aquatic Mammals, Marineland Research Laboratory, St. Augustine, Florida, pp. 1–9.

Crook, J. H., 1970, Social organisation and the environment: Aspects of contemporary social ethology, *Anim. Behav.* **18**:197–209.

Crook, J. H., and Goss-Custard, J.D., 1972, Social ethology, *Annu. Rev. Psychol.* **23**:277–312.

Dudok van Heel, W.H., 1972, Transport of dolphins, *Aquatic Mammals* **1(1)**:1–32.

Eckstein, P., and Zuckerman, S., 1956, The oestrous cycle in the mammalia, in: *Marshall's Physiology of Reproduction* (A.S. Parkes, ed.), pp. 226–396, Longmans, Green, London.

Essapian, F. S., 1963, Observations on abnormalities of parturition in captive bottle-nosed dolphins, *Tursiops truncatus*, and concurrent behavior of other porpoises, *J. Mammal.* **44**:404–414.

Evans, W. E., 1967, Vocalisation among marine mammals, in: *Marine Bioacoustics*, Vol. 2, pp. 159–186, Pergamon Press, Oxford.

Evans, W. E., 1971, Orientation behavior of delphinids: Radio telemetric studies, *Ann. N.Y. Acad. Sci.* **188**:142–160.

Evans, W. E., and Bastian, J., 1969, Marine mammal communication: Social and ecological factors, in: *The Biology of Marine Mammals* (H. T. Andersen, ed.), pp. 425–475, Academic Press, New York.

Gaskin, D. E., 1972, *Whales, Dolphins and Seals*, Heinemann, London.

Goodall, J., 1965, Chimpanzees of the Gombe Stream Reserve, in: *Primate Behavior* (I. DeVore, ed.), pp. 425–473, Holt, Rinehart and Winston, New York.

Hoese, H. D., 1971, Dolphin feeding out of water in a salt marsh, *J. Mammal.* **52**:222–223.

Hussain, F., 1973, Whatever happened to dolphins? *New Sci.* **57**:182–184.

Irvine, B., and Wells, R. S., 1972, Results of attempts to tag Atlantic bottlenosed dolphins (*Tursiops truncatus*), *Cetology* **13**:1–5.

Kasuya, T., 1972, Growth and reproduction of *Stenella caeruleoalba* based on the age determination by means of dentinal growth layers, *Sci. Rep. Whales Res. Inst.* **24**:57–79.

Kellogg, W. N., 1961, *Porpoises and Sonar*, University of Chicago Press, Chicago.

Klingel, H., 1972, Social behaviour of African Equidae, *Zool. Afr.* **7**:175–185.

Lawrence, B., and Schevill, W. E., 1954, *Tursiops* as an experimental subject, *J. Mammal.*

Layne, J. N., 1958, Observations on freshwater dolphins in the upper Amazon, *J. Mammal.* **39**:1–22.

Martinez, D. R., and Klinghammer, E., 1970, The behavior of the whale *Orcinus orca*: A review of the literature. *Z. Tierpsychol.* **27**:828–839.

McBride, A. F., and Hebb, D. O., 1948, Behavior of the captive bottlenose dolphin, *Tursiops truncatus*, *J. Comp. Physiol. Psychol.* **41**:111–123.

McBride, A. F., and Kritzler, H., 1951, Observations on pregnancy, parturition, and postnatal behavior in the bottlenose dolphin, *J. Mammal.* **32**:251–266.

Morozov, D. A., 1970, Dolphins hunting, *Rybnoe Khoziaistvo* **46**:16–17;U.S. Department of Commerce, NOAA, National Marine Fisheries Service, Fishery-Oceanography Center, La Jolla.

Nishida, T., 1968, The social group of wild chimpanzees in the Mahali mountains, *Primates* **9**:167–224.

Nishiwaki, M., 1972, General biology, in: *Mammals of the Sea, Biology and Medicine* (S. H. Ridgway, ed.), pp. 1–204, Charles C. Thomas, Springfield, Illinois.

Norris, K. S., 1966, Some observations on the migration and orientation of marine mammals, in: *Animal Orientation and Navigation* (R. M. Storm, ed.), pp. 101–124, Oregon State University Press, Corvallis.

Penrith, M. J., 1972, The behaviour of reef-dwelling sparid fishes, *Zool. Afr.* **7**:43–48.

Perrin, W. F., 1970, The problem of porpoise mortality in the U.S. tropical tuna fishery, Proceedings of the Sixth Annual Conference on Biological Sonar and Diving Mammals, Stanford Research Institute, Menlo Park, California, pp. 45–48.

Pilleri, G., 1972. Field observations carried out on the Indus dolphin *Platanista indi* in the winter of 1972, *Investigations on Cetacea* **4**:23–29.

Pilleri, G., 1973, *Cetelogische Expedition zum Indus und Persischen Golf und nach Goa und Thailand im Jahre 1973*, Verlag Hirnanatomisches Institut, Waldau-Bern, Switzerland.

Pilleri, G., and Knuckey, J., 1969, Behaviour patterns of some Delphinidae observed in the Western Mediterranean, *Z. Tierpsychol.* **26**:48–72.

Powell, B. A., 1966, Periodicity of vocal activity of captive Atlantic bottlenose dolphins (*Tursiops truncatus*), *Bull. 5th. Calif. Acad. Sci.*, **65**:237–244.

Reynolds, V., and Reynolds, F., 1965, Chimpanzees of the Budongo Forest, in: *Primate Behavior* (I. DeVore, ed.), pp. 368–424, Holt, Rinehart and Winston, New York.

Ridgway, S. H., 1972. Homeostasis in the aquatic environment, in: *Mammals of the Sea, Biology and Medicine* (S. H. Ridgway, ed.), pp. 590–747, Charles C. Thomas, Springfield, Illinois.

Ross, G. J. B., 1973, The Taxonomy of Bottlenosed Dolphins *Tursiops* Species in South African Waters, with Notes on their Biology, pp. 1–113, MS thesis (submitted), University of Port Elizabeth.

Rowell, T. E., 1967, Variability in the social organisation of primates, in: *Primate Ethology* (D. Morris, ed.), pp. 219–235, Weidenfeld and Nicolson, London.

Saayman, G. S., and Tayler, C. K., 1973a, Social organisation of inshore dolphins (*Tursiops aduncus* and *Sousa*) in the Indian Ocean, *J. Mammal.* **54**:993–996.

Saayman, G. S., and Tayler, C. K., 1973b, Some behaviour patterns of the southern right whale *Eubalaena australis*, *Z. Säugetierek.* **38**:172–183.

Saayman, G. S., Bower, D., and Tayler, C. K., 1972, Observations on inshore and pelagic dolphins on the south-eastern Cape coast of South Africa, *Koedoe* **15**:1–24.

Saayman, G. S., Tayler, C. K., and Bower, D., 1973, Diurnal activity cycles in captive and free-ranging Indian Ocean bottlenose dolphins (*Tursiops aduncus* Ehrenburg), *Behaviour* **44**:212–233.

Schaller, G. B., 1972, *The Serengeti Lion*, University of Chicago Press, Chicago.

Schaller, G. B., and Lowther, G. R., 1969, The relevance of carnivore behavior to the study of early hominids, *Southwest. J. Anthropol.* **25**:307–341.

Sergeant, D. E., 1962, The biology of the pilot or pothead whale *Globicephala melaena* (Traill) in Newfoundland waters, *Bull. Fish. Res. Bd. Can.* **132**:1–84.

Sergeant, D. E., 1969, Feeding rates of cetacea, *Fiskeri dir. Skr. Ser. Hav Unders.* **15**:246–258.

Stoltz, L. P., and Saayman, G. S., 1970. Ecology and behaviour of baboons in the Northern Transvaal, *Ann. Transv. Mus.* **26**:99–143.

Tavolga, M. C., 1966, Behavior of the bottlenose dolphin (*Tursiops truncatus*): Social interactions in a captive colony, in: *Whales, Dolphins and Porpoises* (K. S. Norris, ed.), pp. 718–730, University of California Press, Berkeley.

Tavolga, M. C., and Essapian, F. S., 1957, The behavior of the bottlenosed dolphin (*Tursiops truncatus*): Mating, pregnancy, parturition and mother–infant behavior, *Zoologica N.Y.* **42**:11–31.

Tayler, C. K., and Saayman, G. S., 1972, The social organisation and behavior of dolphins (*Tursiops aduncus*) and baboons (*Papio ursinus*): Some comparisons and assessments, *Ann. Cape Prov. Mus. (Nat. Hist.)* **9**:11–49.

Tayler, C. K., and Saayman, G. S., 1973a, Imitative behaviour by Indian Ocean bottlenose dolphins (*Tursiops aduncus*) in captivity, *Behaviour* **44**:286–298.

Tayler, C. K., and Saayman, G. S., 1973b, Techniques for the capture and maintenance of dolphins in South Africa, *J. Sth. Afr. Wildl. Manage. Assoc.* **3**:89–94.

Tayler, C. K., Saayman, G. S., and Connett, V. L., 1970, Behaviour of captive Indian Ocean bottlenose dolphins *Tursiops aduncus*, color film, Port Elizabeth Oceanarium.

Tietz, R. M., 1963, A record of the speckled dolphin from the south-east coast of South Africa, *Ann. Cape Prov. Mus.* **3**:68–74.

Tobias, P., 1964, Bushman hunter-gatherers: A study in human ecology, in: *Ecological Studies in Southern Africa* (D. Davis, ed.), pp. 67–86, W. Junk, The Hague.

Tomilin, A. G., 1967, Mammals of the U.S.S.R. and Adjacent Countries, Vol. IX, Cetacea, Israel Program for Scientific Translations, Jerusalem.

Townsend, C. H., 1914, The porpoise in captivity, *Zoologica N.Y.* **1**:289–299.

van Lawick-Goodall, J., 1968, The behaviour of free-living chimpanzees in the Gombe Stream Reserve, *Anim. Behav. Monogr.* **1**:161–311.

Winer, B. J., 1962, *Statistical Principles in Experimental Design*, McGraw-Hill, New York.

Wood, F. G., Caldwell, D. K., and Caldwell, M. C., (1970), Behavioral interactions between porpoises and sharks, *Investigations on Cetacea* **2**:264–277.

Chapter 7

SOCIAL ORGANIZATION IN SPERM WHALES, *Physeter macrocephalus*

Peter B. Best

39 Beresford Road
Woodstock, Cape Town, South Africa

I. INTRODUCTION

A. Review of Previous Work

> The cows with their young give from nothing up to 35 barrels, and seem to go in schools together, and we frequently see from twenty-five to fifty and sometimes one hundred or more in a school, with occasionally a large bull among them, and at times, though seldom, we find all sizes together. The male or bull whales seem to separate from the cows and calves when about the size of 35 barrels, as we seldom get them in the schools of the mother and its young to make more oil than that, and we find the young bulls in pods or schools beyond that size; we find them in what we call 40-barrel bulls, where they generally go in larger numbers than they do as they increase in size; we find them again in smaller schools of about the size of 50 barrels, and again about 60 barrels, where we sometimes see eight or ten together, and 70 barrels four or five, and beyond that one, two, and three . . . .

These remarks by Captain H.W. Seabury of New Bedford (Clark, 1887) indicate that the organization of sperm whales into schools consisting predominantly of a particular sex or size group was a fact recognized by experienced whalermen in the eighteenth and nineteenth century sperm whale fishery. Those engaged in open-boat whaling had an excellent opportunity to observe sperm whale behavior at close quarters, and several texts including observations on the behavior and social organization of the species were published by men with first-hand knowledge of this form of whaling (e.g., Beale, 1839; Bennett, 1840). These reports have been summarized by Caldwell *et al*. (1966), whose review has been referred to here when the original publications were not available. Most of the data relevant to sperm whale social organization published prior to 1964 has in fact been included by Caldwell *et al.* (1966), and no attempt will be made here to repeat their detailed review of earlier observations.

Due to the limited processing of the carcass in open-boat whaling, there was little opportunity to determine the exact length, age, or reproductive condition of the animals that were killed. With the advent of modern whaling methods involving processing of the whole carcass, it became feasible to investigate the biology of sperm whales in more detail. Conversely, however, the rapid and more intensive catching methods using noisy, propeller-driven catchers gave less time and opportunity for observations of the undisturbed behavior of sperm whales and their school composition. Furthermore, the main arena for such whaling became the Antarctic zone instead of the tropical and subtropical waters fished by open-boat whalers, the principal quarry being baleen whales or large bachelor sperm whales, and the greater variety of sperm whale groupings available in warmer waters were rarely sampled. For a long time after the end of open-boat whaling, therefore, first-hand observations and new data on social behavior in this species were slow in coming.

Matthews (1938) provided some information on the distribution and migration of the two sexes, based on commercial catch returns, and proposed limits for the migration of females and immature animals into higher latitudes of 40°N and 40°S. He also made the first attempt to correlate the stage of male sexual maturation with social behavior. Christensen (1926) and Schubert (1951) proposed a segregation of sexes and size groups from the South American coast, males being found closer to shore than females, and solitary males being the farthest offshore, but Clarke (1962) was unable to substantiate these findings fully.

The first attempt to investigate schooling behavior quantitatively was that of Clarke (1956), who was able to demonstrate off the Azores that males may be solitary or in schools, but females are invariably gregarious. His conclusion that male schooling behavior may be seasonal failed to take into account the different size classes involved, a significant factor as will be seen later. Other general observations in his paper (the occurrence of female schools consisting predominantly of animals at one stage of the reproductive cycle, and the incidence of mixed schools of juvenile males and females) were largely unsubstantiated by statistical series. Later he presented some data on school sizes for sperm whales observed off the western coast of South America (Clarke, 1962).

Gilmore (1959), in a review of mass strandings of sperm whales, confirmed the existence of bachelor schools and mixed (harem) schools, and considered that the latter were confined to waters with a surface temperature of 20°C or more.

Prior to 1960, therefore, knowledge of the composition and development of sperm whale schools had not advanced greatly beyond the observations of open-boat whalers of the previous century.

In 1958, however, a new method of age determination for sperm whales was discovered, using laminae present in the teeth (Nishiwaki *et al.*, 1958), and this method was later quantified correctly (Ohsumi et al., 1963). This led the way for more detailed investigations of the growth, maturation, and reproductive cycle of the species, which by the 1962–1963 season had become the single most important species numerically in the world's whale catch for the first time since modern whaling started (apart from the years during the second world war). This in turn caused the sperm whale to become the target of increased research effort, and it was clear that rational exploitation of the stocks would be impossible without more knowledge of the social organization of the species (Gambell, 1967; Ohsumi, 1971).

Several different approaches to this problem were used. Aerial observations of sperm whale behavior could obviously yield most interesting information on the social behavior and organization of undisturbed animals (Nishiwaki, 1962; Gambell, 1968), and data from aerial spotting operations for the whale fishery off Durban were used to provide estimates of the number of animals in 1222 schools and their probable size composition (Gambell, 1972). The accuracy of counts of animals in a school from the air is likely to be much greater than counts from a ship, and this large sample of school size estimates is extremely valuable. However, the experience of this author has been that estimates of the size of animals made from an aircraft without some means of calibration must be considered very approximate, and hence field classifications into "small," "medium," or "big" whales as used by the commercial spotters off Durban must be considered tentative. The actual sex and size compositions of 140 schools, in which half or more of the estimated number present were landed at the whaling station, were also examined by Gambell (1972). From these the rate of segregation of small males below 39 ft (11.9 m) in length from female schools was considered to be relatively low, a conclusion which will be contested later in this paper. The number of mature females per "harem" (breeding) bull was calculated using the number of small whales seen per large bull in mixed schools, and a sex ratio was derived from the total catch records or from school composition data for small whales. Clarke's (1956) contention that female schools could consist of animals predominantly at one stage of the reproductive cycle was not fully borne out by Gambell's data, which indicated that usually animals in all three main phases of the cycle, pregnant, lactating and resting, were found in the same school.

Other workers attempted to estimate the number of animals in schools and their size composition from sightings at sea (Gaskin, 1970; Ohsumi, 1971). Gaskin (1970) has described methods of distinguishing between female and small male sperm whales at sea using morphological criteria, but in the experience of this author such a distinction can be most difficult unless a small calf or a female with a conspicuous callus on the dorsal fin is present. Both Gaskin (1970) and Ohsumi (1971) have drawn attention to the tendency for the larger whales to be found in smaller schools, and Gaskin (1970) has produced a linear regression of body length on school size to express this quantitatively for males. Gaskin (1970) also presented data on testis weight for 155 whales and on the histological appearance of the testis for 55 whales, showing that the progression from bachelor schools to pairs to solitary males corresponded to increasing maturity, the majority of animals in bachelor schools being pubertal.

The segregation and formation of all-male groups have been studied by Ohsumi (1966) and Tarasevich (1967). From a comparison of male and female age distributions from August to October, Ohsumi (1966) was able to construct a theoretical age distribution for the males not present at this time of year off Japan, on the assumption that the age distributions for the two sexes should be similar. This theoretical age distribution fitted well with the observed distribution of males in Aleutian waters from June to August. On this basis a mathematical model for the rate of segregation of males to higher latitudes was erected. Tarasevich (1967) examined male sperm whales caught in the Bering Sea, Gulf of Alaska, and Aleutian waters and concluded that they aggregated on a size basis which might or might not result in age uniformity. She also discussed the seasonality of male schooling behavior, and proposed that most sexually mature males (apparently from about 11.6 m upwards in size) remained in mixed schools with females during the winter but left them in summer to migrate to higher latitudes, the sequence of this segregation varying with the size of the animal.

Best (1969a,b) provided some indirect evidence of the size and age at which male sperm whales enter the Antarctic for the first time, based on cyamid infestation and the presence of diatom film. The inference was made that this corresponded to the stage at which males left their parent schools for the first time, an incorrect assumption as will be shown later in this paper.

Tormosov (1970) studied the geographical distribution of male sperm whales within the Antarctic, and found an apparent homogeneity of age for a particular latitudinal zone. He also speculated on the reasons for age and sex segregation in this species.

In a later paper (1975), the same author summarized his findings from the examination of 205 sperm whale groupings and identified six main school types, three of mixed sexes in which females predominated and three consisting of males only.

Investigation of the structure and formation of mixed schools has proved more difficult than for all-male groupings, principally because many mixed school members are well below the official minimum length and so are rarely sampled. It has only been possible to obtain comprehensive data on the composition of such schools if a mixed school should strand in its entirety or if attempts were made to capture a whole school under special permit. Robson and Van Bree (1971) and Stephenson (1975) were able to obtain length and sex data from mixed schools (of 59 and 72 sperm whales, respectively) that stranded on the New Zealand coast, while Ohsumi (1971) examined in more detail members of three mixed schools that were sampled for scientific purposes, from which about 30%, 77%, and 81%, respectively, of the animals present were captured.

Further information on 15 mixed schools of which 20–110% (sic) of the members were caught has been presented in an unpublished paper by Masaki *et al.* Both Japanese studies give extensive data on the sex ratio, size, and age distribution and on the reproductive and physical maturity status of members of these schools. Ohsumi (1971) also presented data from whale-marking which indicated some degree of continuity for the female components of mixed schools.

Finally, Gaskin (1970) and Ohsumi (1971) have speculated on the ontogeny of social groupings in sperm whales. Both authors recognize six basic groupings: nursery schools, harem (mixed) schools, juvenile (or immature) schools, bachelor schools, bull schools (or pairs), and solitary bulls. Harem and nursery schools were considered to be the same social unit at differing times of the year. Gaskin (1970) proposed that the largest bachelor schools were derived from juvenile schools, although both authors figured diagrammatically a contribution to bachelor schools both directly from harem schools and indirectly from juvenile mixed-sex schools. Although both authors tended to favor the theory that breeding was chiefly performed by large harem bulls that rejoined the nursery schools, they referred to the doubts expressed by Rice (in Caldwell *et al.*, 1966), who proposed that breeding was done by younger bulls that remained with the nursery schools for most of the year. This topic will be discussed more fully below.

Despite the recent improvements in our knowledge of sperm whale social organization, it should be stressed that nearly all the observations made so far have been based on the examination of whole schools of whales (or members from them) at one moment of time (normally after death). Such synoptic observations can provide very little information on inter- and intraschool relationships, or on social behavior under different circumstances such as feeding, calving, or mating. The acquisition of such data would require systematic and protracted observations of individual schools and whales in the wild, something that so far has not proved possible for sperm whales. The conclusions reached in this paper should therefore be considered as only representing an outline of the species' social organization, a skeleton in fact that needs "fleshing-out" with direct field observations of social-behavior.

B. Scope of the Present Work

New data on the composition of sperm whale schools is presented in this paper, based on the analysis of observations from the whale-spotting aircraft off Donkergat, Saldanha Bay, South Africa (33°S, 18°E), combined with reports from catching vessels and the postmortem examination of school members on the flensing platform at the Donkergat land station. Estimates of the length or sex of animals at sea or from the air have not been used.

To avoid confusion, the following terminology has been adopted for sperm whale social assemblages in this paper (though not when direct reference is made to the findings of other authors).

"Group" or "Grouping." A general term including either a single whale or an association of whales that might be a school or an aggregation.

"School." An association of whales that from their behavior appear to be a cohesive social unit, i.e., their swimming and respiratory patterns seem coordinated irrespective of the distance between individuals. Some whales might be about one body length apart (when schooling would be termed "tight") or separated by as much as one half to one mile (when they might be termed "singles"). These groups are sometimes called "flocks" by those engaged in whaling off South Africa—presumably from the Norwegian word "flokk" for a gathering of animals.

"Aggregation." More than one school, probably in temporary association. Swimming and respiratory patterns are not necessarily coordinated and animals frequently are spread out over a wide area.

During the 1963 whaling season at Donkergat special attention was paid to the composition of sperm whale groupings as they were encountered by the spotter aircraft or catching vessels. The pilot of the spotter aircraft, being in the best position to make accurate observations, was asked to record the geographical position of each school, the number of animals present, some idea of their size composition (e.g., "all small," "bachelors," "big bull," "calves present") and their behavior ("direction of movement," "resting," "mating," "playing"). Certain of these aspects were checked when possible with the captains of the catching vessels, especially on days when the boats found the group before the aircraft or when the spotter aircraft was grounded. On many occasions the deployment of whales, catchers, and spotter aircraft was too widespread in time and space to make identification of individual groupings possible. However, on 27 occasions during the season it was possible to obtain a fairly accurate estimate of the number and size composition of a school of sperm whales before hunting started, and subsequently to identify animals killed from this school when they were landed at the whaling station. On most of these occasions only a single school was encountered during the day.

When sperm whales were landed at the station they were examined and material for the determination of age and reproductive status was collected and prepared as described in earlier reports (Best, 1967, 1969a, 1970). Individual whales killed from a particular grouping were recognized from the catcher and serial numbers cut into their flukes. The 27 groupings described below involved the examination of 358 animals.

These 358 animals represented 42% of all sperm whales landed at Donkergat in 1963. How representative these whales were of the population at large is problematical, but it is considered likely that the majority of groupings present in the whaling grounds were sampled, although not necessarily in proportion to their abundance. The South African Department of Industries issued a special permit under Article VIII of the International Convention for the Regulation of Whaling for the taking of up to 150 undersized sperm whales in the 1963 season at Donkergat, and this to some extent assisted in obtaining better coverage of the smaller size and younger age groups. However, one school type possibly not represented at all due to size selection was the school of immature whales of both sexes.

Using these data and published results, the composition of sperm whale groupings and their possible ontogeny are discussed. Considerable attention is paid to mixed school structure and the number and identity of breeding bulls. The stage at which males leave their parent school is investigated and compared with the age at which this sex first migrates to the Antarctic. In order to understand the phenomenon of male segregation better, the latitudinal distribution of males and females in higher latitudes in summer has been derived from catch records and is discussed in some detail in relation to possible mechanisms behind the pattern of segregation.

Findings on the migrations and social organization of the species are summarized, and the type of organization shown by the sperm whale is considered, with a discussion on how this might have evolved. The sperm whale is now commercially the most important whale species, and because of its apparently complex social organization, the possible effects of different regimes and levels of exploitation on its social behavior are discussed in a final section.

II. COMPOSITION OF POPULATION

If the social organization of a species is to be correctly interpreted, it is essential to be familiar with the biology and population structure of the species as a whole.

Sperm whales are the most sexually dimorphic in size of all living cetaceans. A physically mature male 52 ft (15.8 m) long is 1.44 times as long and at 43.5 m tons weighs 3.22 times as much as a physically mature female (weight data based on Lockyer, 1976). A difference between the size of the sexes at birth has not yet been demonstrated, and the growth curves for the two species only start to diverge substantially from about five years of age (Nishiwaki *et al.*, 1963). An inflection point in the male growth curve occurs at about 19–20 years of age, or perhaps slightly earlier (Best, 1970), when growth accelerates briefly, apparently in association with the onset of puberty. Males reach physical maturity considerably later in life than females (Best, 1970).

This dimorphism in size is accompanied by changes in body shape (Fig. 1). In particular the head is proportionately larger in the male, and the spermaceti organ grows forward to project further beyond the tip of the skull (Nishiwaki *et al.*, 1963). The significance of the greater development of the spermaceti organ in the male will be discussed later in this paper.

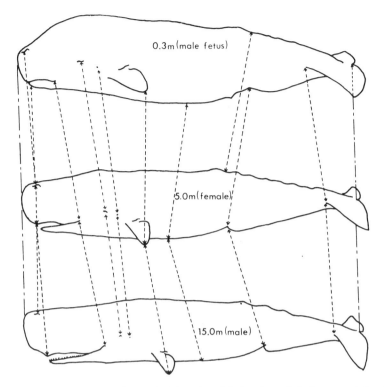

Fig. 1. Comparison of body shapes of adult male, adult female, and fetal sperm whales. (From Nishiwaki *et al.*, 1963.)

It is inferred from fetal sex ratios that equal numbers of sperm whales of either sex are born (Matthews, 1938; Clarke, 1956; Ohsumi, 1965; Best, 1968). In an unexploited population females reach sexual maturity at 9 years and males at 25–27 years of age, and on average a female has a calf once every 5 years (Best, 1974).

As a background to the ensuing discussion, an attempt has been made here to draw up a life table for the sperm whale population. Adult natural mortality rates (instantaneous) for both sexes are estimated to be the same at 0.055. While there are no actual measurements of the juvenile natural mortality rate, it has previously been calculated as an average of 0.07 for the period between birth and age nine, assuming an equilibrium, unexploited population (Best, 1974). However, as natural mortality is likely to be much higher in the youngest animals than in those close to puberty, the rate has been taken here as decreasing in a linear fashion from 0.085 in year one to 0.055 at age nine.

Using these parameters it is possible to construct a theoretical age composition for a stable, unexploited population of sperm whales. This has been converted into a cumulative percentage age composition to clarify the relative contributions of different age groups: a single graph only has been drawn for both sexes because of the similarity between their natural mortality rates (Fig. 2).

One point of interest that becomes apparent is the low reproductive rate: animals at age 0 (birth) comprise only 6% of the total population. A second point is the abundance in the population of subadult males: these comprise about 77% of the total male population, or 38.5% of the population of both sexes. From Fig. 2, the proportion of sexually mature males to sexually mature females in the population can be calculated as about 1 : 2.6.

The full significance of all of these factors will become apparent later in the paper, but initially the most important characteristic to consider is the relative abundance of reproductively idle males in the population. Such a situation predisposes the formation of all-male groupings.

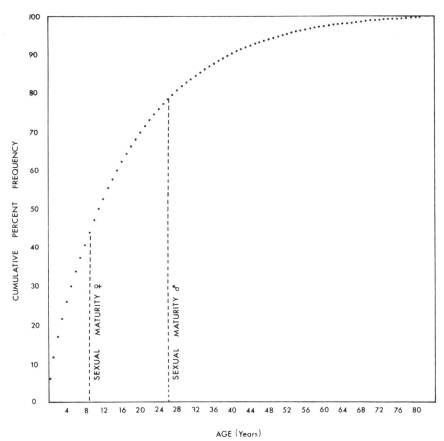

Fig. 2. Theoretical cumulative age composition of sperm whale population.

Table I. Composition of Sperm Whale Groupings (Donkergat, 1963)

Group identification number	Date killed	Number seen by plane	Number seen by boats	Number landed	Number landed as proportion of number seen	Males Small	Males Medium-sized	Males Large	Females
1	March 15	30		25	0.83	4			21
2	19	20		11	0.55	7			4
3	25	3		5	1.67	5			
4	April 1		8	8	1.00		8		
5	11	43		24	0.56	24			
6	12	25		8	0.32	8			
7	13	25		18	0.72	17	1		
8	17		4	4	1.00	1	3		
9	24	13		9	0.69	9			
10	28		2	2	1.00				
11	May 7	12		11	0.92		1	1	
12	13		6	5	0.83		11		
13	19	4		3	0.75				
14	28		100+	25	±0.25	2	1	5	12
15	June 19	40		23	0.58	9	4		21
16	20		12	10	0.83	2			
17	July 6		1	1	1.00	10			
18	August 2		1	1	1.00	1			
19	17	43		34	0.79	29			
20	18	26		18	0.69		5	1	18
21	September 6		30	14	0.47	3		1	10
22	15		25	5	0.20				
23	28		15	13	0.87	13			
24	29		50	32	0.64	31	1		
25	30	28		14	0.50	14			4
26	October 22	20		9	0.45	9			
27	29	27		26	0.96	26			

III. COMPOSITION OF GROUPINGS

In Table I the details are set out of the 27 groupings sampled at Donkergat. On 22 occasions half or more of the group, and on 14 occasions three quarters or more of the group were estimated to have been killed, and the information gathered from these whales is considered to be fairly representative of the group as a whole. In the case of groups containing females and small males, however, the smaller animals below the legal size limit of 35 ft (10.7 m) are poorly represented, and calves of course are completely absent. The information gathered about the composition of mixed schools must therefore be accepted with reservation.

As a matter of interest, the proportion killed of the animals in each group also provides a useful measure of fishing intensity. Mixed groups (i.e., those containing either females or both males and females) were subject to a fishing intensity of from 0.20 to 0.83 with a mean of 0.51. Groups of small bachelors (for definition see below) underwent a fishing intensity of 0.32–1.67 (mean 0.77), medium-sized bachelors a fishing intensity of 0.92–1.00 (mean 0.97) and large bachelors a fishing intensity of 0.83–1.00 (mean 0.94). The trend provided by these values is in fair agreement with fishing mortality rates obtained from catch curves (Best, 1970), where females had the lowest value (0.017–0.034), males less than 25 years old a slightly higher value (0.057–0.074), but sperm whales over 25 years old much the highest fishing mortality (0.157–0.184).

A. All-Male Groups

On 20 occasions all the animals killed from a grouping were males. In most of these instances a high proportion of the animals seen was killed, so it seems probable (on consideration of earlier workers' findings) that the groups originally consisted entirely of males. Such groups have been termed "bachelors."

Sperm whales in these schools tended to occupy a restricted size range (Table II). There was a difference of 6 ft (1.8 m) or less in length between the smallest and largest whales captured from all but two bachelor groups. The variability of body size in each group was calculated from the equation

$$V = \frac{100\, s}{\overline{X}}$$

where $\overline{X}$ = the mean body length in ft, V = coefficient of variability, and s = the standard deviation of the mean length (Simpson *et al.*, 1960). The values of V obtained for each bachelor group are shown in Table III, and varied from 1.1 to 6.5 with a mean of 3.2.

The age composition of bachelor groups examined is given in Table IV. The variability of age within each group was calculated in a similar manner as

Table II. Size Composition of Bachelor Groups

Body length (ft)	3	4	5	6	7	8	9	10	11	12	13	16	17	18	19	23	24	25	26	27	Type
31																	1				Small
32																	2				
33			4	1													1				
34			2	1			1					1			1	2	1	1	2		
35			10	4			4					4			8	5	9	10	4	11	
36	1		5	1	3		3					5			9	5	10	2	1	5	
37	2		2	1	9		1				2				7	1	6	1	2	7	
38	2		1		5									1			1			3	
39						1									4						
40					1	1			1		1				3		1				Medium-sized
41		1							3						1						
42		4							2						1						
43		2				1			2												
44		1				1															
45								1	3												
46								1		2	1										Large
47										1	1										
48										1											
49													1								
50										1											
Total	5	8	24	8	18	4	9	2	11	5	3	10	1	1	34	13	32	14	9	26	

Table III. Coefficient of Variation of Age and Length in Bachelor Groups

Group identification number	Length (ft)			Age		
	s	$\bar{X}$	V	s	$\bar{X}$	V
3	0.75	37.2	2.0	3.56	18.4	19.3
4	1.17	42.4	2.8	0.40	20.3	2.0
5	0.70	35.1	2.0	1.95	15.2	12.8
6	1.12	35.0	3.2	2.35	15.7	15.0
7	0.89	37.3	2.4	2.94	16.9	17.4
8	2.06	41.5	5.0	4.30	23.0	18.7
9	1.01	35.4	2.9	0.35	14.9	2.3
10	0.50	45.5	1.1	0.50	23.5	2.1
11	2.62	42.5	6.2	4.37	22.1	19.8
12	1.50	47.4	3.2	2.55	24.0	10.6
13	1.41	38.0	3.7	1.43	17.3	8.3
16	0.66	35.4	1.9	2.27	15.2	14.9
17				single animal		
18				single animal		
19	0.90	37.7	2.4	2.53	17.8	14.2
23	0.62	35.4	1.8	2.97	15.8	18.8
24	2.30	35.5	6.5	2.50	15.3	16.3
25	1.21	35.2	3.4	2.79	14.1	19.8
26	1.86	35.3	5.3	2.26	15.1	15.0
27	0.72	36.1	2.0	3.19	17.2	18.5

for length, using dentinal growth layers as an indication of age under the assumption that one layer accumulates per year (Best, 1970). The values of V obtained in this instance varied from 2.0 to 19.8 with a mean of 13.7 (Table III). In only two instances was the variability of age less than that of size, and then by only a small amount (group numbers 4 and 9). In the majority of cases, the V values for age were far higher than those for size, although the values for both characters generally seemed to fluctuate in a similar fashion, high V values for age being associated with high values for size. It is difficult to compare the variability of age directly with that of size because of intrinsic differences in the methods of measurement used, and at least part of the higher variability of age is almost certainly due to the difficulties attached to reading growth layers accurately. Nevertheless the degree of disparity between the variability of age and size suggests that bachelor schools are actually more homogeneous for size than for age.

Tarasevich (1967) has reached a similar general conclusion for sperm whales in the North Pacific, stating that male sperm whales aggregate on a size basis which may or may not result in age uniformity. Nevertheless she also concludes that ages tend to be similar within groupings of small whales but that there is more age variation among groupings of large whales. This point will be discussed below.

Besides the similarity in size among individual bachelor whales within the

Table IV. Age Composition of Bachelor Groups. (Donkergat 1963)

Age growth layers	3	4	5	6	7	8	9	10	11	12	13	16	17	18	19	23	24	25	26	27	Small	Medium-sized	Large
10			1																		3		
11				1														1			3		
12			5		1		1					2						1	1	1	11		
13			3	1	4		1								2		4	2		1	18		
14			3		1		1	1	1			3				2	4	3	2	2	23		
15	1		5	1	2		4	1				1			6	2	4	1	1	6	32	1	
16	1		3	3	2		2	1			1				2	2	5	2	2	1	26		
17	1		3		2	1						1		1	7	3	4			2	25		
18		1		1	2				1		2				3	1	2	1		6	22	3	
19	1	2	2	1	1				3			3			3	1	1		3	3	13	5	
20		1	2		2	1									3	2	2			3	13	2	
21		3		1	2										3		1				5	3	
22					1					1					1				1	3	2		1
23		1					1	1	1	1					2						2	1	
24							1	1	1	1												1	2
25	1					1			1												1	2	2
26																							
27									2													2	
28									1	1												1	
29						1																1	1
30																				1	1		
42													1										1
Total	5	8	23	7	17	4	8	2	10	4	3	10	1	1	32	13	32	14	9	26	200	22	7

Table V. Mean Length of Bachelor Groups of Sperm Whales

Mean body length (ft)	Donkergat, 1963	East of New Zealand (Gaskin, 1970)	Total
35	8		8
36	1		1
37	3	1	4
38	2		2
39			
40		1	1
41	1		1
42	2	3	5
43		1	1
44		5	5
45	1	1	2
46			
47	1		1
48		1	1
49	1	1	2
Total	20	14	34

same group, there was a similarity between the size compositions of some groups. Fourteen bachelor groups included whales ranging from about 33 or 34 ft to 39 or 40 ft (10.1 or 10.4 to 11.9 or 12.2 m), while three schools contained whales between about 39 or 40 ft and 44 or 45 ft (11.9 or 12.2 to 13.4 or 13.7 m), and a further three groups contained whales 45 ft (13.7 m) or more in length (Table II). The mean body lengths of each group are combined in Table V with similar data for 14 bachelor schools sighted east of New Zealand (from Gaskin, 1970). The size distribution of these 34 groups seems to fall into two, possibly three, sets: one with a mean body length between 35 and 38 ft (10.7 and 11.6 m), one with a mean body length between 40 and 45 ft (12.2 and 13.7 m), and possibly a third with a mean body length of over 45 ft (13.7 m). The discontinuity between the first and second sets suggests that there is a change in schooling behavior at this stage, and this will be discussed more fully below. For the present, however, it seems sufficient to recognize the similarity in size composition within the sets by classifying those groups with a mean body length between 35 and 38 ft (10.7 and 11.6 m) as "small bachelors," those with a mean length between 40 and 45 ft (12.2 and 13.7 m) as "medium-sized bachelors," and those with a mean length over 45 ft (13.7 m) as "large bachelors."

Norwegian whalermen at Durban have recognized these three classes of whales; small males between about 33 and 37 ft (10.1 and 11.3 m) being called "hannpeis," males between about 37 and 43 ft (11.3 and 13.1 m) "kult," and large males simply "støre sperm" (Capt. T. Haakestadt, personal communication).

The variation in age (expressed as the coefficient of variability V) among members of groups of small bachelors ranged from 2.3 to 19.8 with a mean of 14.8 (Table III). The variability among the three groups of medium-sized bachelors examined ranged from 2.0 to 19.8 with a mean of 13.5, and in two groups of large bachelors was 2.1 and 10.6. There is no sign from these data of an increase in the age variation within schools of larger males as suggested by Tarasevich (1967), but very few schools of medium-sized and large bachelors were examined.

1. Small Bachelors

Fourteen groupings classified as small bachelors were examined containing animals from 31 to 42 ft (9.4 to 12.8 m) in length and from 10 to 29 years of age (group nos. 3, 5–7, 9, 13, 16, 18, 19, 23–27 in Tables II and IV).

Such groups were reported as containing 1–50 animals (with a mean of 22) when sighted from the spotter aircraft or catcher (Table I). However, with the exception of three groupings containing one, four, and five whales, respectively, there seemed to be a tendency for small males to be found in groups of 12–15 or a multiple thereof. There were three groups containing 12–15 whales, four containing 25–28 whales, and two containing 43 whales, while two further groups consisted of 20 and 50 whales. In the case of one group of 43 small bachelors, the pilot of the spotter aircraft reported that there were three main "flocks" with several subsidiary groups, and a group of 27 small bachelors was reported as consisting of two subgroups (of 9 and 17 whales) with one single individual. It seems possible from this that the larger groupings of small bachelors may comprise a number of basic units, perhaps of 12–15 animals each, here called schools.

This observation is interesting when compared with Gambell's (1972) finding that school sizes of small sperm whales (estimated to be less than 39 ft (11.9 m) in length) had a geometric mean of 10.5 or 11.8 whales when unaccompanied by big whales (over 45 ft (13.7 m) long), but a much greater value (a geometric mean of 21.7 or 22.9) when one or more big whales were present. Although the sexes of the animals concerned could not be distinguished from the air, it is probable that the schools of small whales containing big animals represented mixed schools, while the other schools could have been all-male or of mixed sex.

A further characteristic of large aggregations of small bachelors was that they were frequently found distributed over a wide area. Two aggregations of 25 animals each were found over a radius of 15 and 20 miles, respectively, and two of 43 animals each were found over a radius of 8 and 12 miles, respectively. Although one of the aggregations of 43 small bachelors was grouped into three main schools (as described above), the second aggregation of 43 animals was reported by the pilot of the spotter plane as being mostly spread out in singles or pairs. It was also reported from the catchers that the whales from an aggre-

Table VI. Reproductive Status of Members of Bachelor Groups

Type	Number of whales examined	Immature		Maturing		Mature	
		Number	Percent	Number	Percent	Number	Percent
Small	198	124	62.6	69	34.8	5	2.5
Medium-sized	22	9	40.9	10	45.5	3	13.6
Large	8	—	—	2	25.0	6	75.0

gation of 25 small bachelors were spread out over a 20-mile radius. Hence it seems that the basic units comprising large groupings of small bachelors may frequently split up and rejoin over a wide area. Whether in doing so they retain their original identity is at present unknown.

Rice (in Caldwell *et al.*, 1966) also described loose associations of young adult males (10.5–14 m in length) off the California coast that might be scattered over an area 5–10 km square, and although these animals did not gather into tight schools, they apparently moved from area to area more or less as a unit.

According to Norwegian whalermen at Durban, the behavior of small bachelors is typically unpredictable or "mad"—the school may split up and rejoin frequently, take strong evasive action in turning very acutely when pursued, and not follow a regular respiratory pattern. Other schools (i.e., of mixed sex or larger bachelors) are far more predictable in their behavior under similar circumstances. Despite this behavior, and the apparent absence of epimeletic behavior among juvenile males (Caldwell and Caldwell, 1966), there appears to be a fair amount of social cohesion and facilitation in schools of small bachelors (Caldwell *et al.*, 1966).

Male sperm whales have been classed on histological grounds as sexually immature, maturing, or mature (Best, 1969*a*). The majority (nearly two thirds) of males from groups of small bachelors were classed as sexually immature, and very few as mature (Table VI). The segregation of males from mixed schools must therefore begin some time before the onset of sexual maturity.

Schools of small bachelors appear equivalent to the school-type IV of Tormosov (1975), in which the majority of animals are said to be pubertal males, and which normally occurs in tropical and subtropical zones.

2. Medium-Sized Bachelors

The three schools of medium-sized bachelors examined at Donkergat (numbered 4, 8, and 11 in Table I) contained males from 39 to 45 ft (11.9 to 13.7 m) in length and from 15 to 29 years of age (Tables II and IV). An analysis of their sexual condition shows that most were maturing, although a large number were still immature (Table VI). These animals seem to be closest to the "fortybarrel bulls" of old whalermen, typified as ". . . the band of young vigorous males, probably just sexually mature but not yet large enough to secure harems . . . ." (Clarke, 1956). They also appear equivalent to the school-type V of Tormosov (1975), in which the majority of animals are said to be males that have reached maturity shortly before, and which normally is found in temperate zones.

While Gaskin (1970) found bachelor schools with a mean length of 40–45 ft (12.2 to 13.7 m) to contain 3–15 whales, the number of animals in these three schools was 4, 8, and 12, so that medium-sized bachelors seem to form much smaller groups on average than small bachelors. Ohsumi (1971) found that nearly 98% of sightings of schools containing animals estimated to be 40 ft (12.2 m) or more in length consisted of 10 animals or less in number. Observations on schooling behavior are confined to the group of 12 animals, which was reported by the pilot of the spotter aircraft as a single "flock" but in which the whales were spaced as "singles," i.e., there was no tight schooling behavior. This is in agreement with Rice's observations off California (in Caldwell et al., 1966).

3. Large Bachelors

Groups of large males ranging from 45 to 50 ft (13.7 to 15.2 m) in length and from 21 to 42 years of age were only examined on three occasions at Donkergat (those numbered 10, 12, and 17 in Table I). One was a solitary 49 ft (14.9 m) male and a second a pair of males 45 and 46 ft (13.7 and 14 m) in length. The latter were reported as being spaced about half a mile apart. In combination with the behavior recorded for small and medium-sized bachelors, this suggests that tight schooling behavior lessens with increasing body size. However, the third group of six large animals was recorded as "mainly swimming together" when seen from the catchers, indicating that large bulls associate as distinct schools on occasion (see also Caldwell et al., 1966; Gambell, 1972).

The three groups of large bachelors sampled consisted of fewer animals on average than the schools of medium-sized males, continuing the trend seen between schools of small and medium-sized males. This relationship between decreasing school size and increasing individual size has already been described by several previous authors (see Introduction), and will be referred to again below. Ohsumi (1971) found that sightings of schools consisting of whales estimated to be more than 45 ft (13.7 m) long ranged from solitary animals to 6–10 animals, with the majority (nearly 90%) singles or pairs.

Three quarters of the large bachelors examined were sexually mature (Table VI) but by no means were they all old animals. In group number 12, three of the males were still physically immature and only one mature. The solitary whale (number 17 in Table I), however, was physically mature and had 42 dentinal growth layers, being the oldest animal among 849 males examined at Donkergat in 1962 or 1963. This animal had an extremely white head whorl and a large number of scars and scratches on the side of its head. Five barnacles (*Coronula* cf. *reginae*) were found in pits on the front of the head (this being the only record for 1272 sperm whales examined at Donkergat). Although the testes were still large (combined weight 9.8 kg) with spermatozoa present in some seminiferous tubules, there were signs of regression in other tubules with little material present except a germinal epithelium. The mean tubule diameter was the lowest for the 80 mature males examined (112 μm), and fell within the range of immature animals (Best, 1969a). It is possible that this animal was sexually effete, although there is insufficient evidence from males of advanced age to confirm or deny whether a decline in reproductive capability normally occurs, similar to that experienced by females (Best, 1967).

In the school-type VI of Tormosov (1975), the majority of animals are said to be older males, more often observed in high latitudes of the Antarctic. This could be equivalent to the school of large bachelors described above.

B. Groups of Mixed Sexes

Seven groups including female sperm whales were examined at Donkergat, and six of these also included males. The one group from which only females were taken was not heavily fished, 69% of the total number seen being killed, so the possibility of there being small males present cannot be discounted. For present purposes all seven groupings (numbered 1, 2, 14, 15, 20–22 in Table I) have been combined under the heading "mixed groups."

The term "harem" has been avoided where possible when describing these groupings, as some ethologists consider that this term should refer only to a social unit consisting of a stable group of one dominant male and several females that persists throughout at least one breeding season (K. Ralls, personal communication). Evidence given below will show that this situation does not normally occur in sperm whales.

Six of the mixed groups appeared to be discrete schools when sighted, and there was no mention of their being subdivided or scattered over a wide area as recorded for bachelors. The total number of animals seen in each of these cases was remarkably similar, varying from 20 to 40 with an arithmetic mean of 28. The seventh group (numbered 14 in Table I) was described as an extremely large aggregation with an estimated total of over 100 individuals, but the catchers reported that most animals were traveling in "flocks" of up to 20 animals. Hence the usual size of a mixed school is taken to be 20–40 individuals.

Ohsumi (1971) has calculated the average size of a "harem" as 27.1 whales, while Masaki *et al.* have unpublished data giving a mean school size of 21.6 (geometric) or 26.9 (arithmetic). The latter authors, however, considered that these might be underestimates due to previous fishing of the stock and/or undercounting of the number of whales (especially calves) present. In a summary they used a total school size of 29.8. Gambell's (1972) data for schools classified as "small+big" or "small+big+medium" (presumably mixed schools) give a geometric mean of 21.7 or 22.9 whales per school. A median figure from all these estimates might be 25 whales.

The tight schooling behavior of individual mixed schools is quite characteristic (see Caldwell *et al.*, 1966). Such schools will split up into smaller units upon being chased but will usually reform very shortly after harassment has ceased. They therefore exhibit considerable social cohesion. "Standing-by" or epimeletic behavior has been recorded frequently for female sperm whales when their calves are injured, although this response can vary. Similar behavior between adult females has also been recorded (Caldwell and Caldwell, 1966), as has cooperative behavior between school members during probable calving (Gambell *et al.*, 1973; Pervushin, 1966).

When group number 20 was first encountered, the pilot of the spotter plane reported that two animals were swimming in front of the remainder of the school and that they appeared considerably larger than the rest. All the whales killed from this school were females, the largest being 36 and 35 ft (11 and 10.7 m) long and the remainder 33 ft (10 m) or less. The two largest females were also the oldest, bearing 12 and 14 ovarian corpora. It seems possible from this description that old females may, occasionally at least, lead mixed schools. Ohsumi (1971) has proposed a matriarchal organization for the sperm whale. Following Ralls (1976), however, this type of organization should perhaps be more strictly termed "extended matricentric family" or "extended mother-family," as there is no evidence that females are dominant to all males (including adult bulls) when these are present in the school.

1. Female Component

Despite the sexual dimorphism of sperm whales, which would tend to promote the killing of the larger males, the majority of animals killed from mixed groups at Donkergat were usually females. In six schools they formed 36%, 71%, 80%, 84%, 91%, and 100%, respectively of the catch, giving a mean of 77%. In a seventh instance the proportion of females was 48%, but this concerned a large aggregation of over 100 animals of which only 25 were taken and in which it was reported that large numbers of small calves were present, so that selection against lactating females may have affected the sex ratio of the catch.

Previous estimates of the proportion of females in a mixed school have been 71.8–84.2% with a mean of 77.1% (Ohsumi, 1971), 75% (Stephenson, 1975), 78% (Robson and Van Bree, 1971), and 82.6% (Gambell, 1972), while Masaki *et al.* have unpublished data for 15 schools ranging from 50% to 100% with a mean of 78%. The latter authors have readjusted this figure to 69.8% to account for whales that escaped, but in doing so they have assumed that calves must all be 20 ft (6.1 m) or shorter, whereas male suckling sperm whales have been found up to 31 ft (9.4 m) in length at Durban (unpublished data in the author's possession). Hence the proportion of males may have been overestimated in the adjustments made to the school composition. For the present it is concluded that mixed schools of sperm whales contain on average about 78% females.

The proportion of these that are sexually mature is difficult to estimate because of legal size restrictions on the catch, the mean length at sexual maturity (28 ft. or 8.5 m, Best, 1968) being below the minimum length prevailing at the time (35 ft or 10.7 m). From body length alone, the percentage of mature animals among a stranded school containing 46 females was calculated as 78.3% (Robson and Van Bree, 1971); and in a stranded school containing 54 females as 76% (Stephenson, 1975). In the females from three schools sampled by Japanese whalers, the percentage of mature animals varied from 68.8% to 85.7% with an average of 79.6% (Ohsumi, 1971), while the percentage of mature animals among the females caught from 15 mixed schools off the Japanese coast varied from 42.9% to 100% with a mean of 79.4% (Masaki et al., unpublished data). Masaki et al. adjusted this figure to account for whales that escaped the catchers, giving a value of 15.5/20.8 or 74.5%. A figure of 75% maturity has therefore been adopted.

No justification has been found from these data for Tormosov's (1975) recognition of three types of school containing both sexes i.e. type I, in which females greatly predominate and in which the percentage of young immature females is high; type II, in which females only slightly exceed males in number and in which the percentage of immature females is low; and type III, in which females greatly predominate but in which the percentage of immature females is low.

The reproductive condition of females landed from each mixed group is shown in Table VII, each mature animal being classified as pregnant, lactating, or resting, as described in an earlier report (Best, 1968). Clarke (1956) has suggested that there may be segregation of female sperm whales off the Azores into schools consisting predominantly of lactating or pregnant animals, and Tormosov (1975) believed that female sperm whales formed concentrations of a

Table VII. Reproductive Condition of Females in Mixed Groups

Group identification number	Number immature	Number mature			Total
		Pregnant	Lactating	Resting	
1		8	7	6	21
2			3	1[a]	4
14			8	4	12
15		4[b]	10	7	21
20	1	1	8	8	18
21	1	1	5[a]	3[c]	10
22		2	1	1	4

[a] One also ovulating.
[b] One doubtful record.
[c] Two also ovulating.

particular type for a certain period according to their physiological state and biological rhythm. This does not seem to be the case off the west coast of South Africa. Only three mixed groups examined (numbered 1, 15, and 20 in Table I) are considered to be represented by a sample large enough for analysis, and the proportions of pregnant, lactating, and resting females in each group have been compared with the ratios expected from a study of the reproductive cycle. These are 1.27 pregnant: 2 lactating: 0.73 resting for group number 1, and 1 pregnant: 2 lactating : 1 resting for groups 15 and 20 (see Best, 1968). In none of the groups, however, were the observed proportions significantly different at the 5% level from the expected proportions (chi-square test). These limited data suggest that segregation of a particular reproductive class of females is not a general occurrence off Donkergat. In fact, most of the schools examined contained females in all three main stages of the reproductive cycle, as was the case for the schools examined by Ohsumi (1971) and Gambell (1972).

Due to the difficulties attached to reading maxillary teeth from female sperm whales, the accumulation of ovarian corpora has been used as an alternative method of age determination (Best, 1970). From the age composition of the females landed for each mixed group (Table VIII), it is immediately clear that, in contrast to most bachelor groups, the animals present are not restricted to a narrow age range in each group. Bearing in mind that the age of female sperm whales when fully recruited to the Donkergat catch is equivalent to 5 or 6 corpora

Table VIII. Age Composition of Females in Mixed Groups

Age (corpora count)	Group identification number							Total
	1	2	14	15	20	21	22	
0					1	1		2
1					1		1	2
2		1			2		1	4
3	1			1	1	1		4
4	3		2	1	4	1		11
5	2		2	2	3			9
6	3	1	1	2	1	1	2	11
7	3		1		1	2		7
8		1		2	1	3		7
9	2			6				8
10		1	3	3	1			8
11	1		1	1				3
12				1	1			2
13	3							3
14					1	1		2
16	1		1					2
17	2		1					3
Total	21	4	12	19[a]	18	10	4	88

[a] Impossible to determine age of two females.

(because of the minimum size limit), and that no animal was ever found with more than 22 ovarian corpora (Best, 1970), these groups seem to comprise a random selection of females at all ages. Ohsumi (1971) found females ranging from 4–7 to 40–43 years of age in whales landed from three mixed schools.

2. Male Component

Three of the mixed groups examined at Donkergat included small males, i.e., those of 39 ft (11.9 m) or less in length. In the case of groups 1 and 15 these were all sexually immature, while in group 2 there were three immature and four maturing individuals. A fourth group (number 14) included four medium-sized (40–45 ft (12.2–13.7 m) long) as well as nine small males, but as these animals were part of a large aggregation of over 100 animals, their exact relationship to the female schools cannot be determined. Five of these males were sexually immature and eight maturing. All these groups were encountered during the period from late March to late June, which is outside the normal breeding season (Best, 1968).

The remaining two groups (numbers 21 and 22) are of particular interest, for they were found in September, a month at the beginning of the breeding season in which about 9% of conceptions normally take place (Best, 1968): group 21 actually included three ovulating or recently ovulated animals among a total of nine mature females examined. Both groups were accompanied by a large male, one 48 ft (14.6 m) and the other 49 ft (14.9 m) in length, although three small, immature males were also present in group 21. The large male associated with group 21 was sexually mature but that with group 22 was classified on histological grounds as maturing, although its combined testis weight of 8.0 kg was actually greater than that of the large male with group 21 (6.9 kg). Neither bull was an old individual, the animal with group 21 having 26 dentinal growth layers, while the animal with group 22 had about 29 growth layers (tooth twisted at the root). Both animals were physically immature, the epiphyses of the caudal vertebrae being unfused.

The timing of the capture of these two schools, the presence of sexually active females in at least one school, and the presence of a single large mature (or maturing) male with each suggest that both schools represented active breeding units. The main breeding force in each of these schools was the single large bull, here called the "schoolmaster," and the absence of mature males from female schools outside the breeding season supports the contention made earlier (Best, 1969a) that schoolmasters join the female schools for the breeding season and are not necessarily found with them outside.

Evidence in support of this hypothesis comes from an incident on a recent whale-marking cruise. On January 15, 1974, a school of 12 sperm, composed of one medium-sized to big male (estimated at 45 ft (13.7 m) in length) and the rest either small males or females was encountered at 30°56'S, 42°17'E. Seven whales from this school, including the large male, were marked. The large male was seen twice, once at the outset of marking and the second time about 2 hr later, when it was still with a number of the rest of the school. During the 1974 whaling season, five of these seven marked whales were killed and landed at Durban, one on March 14, and the others on April 18. All were females. On neither occasion was a large male killed, although such an animal would have been the first target of the whalers. The breeding season off Durban extends from September to May, with about 80% of conceptions occurring from November to January (Gambell, 1972). The first encounter with the school therefore coincided with the peak of the breeding season, but 2–3 months later, at the end of the breeding season, the schoolmaster bull had already left this particular female school.

Further evidence of the transitory nature of the relationship between schoolmaster and mixed school is available from data on cyamid infestation. It has been established that female and small male sperm whales are almost exclusively infested with *Neocyamus physeteris*, whereas the majority of medium-sized and large males are infested with *Cyamus catodontis*, presumably due to the different migratory habits of the two groups of whales (Best, 1969*a*). On three occasions at Donkergat, a medium-sized or large male infested solely with *C. catodontis* was found associated with females and small males that were solely infested with *N. physeteris*. On September 25, 1962, a 43-ft (13.1 m) long male infested with *C. catodontis* was killed in the same position as three small males and three females infested with *N. physeteris*. On June 17, 1963, a 47-ft (14.3 m) long male bearing *C. catodontis* was killed while lying in the middle of a group of small sperm of which several also were killed. Unfortunately at least two schools were fished this day, so it is not possible to identify positively the animals associated with this male, but 11 females from the day's catch were infested solely with *N. physeteris* and none with *C. catodontis*. And on September 6, 1963, a 48-ft (14.6 m) long male carrying *C. catodontis* was found lying in the center of a school of 30 animals (group number 21 in Table I); five females killed from this school bore only *N. physeteris*. This evidence suggests that in each case the medium-sized or large male had only recently joined the mixed school, for cyamids are capable of moving from one host to another within a very short space of time from the initial contact: a newborn gray whale with a bleeding umbilical stalk (Eberhardt and Norris, 1964) and a sperm whale calf with an unhealed umbilical scar (personal observation) were both already infested with numerous cyamids. Cross-infestation may have taken place in group 22, where two females carried *C. catodontis* only and two a mixture of *C. catodontis* and *N. physeteris*, although in this instance no cyamids were found on the large male (49 ft or 14.9 m long) present.

At this stage it is worth discussing the contention of Rice (in Caldwell *et al.*, 1966) that "the widely held harem concept may be erroneous. There may be no males, or one or more large adult males, with each school of females, and usually several smaller, but sexually mature, males accompany each school of females." This statement was apparently based on observations of sperm whale groupings off California. Later the same authors state, "The generally held assumption that the solitary old males join the herds and engage in breeding may be erroneous. Some whalers believe that such males are more or less 'outcasts,' that they live alone the year around, and that most of the breeding is done by the younger, adult males that remain with the pods of females."

The unlikelihood of the latter males being fully fertile is shown by the low density of spermatozoa in the vas deferens fluid of males less than 43–44 ft (13.1–13.4 m) in length (Best, 1974). The seasonal nature of the relationship between schoolmaster and mixed school is more likely to be responsible for the observed absence of large males from female schools during a large part of the year. Tormosov and Sazhinov (1974) have made direct observations of mating behavior during the southern breeding season (December) between a male 14–15 m long and a 10-m animal from a school of putative females and young males. They conclude that males 13.5–15 m long normally play the predominant role in reproduction.

That considerable competition may take place for "possession" of a mixed school and that this competition is restricted to large males, is intimated by the presence of wounds and scars attributable to intraspecific fighting on males above a certain size (Fig. 3). Such scars are parallel lines, normally in pairs but sometimes three or more (up to seven or eight) together, and are typically found on the head region, but they can be found farther back on the body, even around the region of the dorsal fin and flipper (Fig. 3a). Such scars could not be caused by any cephalopod species, even those with hooks on the arms or tentacles, for it would seem impossible for them to make such exactly parallel lines for any distance. The maximum spacing between the parallel lines was measured in five whales at Durban, with the following results: 10, 11, 12–14, 14, and 12–16 cm. The maximum interdental spacing of male sperm whales over 13 m in length varies from 7.5 to 15 cm, and in whales over 14.6 m in length is normally 11 cm or more (Tomilin, 1957). This seems to confirm that such marks must be caused by large male sperm whales, refuting Rosenblum's (1962) suggestion that they were made by either females or young males. Of 50 males examined at Durban in August and September 1971, only those 43 ft (13.1 m) or more in length were found bearing these scars (Table IX), indicating that serious fighting is confined to large males. Most scars were already healed when seen, but a 47-ft (14.3-m) male bore fresh wounds 2.0–2.5 cm wide and 1.5 cm deep (Fig. 3c).

Matthews (1938) noticed the difference in the degree of such scarring between males and females but attributed it to females not attacking such large squid as males.

Although eyewitness accounts of battles between individual sperm whales exist in the literature of open-boat whaling (see Caldwell *et al*., 1966), only one modern account of such behavior exists (Zenkovich, 1962). This may be because the presence of screw-driven vessels can be detected by sperm whales at a distance of up to eight miles, when their behavior usually changes markedly

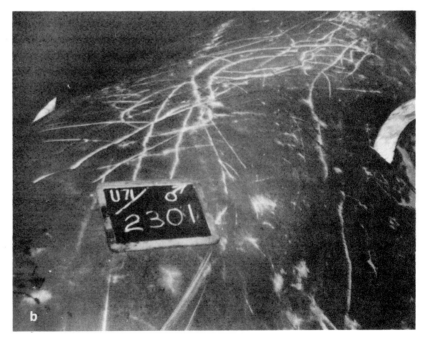
Fig. 3. Scarring of male sperm whales killed off Durban, August–September, 1971:

Social Organization in Sperm Whales

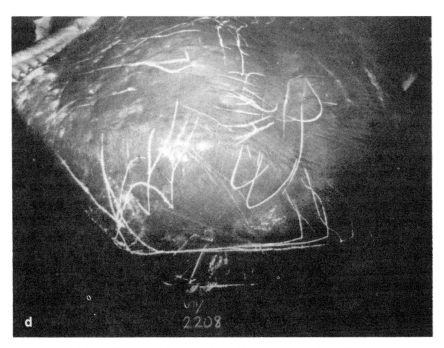

(a) 45 ft (13.7 m), (b) 46 ft (14 m), (c) 47 ft (14.3 m), (d) 45 ft (13.7 m).

Table IX. Incidence of Parallel Scars on Male Sperm Whales, Durban, August/September, 1971

Body length (ft)	No scars seen	Scars present
28	1	
30	1	
32	1	
33	4	
35	2	
36	1	
37	2	
38	2	
39	6	
40	1	
41	2	
42	7	
43	1	1
45	3	1
46	2	2
47	1	1
48		1
49	3	
50	1	
52	1	2
Total	42	8

(Gambell, 1968). It is surprising, however, that such fights have never been reported by the whale-spotters flying out of Durban during the last seven years in which observations of behavior have been specifically recorded.

When first seen, the schoolmaster of group 21 was in the center of the school, whereas the schoolmaster of group 22 was found half a mile away from the main school. It is not clear if this difference in behavior is in any way connected with the fact that several ovulating females were found in group 21, but none in the very few animals caught from group 22. According to earlier published reports, adult bulls may occasionally withdraw a short distance from the main body of females (Caldwell *et al.*, 1966).

C. Effective Sex Ratio in the Population

In an earlier report (Best, 1969*a*) mention was made of the importance for stock assessment purposes of determining the number of mature females served by each schoolmaster during a breeding season. Ohsumi (1966) calculated the mean harem size to be 16 females from a comparison of male and female age compositions, and 14 females from the examination of mixed schools killed under special permit (Ohsumi, 1971). Extremes of 10-20 females per schoolmaster were adopted in calculations of the maximum sustainable yield of socially mature males in the southeast Atlantic (Best, 1970).

The number of mature females in a mixed school can be calculated from the data on the "typical" sex and age composition of a mixed school reviewed above. With a range of school sizes of 20-40 animals and a median of 25, of which 78% will be females, there will be 15.6-31.2 females (mean 19.5) per mixed school. As about 75% of these will be mature, the number of mature females in a mixed school will range from 11.7 to 23.4 with a mean of 14.6, i.e., 60% of the total school size (cf. 50% as stated by Ohsumi, 1971). This would be equivalent to the number of mature females available to a schoolmaster, if there was only one such bull per school and his tenure lasted the breeding season.

However, although only single schoolmasters were found with groups 21 and 22, there are earlier reports of up to four adult bulls per school (see Caldwell et al., 1966). Data are available from Donkergat for 29 instances during the period August to October, 1958 to 1967, when sexually mature bulls [45 ft (13.7 m) or more in length] were killed on the same day as females, and assumed therefore to be in association with female schools (Best, 1970). The frequency distribution of numbers of bulls per school is shown in Table X, the periods 1958-1962 and 1963-1967 being given separately as there was a striking difference between the size of schoolmasters in these two periods. Up to five bulls per school of females were found, although on the majority (62%) of occasions only one bull was present. The geometric mean of the number of mature males per school was 1.48 in 1958-1962 and 1.66 in 1963-1967.

If allowance is made for more than one schoolmaster per mixed school, it is clear that the actual number of mature females served per bull might be rather less than the values given previously. If the 1958-1962 figure of 1.48 schoolmasters per school (considered to be nearer the unexploited stock condition) is adopted, the ratio will then become one bull per 7.9-15.8 females, with a mean of 9.9.

Table X. Number of Mature Males (45 ft (13.7 m) or More In Length) per School of Females

Number of bulls	Donkergat		Total
	1958-1962	1963-1967	
1	11	7	18
2	3	0	3
3	0	3	3
4	1	2	3
5	1	1	2
Geometric mean	1.48	1.66	

The length of a schoolmaster's tenure with a mixed school is unknown, although from the evidence given above it would not seem to persist after the breeding season has closed. Whether a particular bull would be replaced by others during the same breeding season, so effectively reducing the number of mature females served per bull, is also not known.

Distinction should be made at this stage between the number of schoolmasters attending a mixed school and the number of such males (both reproductively active and idle) needed in the population to achieve maximum breeding efficiency, here called the effective sex ratio in the population. It is possible that a surplus of males is necessary to create competition for females, so reinforcing sexual behavior and ensuring maximum breeding success. It is also possible that a surplus of males is required to maintain, through adequate selection, those genetic characters carried by the male that are optimal for the species.

It is of course exceedingly difficult to ascertain the actual effective sex ratio for the sperm whale population, but it is possible to examine the ratio of mature males to mature females in the population as a whole, assuming that this must represent the upper limit of the effective sex ratio. Under the assumptions of a stable population, identical natural mortality rates for the two sexes, and with the ages at sexual maturity of 9 for females and 26 for males, the ratio can be calculated as about one mature male per 2.6 mature females (see Fig. 2).

This means that for each mixed school (containing on average about 15 mature females), 5.8 mature males will be available, and the ratio of "active" to "idle" bulls in the population is therefore about 1 to 3.9.

This ratio is not inconsistent with the degree of segregation of large males (over 25 years of age) to the Antarctic in summer, which is calculated as 75–90% (see below). These animals might therefore be the idle bulls that have moved to higher latitudes for feeding after failing to secure a mixed school.

IV. ONTOGENY OF SCHOOLING BEHAVIOR

A. Segregation of Males from Mixed Schools

The basic social grouping of sperm whales is the mixed school, the active breeding unit. The tight schooling behavior of mixed schools has been mentioned previously and when the random age and reproductive compositions of the females are taken into account it seems likely that these schools are stable units in which there is very little permanent subdivision, at least among the mature female component. Ohsumi (1971) has given some evidence in support of this from mark returns in the North Pacific. There have been no recorded instances of whales being marked in the same school but recaptured in different schools in the same season as marking. On the other hand, two females were marked in a school and recaptured together in a school five years later, and on two occasions two females marked in one school have been recaptured together in a school ten years later. There is a further, not so convincing, example of two marked females being recaptured together again after eight years. It is significant that all four records only concern females, and it seems quite possible that this sex maintains close associations throughout life. In fact recent morphological studies suggest close family relationships within sperm whale groupings (Veinger, in Berzin, 1971).

Nevertheless, from the low proportion of males within mixed schools, it is clear that these schools cannot retain all the young of this sex born each year. The exact stage at which males leave the mixed schools is unknown. Among 31 males from two mixed schools stranded on the New Zealand coast, there were only three males over 12 m in length, one of which was a large bull 16.31 m long (Robson and Van Bree, 1971; Stephenson, 1975): the method of length measurement used for one of these schools, however, would tend to produce greater sizes than if the standard method had been used (Stephenson, 1975). Among 16 males from the three mixed schools sampled by Japanese whalers, there were no animals over 31 ft (9.4 m) in length and none older than ten years (Ohsumi, 1971). None of the animals that escaped from these schools was estimated to be larger than 38 ft (11.6 m). Ohsumi concluded that all males gradually leave the nursery school after reaching puberty (nine years) but that some must leave after weaning (at about two years). Rice (in Caldwell *et al.*, 1966) found that the size of males encountered with females off the California coast ranged up to 11 m, while those found in exclusively male company varied from 10.5 m upward.

On theoretical grounds it is possible to estimate the age at which this segregation takes place from the cumulative age distributions of males and females (in Fig. 2), assuming that female calves remain in the same school throughout life.

If the proportion of males in the school is 0.22, then the relative proportion of males to females will be 0.22/0.78 or 0.28. This level is reached at an age of four to five years. This calculation assumes knife-edge male segregation, whereas it is probably gradual; however, if segregation occurs at a constant rate, the result can be considered equivalent to the mean age at which males leave mixed schools.

The result of this calculation is naturally very sensitive to changes in the observed proportion of males to females in the school. For instance, if the proportion of males is 30% rather than 22%, as is suggested by the adjusted school composition of Masaki *et al.*, the mean age at segregation becomes 8–9 years. As there must be some doubt about the accuracy of the observed percentage of males due to size selection and incomplete sampling of the schools, the estimate 4–5 years for the mean age at segregation must be accepted with reservation.

In six groups of small bachelors off Donkergat the modal length of the animals caught was 35 ft (10.7 m), in two groups 36 ft (11 m), and in a further two groups 37 ft (11.3 m) (Table II). The remaining small bachelor groups had an ill-defined modal length. Unfortunately these modes correspond closely to the prevailing minimum legal size, so it is possible that smaller animals may be underrepresented due to selection by the gunners. Nevertheless, in some instances a high proportion of the original group was estimated to have been killed (0.79 in group 19, 0.83 in group 16, and 0.96 in group 27), so any size selection could not have seriously affected the modal length of the catch. On this evidence, the segregation of males from mixed schools seems to be completed by a body length of 35 ft (10.7 m). From the combined age compositions of all 14 small bachelor groups, this stage corresponds to a modal age of 15 years (Table IV).

To summarize, the segregation of male sperm whales from female schools seems to finish at a length of 35 ft (10.7 m) and an age of 15 years. The mean age at which males depart is less certain but may be as low as 4–5 years when they would be about 25–26 ft (7.6–7.9 m) long (Best, 1970).

If a significant proportion of juvenile females should also leave their parent school, then the mean age at which males segregate would actually be younger than the 4–5 years calculated above.

Clarke (1956) has mentioned a type of mixed school consisting of juvenile males and females, weaned but still quite young, which associate ". . . as boys and girls go to school together . . . " and which may later separate when one or both sexes becomes mature. Gaskin (1970) has recorded sightings of immature, mixed-sex schools off New Zealand, although no animals from these were killed and examined. Murphy (1947) encountered a pod of calves of which one was killed and proved to be a male "not more than 25 feet long" which had recently fed on squid. If such schools of juvenile males and females do exist they would be well below the minimum legal size of 35 ft (10.7 m) in effect for land stations until 1973, and so they would rarely if ever be sampled commercially in modern whaling. Groups of small sperm whales of which all the members were reported to be "too small to shoot" were sometimes encountered off Donkergat (there were three in 1963), but it is possible of course that these represented mixed or bachelor schools with an average size below 35 ft (10.7 m) rather than a collection of juvenile males and females.

Sightings of schools of whales estimated to be less than 30 ft (9.1 m) in length were also reported by Ohsumi (1971) and were regarded as being comprised of juveniles. The size of such schools varied from 3 to 50 with a mode around 6–10 whales. They were comparatively rarely seen, comprising only 1.7% of 594 schools sighted in the North Pacific and southern hemisphere combined.

A segregation of juveniles of both sexes, rather than of males only, is supported by the evidence from mixed schools sampled under special permit. Ohsumi (1971) found 14 males and 12 females between the ages of 4 and 11 years in three such schools, while unpublished data provided by Masaki *et al.* includes 25 males and 37 females between 3 and 10 years old in four schools sampled. Neither of these sex ratios is significantly different from parity (chi-square test, $P > 0.80$ and 0.10, respectively), while if a male exodus from the mixed school had predominated, there would have been a preponderance of females in this age range. Presumably the exodus of females ceases before that of males or is balanced by a reentry of young females from other juvenile schools.

Further evidence of female segregation is provided by the disparity between the observed percentage of mature females in mixed schools (75%) and the expected percentage in the population as a whole (approx. 56%, see Fig. 2). This suggests that 19/44 or about 48% of juvenile females are absent from mixed schools.

If a segregation of the young of both sexes does occur, as is suggested by these data, it must take place at too young an age to be correlated with the onset of puberty (Ohsumi, 1971). Lactation is a prolonged affair in the sperm whale, on average lasting two years (Best, 1968), although evidence is now available indicating that individual male sperm whales may continue to suckle to an apparent age of 13 years and a length of 31 ft (9.4 m)—data from calves collected under special permit at Durban whaling station, 1971 and 1973. It is therefore possible that most animals (at least of the males) leave their parent school once they have been weaned. Lactation in the southern sperm whale has been assumed to end in March or April (Best, 1968; Gambell, 1972), although data for November to January were unavailable. This corresponds with the calving season rather than the breeding season, so that as one set of juveniles is being weaned a new set of calves is being born. It is possible that the mothers of newborn calves (or other members of the school,) may actively reject larger juveniles, as occurs in several social ungulate species (e.g., von Richter, 1971; Rowe-Rowe, 1973), so helping in the process of weaning.

Tarasevich (1967) proposed that there were significant seasonal changes in the segregation of male sperm whales from females. She stated that most sexually mature males remained in mixed schools with the females during the winter in the North Pacific, but that in spring first small males (11.6–13 m long) and then larger males left the female schools and segregated themselves in higher latitudes. Males apparently rejoined female schools in autumn.

The data obtained at Donkergat are not in complete agreement with this. Groups of small bachelors were found in every month from March to October except July, and there was little difference between the proportion of small males in mixed groups in March, May/June or August/September (Table I). On this evidence the segregation of small males from schools containing females seemed to be permanent, at least during eight months of the year (including winter). Only three schools of medium-sized bachelors were examined in this report, all of which were encountered in April or May, so it is not possible to speculate on the permanency of their segregation. Large males, however, did seem to show a seasonality in their relationship with mixed schools, for evidence given previously (Best, 1969a) and in the present report indicated that these animals were found more often with mixed schools in spring than in autumn or winter. This is connected with the timing of the female breeding season and may be associated with a cycle of Leydig cell activity (Best, 1969a). The number of large males involved in this activity, however, is small, and those unable to secure a mixed school may remain segregated for the remainder of the spring and summer.

If the majority of bachelors are permanently segregated from mixed schools between their initial departure as juveniles and their return as schoolmasters, it is interesting to speculate on whether they form permanent, stable groups much as mature females apparently do.

The data on the relationship between mean body length and school size for bachelor males presented by Gaskin (1970) has been recalculated, giving a regression of the number of males in a school on their mean body length of $y = 117.26 - 2.49x$. From this it can be calculated that the reduction in school size between mean body lengths of 40 and 46 ft (12.2 and 14 m) is from 17.7 whales to 2.7 whales. Using the growth curve presented by Best (1970), these lengths correspond to ages of about 21 and 28 years, respectively, which gives an instantaneous rate of reduction in school size over this age range of 0.27.

Total mortality rates (including fishing mortality) for the same stock vary from 0.05 for the age range 28–39, to 0.13 for the ages 31–45, and 0.125 for animals between 40 and 50 years old (Gaskin and Cawthorn, 1973). As the total mortality rate appears to increase with increasing age, the value for animals younger than 28 years is probably closer to 0.05 than 0.13. Consequently the rate of reduction in school size for males between 40 and 46 ft (12.2 and 14 m) in length is too great to be attributed to the combined effect of fishing and natural mortality alone. Presumably the schools must fragment as the bachelors get older, possibly as a result of increasing antagonism as the animals approach schoolmaster status (as is suggested by the trend for increasing individual distance in schools). It is therefore unlikely that individual bachelors remain in close association with each other for any protracted length of time.

B. Segregation of Males to Higher Latitudes

In an earlier report (Best, 1969a) it was concluded from cyamid infestation that the separation of male sperm whales from mixed schools occurred at a body length of 39–40 ft (11.9–12.2 m). This is now considered unlikely because of the evidence given above. The significance of trends in cyamid infestation, however, is not reduced. The predominance of *Neocyamus physeteris* on male sperm whales below 39 ft (11.9 m) in length indicates that these animals probably remain in the same environment as females, for this is the cyamid characteristic of females. Thus, although schools of small bachelors are segregated from mixed schools, they do not appear to penetrate Antarctic regions. This hypothesis is supported by the relative infrequency of the Antarctic diatom *Cocconeis ceticola*, on male sperm whales below 40 ft (12.2 m) in length, and by the similarity in seasonal migration patterns of females and small males (Best, 1969b).

In contrast to small males, medium-sized and large males landed at Donkergat frequently carried a film of the Antarctic diatom *Cocconeis ceticola*, indicating that they had recently migrated from colder waters. Such animals would be found most often in autumn and early winter (Best 1969b). Males 42 ft (12.8 m) or more in length were also infested exclusively with the cyamid species *Cyamus catodontis*, while females and males smaller than this normally carried the species *Neocyamus physeteris* (Best, 1969a). It seems reasonable to assume that this change in cyamid infestation reflects a change in migratory behavior, although the species of cyamid infesting sperm whales in the Antarctic has never been identified (so far as the author is aware).

Through the kind services of Mr. Sidney Brown, the Discovery Investigations' collection of cyamids from male sperm whales in the Antarctic has been made available to and identified by Mr. Charles Griffiths of the Department of Zoology, University of Cape Town (Table XI). The cyamids from all 12 animals examined [nine of which were 46 ft (14 m) or more in length] proved to be *Cyamus catodontis*, the data covering Areas I, II, and III of the Antarctic.

It therefore seems that *C. catodontis* is a more eurythermous species than *N. physeteris*. Consequently the point at which 50% of males are infested with either cyamid species will represent the stage at which 50% of males enter the Antarctic for the first time. This occurs at a length of 39–40 ft (11.9–12.2 m) (Best, 1969a) which from the growth curve would be equivalent to an age of about 19 years (Best, 1970).

This size and age correspond with the discontinuity in size distribution between groups of small and medium-sized bachelors, the former averaging 35–38 ft (10.7–11.6 m) in length and the latter 40–45 ft (12.2–13.7 m) (see above). They also coincide with the attainment of puberty (Best, 1969a) and the beginning of the period of accelerated growth (Best, 1970).

Although the distribution of the sperm whale sexes in the North Pacific is apparently less correlated with latitude than in the southern hemisphere (Ohsumi and Nasu, unpublished data), it has been calculated that 50% of males are segregated in Aleutian waters at an age of 19 years (Ohsumi, 1966).

The segregation of medium-sized males to the Antarctic means that these (and large males) become migratory between these latitudes and warmer waters. This leads to a quite different seasonal migratory pattern to that of females and small males (Best, 1969b). Evidence presented by Ohsumi (1966) indicates that the degree of segregation to higher latitudes (and hence the strength of the annual migration) increases with age, and it has been calculated that 75–90% of large males migrate to higher latitudes in summer (Ohsumi, 1966; Best, 1974). Ohsumi (1966) has also demonstrated that the rate of segregation falls off again in males over 25 years of age, presumably coincidental with the acquisition of schoolmaster status.

Table XI. Identification of Cyamids from Male Sperm Whales in the Antarctic

Whale no.	Length	Date killed	Locality	Cyamid species
1728	50.5 ft (15.4 m)	March 11, 1928	Grytviken, South Georgia	*Cyamus catodontis*
1730	50.3 ft (15.3 m)	March 11, 1928	Grytviken, South Georgia	*Cyamus catodontis*
—	—	1928	South Georgia	*Cyamus catodontis*
3011	51.2 ft (15.6 m)	January 21, 1930	Grytviken, South Georgia	*Cyamus catodontis*
B23	47 ft	November 27, 1946	57°06'S, 57°54'E	*Cyamus catodontis*
B28	54.3 ft	November 29, 1946	58°37'S, 62°07'E	*Cyamus catodontis*
B32	51.3 ft	November 29, 1946	58°37'S, 62°07'E	*Cyamus catodontis*
—	—	November 29, 1955	56°40'S, 45°47'W	*Cyamus catodontis*
SH169	55 ft	December 14, 1955	65°58'S, 70°49'W	*Cyamus catodontis*
—	54 ft	December 14, 1956	62°40'S, 63°33'W	*Cyamus catodontis*
—	—	1956/57	Antarctic Area I	*Cyamus catodontis*
SGL 517	46 ft	February 1, 1965	52°09'S, 40°15'W	*Cyamus catodontis*

V. GEOGRAPHICAL DISTRIBUTION OF GROUPINGS

In order to further understand the strategy involved in the social organization of sperm whales, it is necessary to have some indication of the geographical dispersion and segregation of the different groupings. This section deals with the southern hemisphere exclusively, as the situation there appears rather more clear-cut than in the North Pacific, apparently due to the less complex oceanographic situation in higher latitudes of the southern hemisphere.

In a diagrammatic representation of the social structure of various sperm whale schools and their approximate distribution in southern latitudes, Gaskin (1970) has intimated that bachelor schools are restricted to latitudes between about 35° and 55°S, and solitary males to latitudes south of 35°S. Supporting evidence for this is not given in the text.

Bachelor schools and solitary males are certainly encountered at the latitude of Durban (30°S), although their frequency of occurrence is definitely seasonal (Gambell, 1972). Size composition data from land stations at Paita (approx. 5°S), Pisco (approx. 14°S), Iquique (approx. 20°S), and Talcahuano (approx. 37°S) on the west coast of South America indicate that males up to 16.0 m length were taken at all stations and that males 12.0 m or more in length comprised 45.7%, 49.7%, 43.5%, and 52%, respectively, of the total number of males examined (Clarke *et al.*, 1968). Unless all these medium-sized and large males were associated with female schools (which seems unlikely from the high proportion of males in the total sample examined—55.5% at Paita, 60% at

Pisco, 59.6% at Iquique, and 71.7% at Talcahuano), it must be concluded that schools of medium-sized and large males do penetrate into tropical and subtropical latitudes, although without data on their relative abundance in different latitudes it is not possible to estimate the strength of this northern migration. The northern limits to the distribution of bachelor schools shown by Gaskin (1970) are therefore unsubstantiated by these data.

Schools of females and juvenile males are generally considered to be based in tropical and subtropical waters, with a southern limit to their distribution of 40°S (Matthews, 1938) or the 20°C isotherm (Gilmore, 1959), although individual schools containing females have been found as far as 56°S (Ivashin and Budylenko, 1970) and 59°S (Ivanov, 1972). The southern limit to their distribution can be investigated in more detail through the examination of catch statistics from the Antarctic.

Mr. Vangstein of the Bureau of International Whaling Statistics (BIWS), Norway, has kindly provided length data by 10° square of latitude and longitude for all male sperm whales killed in the Antarctic each season from 1961/1962 to 1968/1969. These data have been combined for all eight seasons and for each 10° series of latitude, series D, A, B, and C referring to 40–50°, 50–60°, 60–70°, and 70–80°S, respectively. The data have been analyzed separately for nine regions of longitude, corresponding to the divisions of the southern hemisphere adopted by the scientific committee of the International Whaling Commission (IWC) for the purposes of stock analysis (Anon, 1973). These regions are

West Indian	20–60°E
Central Indian	60–90°E
East Indian	90–130°E
East Australian	130–160°E
New Zealand	160°E–170°W
Central Pacific	170–100°W
East Pacific	100–60°W
West Atlantic	60–30°W
East Atlantic	30°W–20°E

The biggest differences in size composition between regions occur in series D, or in the area between 40° and 50°S (Fig. 4). In some regions, notably the New Zealand, East Australian, and Central Indian, the size composition of male catches in series D peaks at 39 ft (11.9 m), whereas in other regions the modal length at this latitude is as high as 43 ft (13.1 m). The size compositions that peak at 39 ft (11.9 m) are markedly skewed to the left, with an exceedingly low percentage of animals less than 38 ft (11.6 m) long. As 38 ft (11.6 m) was the minimum size limit prevailing, this suggests that there has been some over-generous measuring of smaller animals. Nevertheless it indicates that a substantial proportion of males caught in these regions was less than 40 ft (12.2 m) long.

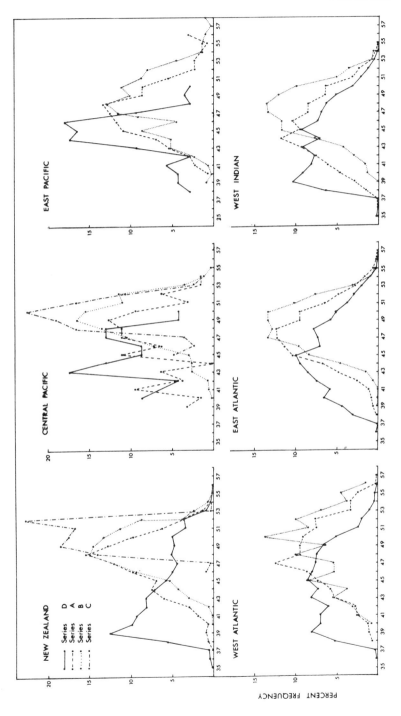

Social Organization in Sperm Whales

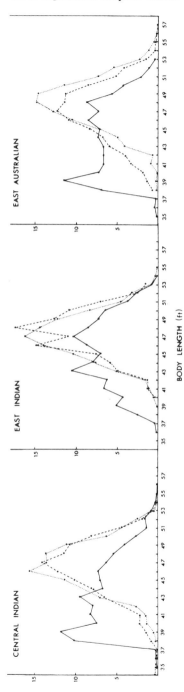

Fig. 4. Size composition of male sperm whale catches by latitudinal series for each division in the Antarctic, seasons 1961/1962 to 1968/1969 combined.

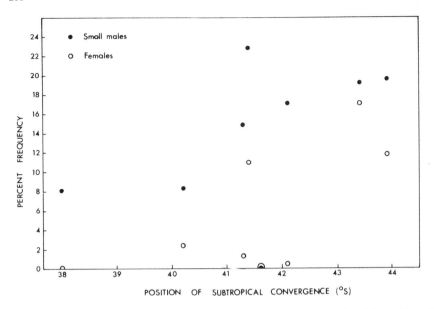

Fig. 5. Proportion of female and small male sperm whales in the series D catch and the position of the subtropical convergence in each division.

In Fig. 5 the percentage of males less than 40 ft (12.2 m) long killed in series D has been plotted against the position of the subtripical convergence for each region. The position of the convergence was estimated from the average of its positions at each 5° of longitude, based on the chart given by Deacon (1937); no data were available for the East Pacific, and those for the Central Pacific were incomplete. The percentage of females in the series D catch (from the same BIWS data) has also been plotted against the position of the subtropical convergence in each region (Fig. 5). In both instances there is a clear correlation between the incidence of whales and the position of the convergence: as the mean position of the convergence shifts farther south, particularly beyond 42°S, the proportion of small males and females in the catch between 40° and 50°S increases markedly.

This circumstantial evidence strongly suggests that the limit to the southward migration of females and small males is related in some way to the position of the subtropical convergence.

A similar heterogeneous distribution, in which adult males are distributed in higher latitudes than females and young outside the breeding season, has been described for three otariid seals in the eastern North Pacific, the California sea lion (*Zalophus californianus*), the Steller sea lion (*Eumetopias jubata*), and the northern fur seal (*Callorhinus ursinus*), and possibly one phocid, the elephant seal (*Mirounga angustirostris*). According to Bartholomew (1970), this pattern of distribution is a correlate of the sexual dimorphism of these animals, the adult males being three to six or more times heavier than the adult females, as well as being fatter. His hypothesis is that other things being equal, large size and subcutaneous fat favor reduced thermal conductance, so that the large, fat males should be better adapted to cold waters than the smaller and less fat females—in effect an intraspecific analogy to Bergmann's rule.

Gambell (1972) has shown that male sperm whales in general have higher growth rates for blubber thickness than females. Thus, although both sexes have similar blubber thicknesses at around 30 ft (9.1 m) in length, the blubber of an adult male (52 ft (15.8 m) in length) is about 1.5 times fatter than that of an adult resting female (36 ft (11 m) in length).

Whales, however, appear physiologically to be many times overinsulated for life in the coldest water (Kanwisher and Sundnes, 1966), so that it is unlikely that water temperature could be a direct limiting factor to sperm whale distribution. In addition, McNab (1971) has pointed out that there is no reduction in the *absolute* amount of heat loss through the surface by a *relative* reduction in surface area, and that large individuals of a species lose more heat via their surface than small individuals, if all other factors are equal, due to their larger surface areas. This means greater energy expenditure, and so a greater requirement for food. We must look elsewhere, therefore, for a rational explanation of the heterogeneous distribution pattern of sperm whales.

The distribution of males within the Antarctic in summer has also been investigated through examination of the catch statistics provided by Mr. Vangstein. The size composition of males for each series, based on the average of all nine regions (except for series C, where the mean was taken of the two regions for which data were available), is shown in Fig. 6.

Even in those regions in which a very small percentage of small males occurs in the series D catch, the modal length of males in series A (50–60°S) is substantially greater than that in series D (see Fig. 4). The difference between the size compositions in series A and B is not so great, although that in series A is skewed to the right and contains a higher percentage of animals less than 48 ft (14.6 m) long. In the two regions for which data exist, however, males in series C (70–80°S) tend to be substantially bigger than those in either D, A, or B. There is therefore a tendency for male sperm whales within the Antarctic region in summer to be stratified by latitude, the bigger animals being found farther south.

Tormosov (1970) has also found that the older male sperm whales are distributed farther south, there being a difference of 8–12 years in mean age between animals taken from 40 to 49°S and from 62 to 64°S.

This distribution pattern would seem to be a logical extension of Bartholomew's (1970) hypothesis, previously mentioned, that larger (and fatter) males should be better adapted physiologically to colder waters. The arguments against this hypothesis have already been made. Tormosov (1970), on the other hand, has concluded that the migration of sexually mature males to higher latitudes is linked with their recuperation from the breeding season, with the cold water

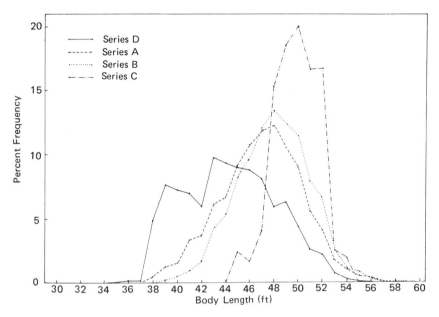

Fig. 6. Average size composition of male sperm whale catch by latitudinal series in the Antarctic, seasons 1961/1962 to 1968/1969 combined.

environment also promoting the process of spermatogenesis for the next breeding season. Studies to date, however, have failed to reveal a seasonal cycle of spermatogenesis in the sperm whale (Best, 1969a; Gambell, 1972).

Budylenko et al. (1970) have correlated the occurrence of female sperm whales at a latitude of 56°S (see Ivashin and Budylenko, 1970) with the penetration of warm, saline water from the north (apparently caused by anticyclonic conditions), as this created favorable conditions for the distribution and feeding of squid.

Unfortunately little is known about the distribution and abundance of the sperm whale's main food item, bathypelagic and mesopelagic squid. The squid fauna for the Antarctic is quite characteristic, 55% of the species and 31.2% of the genera being endemic (Filippova, 1972). Such a high degree of endemism is apparently unusual for the Cephalopoda. However, on the face of it, it is difficult to see why these animals might be less acceptable to female and small male sperm whales than to the larger males.

However, according to Kirpichnikov, as quoted by Berzin (1971), cephalopods are less numerous in the Antarctic as a whole in both species and numbers than in the warm waters of lower latitudes, so that sperm whales are more evenly distributed in the Antarctic and do not form large and dense accumulations. If this is so [and according to Knox (1970) there are no estimates of the standing crop of Antarctic cephalopods, although large concentrations have been reported], this might explain why mixed schools and schools of small bachelors are excluded from the region, their contribution in terms of over-all numbers and in density per unit area being incompatible with the availability of suitable squid.

It has also been suggested that certain food-seeking capabilities in sperm whales (connected with the ability to dive deeper and to catch rapidly moving or large cephalopods) are not only sexually linked but also serve to differentiate between males (Tarasevich, 1967). This difference in diving behavior is supported by the observations of open-boat whalers that females and young males took out only one and a half tubs of line (300 fathoms or 549 m) when harpooned, while bulls could take out three tubs or 600 fathoms (1098 m) (Caldwell *et al.*, 1966). Tarasevich (1968) has shown that in some areas of the North Pacific large male sperm whales (over 13 m long) fed on larger individuals of the same squid species than small males (11.6–13 m long). This she attributed to the young squid being distributed in the upper ocean layers while the adults were found deeper, as observed in some squid species (Akimushkin, as quoted by Tarasevich, 1968). It is interesting to note that one such species is *Onychoteuthis banksii* which, according to Korabelnikov (1959), forms the bulk of sperm whale food throughout the Antarctic. Betesheva (1961) also found that male sperm whales in the Kuriles region of the North Pacific consumed a greater variety of food and larger animals than females. It is therefore possible that the absence of small males and female sperm whales south of the subtropical convergence reflects a relative scarcity of young squid and warm-water forms of squid in the upper layers of the ocean, with the adults and cold-water squid being distributed deeper and out of the reach of the diving range of the smaller sperm whales, as has been proposed for the Pribilof Islands region of the North Pacific (Tarasevich, 1968).

Clarke (1956) has in fact shown that the squid eaten by male sperm whales in the Antarctic (*Moroteuthis robusta*) varied from 0.6 to 2.4 m in mantle length, with a mean of 1.3 m (0.6–1.8 m according to Iukov, 1971), while the squid eaten off the Azores (chiefly *Histioteuthis bonelliana* and *Cucioteuthis unguiculatus*) were smaller, averaging 0.95 m in length. Assuming these species have a similar length/weight relationship, and if their weight is proportional to the cube of their length, this difference in length would be equivalent to an approximate difference in weight of 2.6 times. The apparent greater size of food particles available in the Antarctic should therefore carry with it a relative advantage to adult male sperm whales.

The feeding requirements of sperm whales can be estimated from Sergeant's (1969) calculations based on heart weight. If the ratio of heart weight to body weight equals approximately one tenth of the feeding rate in large whales, an adult female sperm whale 36 ft (11 m) long and weighing about 12 m tons (Omura, 1950) will require roughly 420 kg of food daily, while an adult male 52 ft (15.8 m) long and weighing 39 m tons (Omura, 1950) will require about 1365 kg daily (The body weights used here are uncorrected for fluid loss to be comparable with the relative heart weight data). The maximum weight of squid recorded in a sperm whale stomach appears to be about 300 kg (Rice, in Caldwell et al., 1966). The maximum weight of food found in the stomachs of 277 males and 79 females examined at Donkergat whaling station was 130 kg (both first and second stomachs combined), although the great majority of stomachs contained far less than this. Apart from the stomach mentioned above, which contained 78 whole squid, the maximum number of fresh squid found in the next 13 fullest stomachs ranged from 24 to 56, with a mean of 31. Although some squid may have been regurgitated when the whale was harpooned or when compressed air was forced into its abdomen, this is by no means a general occurrence (personal observation). It therefore seems very likely that even female sperm whales have to make several feeding dives to achieve their daily food requirements. This exercise must involve the capture of many individual squid. At Donkergat in 1962, 38 batches of fairly fresh squid from 15 male sperm whale stomachs were weighed, totaling 211 individuals. Each batch consisted of similar-sized squid. The mean weights of squid in individual batches varied from 0.23 to 82.5 kg, but the majority of squid (96.7%) weighed averaged between 0.23 and 2.73 kg, with a mode between 1 and 2 kg. Assuming a mean squid weight of 1.5 kg, an adult female might have to catch about 300 squid a day and an adult male about 900. Although predation is probably directed to concentrations of squid, feeding must still be a time-consuming affair.

The relatively greater advantage to adult males afforded by the presence of much larger food items in the Antarctic than outside it therefore seems clear. However, there is no evidence available yet to indicate whether within the Antarctic itself there is a gradation in squid size with latitude that could account for the pattern of distribution of medium-sized and large males.

The heterogeneous distribution of the adult males and the females and young of certain pinniped species mentioned above has also been attributed by Sergeant (1973) to the possibility of there being a greater availability of large prey fish in cold waters, so giving adequate food to the large-bodied male seals which need a higher food intake.

McNab (1971), in his critique of Bergmann's rule, has suggested that body size in "hunters" is mainly determined by two factors: (1) the frequency distribution by size of available food particles and (2) the presence of other species that utilize the same food resource. Latitudinal variation in the size of hunters results from latitudinal variation in these factors. McNab states that in high latitudes a large size tends to occur in mammals and birds, and these regions are also marked by a low diversity of species, so suggesting an increased size of available food particles.

VI. MIGRATIONS

The conclusions of the preceding two sections can perhaps be best summarized pictorially (Fig. 7). The relative distribution of the different social groups is shown here in relation to the month of the year, so that the timing of events in relation to the migratory cycle can be better understood.

For the southern hemisphere, data on sperm whale migrations have been summarized by Best (1969b) and Gambell (1972). Mixed schools and schools of small bachelors are believed to have similar distributions and migratory cycles, even though they may be socially segregated from each other. They appear to

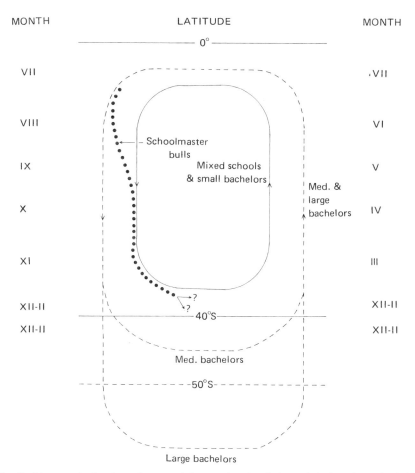

Fig. 7. Diagram of migrations of sperm whale groupings in relation to latitude and month of the year, southern hemisphere.

shift towards the equator in autumn and towards the subtropical convergence in spring. Schools of medium-sized bachelors, on the other hand, move north later in the year and come south earlier, while large males come north last of all and seem to leave earliest for higher latitudes. This pattern would correlate reasonably well with the differing distances that animals have to travel between tropical waters and their "feeding grounds" in higher latitudes. A rendezvous between the medium-sized and large bachelors and the schools containing females probably takes place around mid-winter, presumably in latitudes closer to the equator than to the subtropical convergence (Best, 1969b). It is believed that a selection process occurs in early spring, and that most of those medium-sized and large males unsuccessful in obtaining access to a mixed school then migrate back to the Antarctic. Estimates made above and by Ohsumi (1966) suggest that only 10–25% of large males become schoolmasters each year. The queries shown in Fig. 7 indicate our ignorance of whether schoolmasters leave the mixed schools once breeding is completed and migrate to the Antarctic, or whether they remain north of 40°S for most of the summer.

No accurate figures for the ranges covered by individual schools exist, although it appears that in general males move greater distances than females (Ohsumi and Masaki, 1975), and the maximum distance yet recorded for the movement of a marked individual (a male) was 7400 km (Ivashin, 1967). Mark returns for animals at large for more than one month between marking and recapture have been analyzed for 17 males and 22 females in the southern hemisphere. Approximate measurements of straight-line movements between marking and recapture produce average figures of 850 nautical miles for males and 372 nautical miles for females, with a significant difference between the two (Mann-Whitney U test, $P < 0.0069$). "Home ranges" for mixed schools are therefore probably considerably smaller than those of bachelor schools in general.

Berzin (1971) has proposed that each school has its strictly defined wintering ground, but as the area of summer habitat is vast, the schools cross extensively from one foraging ground to another, returning in autumn to their respective wintering grounds. There is insufficient evidence to test this theory at present. However, there are six recoveries of female sperm whales at Durban that were also marked in the whaling grounds from two to six years previously, and for all of which the calendar date of recovery fell within two months of the date of marking. This suggests that migrating patterns and routes may be consistent, at least for this sex.

A peculiarity about the patterns of migration and reproduction described above is that at the height of the mating season (from November to January; Best, 1968; Gambell, 1972) the majority of males of breeding age appear to be distributed allopatrically to the mixed schools, with large geographical distances between them. This seems a most unusual distributional pattern, as in other animals with a similar reproductive strategy (e.g., fur seals), surplus males of breeding age are distributed close to the periphery of the breeding colony during the height of the mating season. In this way a dominant male who becomes fatigued or dies is immediately replaced, so maintaining reproductive efficiency. The migratory pattern described above does not seem to allow for extensive replacement of schoolmasters once the selection process has been completed. This will be discussed more fully below.

VII. SOCIAL ORGANIZATION AND ITS DEVELOPMENT

A. Form of Social Organization in the Sperm Whale

A pictorial summation of the social organization of the sperm whale as described in this paper is shown in Fig. 8. In effect this differs from Ohsumi's model (1971) only in that recognition is given to three types of bachelor schools, and more detail on age and reproductive composition is included.

The basic social unit of the sperm whale appears to be the mixed school of adult females plus their calves and some juveniles of both sexes, normally numbering 20–40 animals in all. Although the primary unit could be considered as the cow and calf (possibly together with some of its previous offspring), as in the beluga (Kleinenberg *et al.*, 1964), there is evidence for the sperm whale that bonds between females within mixed schools may persist for many years, so the school has been considered as the basic unit. The adult female component of these schools consists of animals of all ages and in all stages of the reproductive cycle, while the male component is mostly sexually immature. Segregation of males (and probably of some females) probably begins shortly after weaning, when schools of juveniles may be formed. Recruitment to schools of small bachelor males may take place from juvenile schools or from the mixed school direct. The female component of juvenile schools presumably returns to a mixed school before attaining puberty. Schools of small bachelors number 12–15 animals, although these may aggregate to form much larger units. These schools appear to fragment as the members grow older and reach sexual maturity, while it is quite possible that (at least at the stage of development between schools of small and medium-sized bachelors), school members are derived from a number of different small bachelor schools. This could arise through a selection process when only males above a certain size migrate to the Antarctic for the first time. Adult (schoolmaster) bulls only join mixed schools during the breeding season, when between 10 and 25% of mature males are involved in breeding activity.

The cohesiveness of mixed schools, the low proportion of males (particularly sexually mature males) within them, and the temporary association of large mature males with mixed schools all strongly indicate that the sperm whale's social organization is basically polygynous. Most previous workers have agreed with this conclusion, apart from some who considered that breeding was

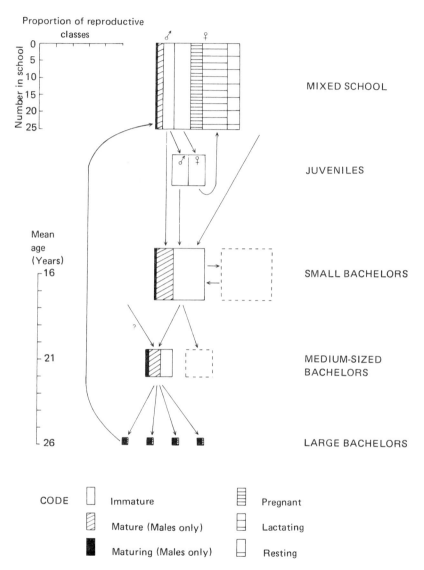

Fig. 8. Diagram of social organization of sperm whale (size of square represents average size of school, extent of shading represents average proportions of different reproductive classes within the school).

performed by much smaller and younger males; this theory has already been criticized (Section III).

According to Yablokov (unpublished data), it has been possible, through the investigation of color phenotypes within sperm whale schools, to determine the male parent of fetuses. There is apparently strong evidence for a single male parent for several fetuses in one harem school, although there were several males included in the group. This is further confirmation of the existence of polygyny in the species.

Polygyny is here defined as the situation where more females than males contribute genetically to the population, and no assumptions have been made on the existence of behavioral pair-bonds, something that would be difficult to establish for the sperm whale with our present ignorance of its social behavior.

The form of polygyny in the sperm whale, however, may not be exactly the same as for other polygynous mammals, for example, pinnipeds. The peculiarity in distribution pattern has already been mentioned, from which it appears that replacement of breeding bulls during the breeding season is difficult due to the geographical divide between mixed schools and surplus mature males. To date there has been no demonstration of a significantly higher natural mortality rate in adult males than adult females (see Section II), and this suggests that the possession of a "harem" may not be as strenuous an exercise as, for example, in some pinnipeds, where the natural mortality rate of harem bulls may be as much as three times that of adult females (Johnson, 1968). The need for natural replacement during the breeding season will therefore be correspondingly less.

Tormosov and Sazhinov (1974) have proposed a form of polygyny for the sperm whale different from that for pinnipeds. Its most characteristic feature is not that one male has a harem of several females but that during the period of reproduction several large adult males become sexual monopolists among several family groups, forming a local accumulation with females prepared to mate. They also conclude that promiscuity without sharp competition between males is a possibility under such circumstances. The description of pinniped polygyny given by Peterson (1968), however, is similar in some respects to that proposed for the sperm whale by Tormosov and Sazhinov (1974). I quote: "The term 'harem' is usually used by zoologists to designate a group of females associated with a single male. None of the species I have described above, nor any other pinniped pattern with which I am familiar, fits this definition very closely. Female pinnipeds seem to aggregate without regard for the males, except that in highly dimorphic species at times of low population density, they aggregate in groups at places not used by bulls for fights and displays. In no study of which I am aware has it been demonstrated that a constant group of females associates with a single male for any prolonged period."

It is difficult to speculate on the exact mode of polygyny in the sperm whale without more information on the behavioral processes that occur between the sexes prior to and during reproduction, e.g., how the rendezvous between breeding males and females occurs, the existence of pair-bonds, and what degree of competition between males takes place. Could the tooth marks found on the head and body of adult males represent the results of disputes within bachelor schools in establishing a dominance hierarchy, rather than in competition for access to females? This appears unlikely given the small number of animals present in bachelor schools at the stage when they reach social maturity and the great individual distances observed within such schools. However, it is an interesting possibility that, even within pairs of adult males, some sort of mating priority could be established through a dominance hierarchy prior to a rendezvous with females, thus avoiding or at least reducing the number of more serious disputes that might arise over access to females. The number of schoolmasters actually found with mixed schools during the breeding season (1–5) is very similar to the number of animals found in a school of large bachelors (see Section III.A). It is therefore possible that a bachelor school, rather than several individual males, makes the contact with females during the breeding season. In this respect the situation would resemble that described for the African lion by Bertrand (1975), although there is no evidence that such males would be related to each other, as in the lion.

One may conclude that the sperm whale does have a polygynous social organization, but its exact form will only be known when more observations have been made of sperm whale behavior during the reproductive cycle.

B. Evolution of Polygyny

Superficially at least there seem to be several similarities between the social organizations of the sperm whale and some of the otariid seals. In view of Bartholomew's (1970) interesting model for the evolution of pinniped polygyny, it is worth speculating on the development of the sperm whale's social organization, although at this stage in our knowledge it is not possible to construct a complete model for this species.

As mentioned by Bartholomew (1970), for pinnipeds any useful model of the evolution of polygyny must account for extreme gregariousness, because in the absence of gregariousness during the breeding season, organized polygamy is impossible. The sperm whale has evolved as a predator of mesopelagic and bathypelagic squid. Two factors arising from this would seem to promote gregariousness. Firstly, their prey is active and must be caught in a time span restricted by the diving limitations of the whale. Daily food requirements also seem to be such that the exploitation of schooling squids would be favored. Under these circumstances cooperation between individuals in the location and capture of their food (as observed for some delphinid species feeding on pelagic schooling fish; Fink, 1959; Morozov, 1970; Tayler and Saayman, 1972) would be mutually advantageous. Secondly, newborn sperm whales do not appear to be as proficient in diving as adult females (personal observation) and must presumably learn to become successful predators on mesopelagic and bathypelagic squid. These two factors would tend to produce a long period of dependence of the calf on its mother, and in fact lactation usually continues for two years in the sperm whale (Best, 1974). The schooling habit might be advantageous to both mother and calf during this period. Survival of the young against attacks by sharks or other predators might be enhanced (see Gambell, 1968, for evidence of the attraction of sharks to calving sperm whales and their protective reaction). The mother of the newborn calf might also be able to leave its offspring for short periods with other members of the school while feeding at depths which the young cannot reach. Although such caregiving behavior by other members of the school has not been directly observed for the sperm whale, Durban whalers claim that sperm whale calves can change their swimming partners within the school when being chased, making it difficult to distinguish the actual mother. The author has seen a newborn calf "stand by" a harpooned female that was sexually immature and so obviously not its mother. Temporary abandonment of the calf with other members of the school has been observed in wild bottlenose dolphins *(Tursiops aduncus)* when the parent enters the surf zone to feed (Tayler and Saayman, 1972).

The gregariousness of sperm whales is probably enhanced by their high mobility, enabling them to cover a potentially large foraging range. The schools containing mature females appear to be year-round units, so an upper limit to school size may be set by a compromise between the degree of gregariousness in their prey, mesopelagic and bathypelagic squid, and the energy demands of a school of sperm whales.

The second group of "operators" that Bartholomew (1970) considered essential for the dynamism of his model of pinniped polygny favors the sexual dimorphism of males. In pinnipeds sexual dimorphism in size is considered to arise by a combination, on the one hand, of large size and subcutaneous fat that favor prolonged territory maintenance due to resistance to starvation and, on the other hand, by a combination of aggressiveness of the males (due to increased testosterone levels) and gregariousness of the females. In the sperm whale a rather different situation exists. The gregariousness of females is apparently permanent, all reproductive classes being present in the harem (i.e., mixed school) with no segregation of receptive females and with the mixed school continuing to feed normally throughout the breeding season. The advantage that large size and thick blubber bestows in a capacity for prolonged fasting does not seem to apply in this situation; it has not been demonstrated that large males cease to feed while present with the mixed school, and significant seasonal variations in blubber thickness do not clearly occur for any sex or class of sperm whale (Berzin, 1971; Gambell, 1972).

Contrary to polygynous pinnipeds, sperm whales have a very low reproductive rate, probably the lowest recorded for any marine mammal to date. The mean pregnancy rate for mature females in an unexploited population is about 20% (Best, 1974), compared to 69–80% for *Callorhinus* (Kenyon *et al.*, 1954). This low reproductive rate is largely a consequence of the prolonged suckling period, which comprises an estimated 40% of the total reproductive cycle. It has two chief consequences. Firstly, the proportion of females coming into estrus annually during the breeding season is only a small percentage of the total number of mature females: taking figures given in Table XIX of Best (1968), and omitting those animals that ovulate outside the main breeding season (and so are probably not impregnated), the proportion of the mature female population ovulating each year can be calculated as 1.388/4 or about 35%. Secondly, for the survival of the stock, it is essential that a high efficiency of fertilization is achieved.

Because such a small proportion of females ovulates annually, there must be an inevitable surplus of males in the population as a whole. The cohesive social unit formed by females also means that several potentially receptive females may be encountered in the same school during a breeding season so that "possession" of an entire school would provide a successful bull with a much better chance of his characteristics as a schoolmaster passing to his progeny. Selective forces therefore exist to favor such characteristics (large size, strength, etc.) through competition between mature males for access to a mixed school. Such competition may be enhanced by seasonal variations in testosterone output, as is suggested by changes in the size of Leydig cells in the testis (Best, 1969a), so increasing aggressiveness.

The forces acting to produce sexual dimorphism in size in sperm whales also have the consequence in doing so of significantly increasing the nutritional demands of the male over the female. This has two effects: firstly, the advantages of the male staying with the female school permanently are reduced, as the feeding requirements of the two sexes diverge and the possibilities of synchronizing feeding and diving behavior patterns recede. Secondly, any adaptation that would enable the male to dive deeper and longer and so reach the larger individual squid believed to occur at greater depths would have a selective advantage.

The sexual dimorphism in size of sperm whales is in fact accompanied by a change in body shape, so that adult males have a relatively larger cephalic and shorter caudal region than adult females, as mentioned in Section II. The distance from the tip of the snout to the center of the eye is about 27% of the body length in a physically mature male, but only 20% of the body length in a physically mature female (Nishiwaki *et al.*, 1963). This distance corresponds roughly to the length of the spermaceti organ, so that the relative increase in the length of this organ between adults of either sex is about 1.35 times. The function of the spermaceti organ is still controversial. Opinion is divided between its use as a buoyancy regulator for deep diving (Clarke, 1970), as a reverberation and sound-focusing chamber for the production of echolocation clicks (Norris and

Harvey, 1972), and as a method of (1) assisting in the evacuation of lungs prior to diving and (2) absorbing nitrogen at extreme pressures from air in the right nasal cavities before it enters the bloodstream (Schenkkan and Purves, 1973).

The circumstantial evidence of a correlation between diving ability and the sexual dimorphism of the head seems to suggest, to this author at least, that the development of the spermaceti organ is somehow concerned with adaptation to prolonged submergence at great depths. This head dimorphism seems to commence between birth and the time of weaning and is usually quite distinct by a length of about 9 m (Nishiwaki *et al.*, 1963), when the females reach puberty. Similar sexual dimorphism in melon development has been described for the pilot whale *(Globicephala melaena)* and false killer whale *(Pseudorca crassidens)*, but its function is still obscure (Mead, 1975).

The evolution of sexual dimorphism in body size is therefore accompanied by sexual dimorphism in the size of the spermaceti organ. This in turn may have the ecological consequence of favoring male segregation from females. Whatever the initial stimulus for male calves leaving their parent school, their early development of a relatively bigger spermaceti organ (and so presumably better diving abilities) means that the advantages of residing in a school of mature females with a relatively restricted feeding strategy are reduced and the advantages of schooling with other males of a similar size and stage in development are increased. Early and permanent segregation of males from female schools could therefore be reinforced in this manner.

Such segregation (and the seasonal dispersal of medium-sized and large males to higher latitudes in summer) has the important ecological effect of increasing dispersion and so reducing intraspecific competition for available space and resources, thus being advantageous to the species as a whole.

In particular, the ability of male sperm whales above an average length of 39–40 ft (11.9–12.2 m) to penetrate latitudes south of the subtropical convergence in summer, so giving them access to a food supply separate from that of mixed schools or small bachelors, must facilitate the spurt of accelerated growth seen in animals of this size.

Bartholomew (1970) has listed several reproductive adaptations of female pinnipeds to polygyny. One possible adaptation of sperm whale reproductive behavior is the length of the gestation period, which is estimated to be 14–16 months (Best, 1974). Nearly all other cetaceans so far investigated tend to have gestation periods of 9–12 months (Harrison, 1969). It may be of some significance that the only other cetacean species in which pregnancy is known to greatly exceed one year are the pilot whale *(Globicephala melaena)*, in which it lasts 15–16 months (Sergeant, 1962), and the beluga *(Delphinapterus leucas)*, in which it lasts 14.5 months (Brodie, 1971). Both species can be considered as polygynous, although the evidence for *D. leucas* is less convincing (Kleinberg *et al.*, 1964). It is difficult to visualize how such a feature could be linked in any way to an annual migratory cycle, but it is possible to consider it in the light of an adaptation to polygyny. By prolonging pregnancy for about three months over the year, effective separation of the breeding and calving seasons is achieved. By the time the peak of the calving season is reached in sperm whales, about 88% of conceptions have already occurred (Best, 1968). It is postulated that in this way interference in the social structure and cohesiveness of the mixed school is minimized during the calving period when the young whale is particularly dependent on its mother, so reducing calf mortality. Caldwell and Caldwell (1972) have reported that in a captive situation female dolphins normally have to protect calves from the aggression of adult males.

A possible male reproductive adaptation to polygyny in the sperm whale is the delayed onset of full sexual maturity. Although competition between adult males for females is considered fundamental to the evolution and maintenance of polygyny, so that there will always be an excess of mature males in the population, such competition could be counterproductive if carried to extreme. Thus intermale strife could reach proportions that would actively interfere with successful mating behavior and might adversely affect the survival of young calves in the same school. In the northern fur seal a limit to the number of surplus males arises from a combination of a greatly increased natural mortality rate in harem bulls and a higher age at sexual maturity in the male. On the other hand there appears to be no difference between the natural mortality rates of adult sperm whales of either sex, but males reach full sexual maturity at an age 2.8 times that of females, and this sets a limit to the number of surplus bulls in the population.

VIII. IMPLICATIONS FOR MANAGEMENT

A study of the sperm whale's social organization and its development is of more than academic interest. Sperm whales are (numerically and by weight) the most important single species in the present world whale catch, approximately 21,000 being taken in 1974–1975. If their stocks are to be managed successfully, it is important to understand the species' social organization fully and to predict how it might be affected by different levels and types of exploitation.

Because of the sperm whale's essentially polygynous organization, it has been proposed that exploitation should be based on the "surplus" males (Bartholomew, 1974), and in practice the total world catch has been predominantly of males since the beginning of this century. It is inevitable, however, that this type of exploitation will lead to a change in the ratio between the sexes in the adult population. How this change might affect the social structure or the reproductive rate of the population as a whole is a matter of conjecture. The "effective sex ratio" in an unexploited population has been discussed above, but the ratio to which this might decline as a result of differential fishing for males and which might still maintain an effective reproductive rate is not known. As present indications from a mathematical population model are that the size of the potentially sustainable catch of males is almost directly related to the "harem" size and the number of reserve males needed for each mixed school (Anon., 1977), it is obvious that more information is needed on this vital question before a management strategy based on predominantly male catches can be safely adopted.

There seem to be only two methods by which this information might be obtained. One of these is by experimental manipulation of a stock, i.e., by removing a certain proportion of the adult male component and observing what effect this has on the reproductive rate. The second is by studying the effects of a similar management policy on another polygynous species and by drawing analogies from the historical record. At this time it is only possible to consider the latter approach.

The northern fur seal is a polygynous species that has (for a considerable part of its recent history) supported an industry based on the harvesting of sub adult males. Fortunately the species has also been extensively studied for many years, including a period during which the population was recovering from prior overexploitation. Coincidentally one of the population parameters that has been monitored for the longest period is the annual count of harem and idle bulls on the rookeries. During a period of sustained population increase (from 1912 to 1924), the mean annual ratio of harem to idle bulls on the Pribilof islands was 6.23 : 1 (Anon., 1962). In the early 1950s, when the total population was at a much higher level, pup production had virtually stabilized, and the rookeries appeared to be overcrowded, the mean annual ratio of harem to idle bulls was much lower and seemed to be approaching parity. A program of deliberately reducing the mature female population then began in 1956. Assuming that the 1950s situation was closer to that prevailing in the unexploited population, a reduction in the harem reserve of about 84% from the "unexploited" condition appeared to have no adverse effect on the reproductive rate. If a direct comparison is made, it should be possible to reduce the active-to-idle bull ratio in the sperm whale from 1 : 3.9 in the unexploited population (see above) to 1 : 0.62 without adversely affecting the reproductive rate.

A major difference between the two management situations, however, is that the northern fur seal harvest was usually restricted to males 2–5 years of age, and older animals were left undisturbed; in the sperm whale exploitation has been principally directed to the largest (and therefore the oldest) animals. The latter procedure is therefore likely to result in major changes in the size and age composition of the adult male component of the population; such changes in mean size have been observed in at least one region (Jonsgård, 1960). In the case of the northern fur seal, however, the harvesting of only subadult animals should, after an initial period of stabilization following the onset of exploitation, result in an adult male component that is reduced in numbers but whose size and age structure are unchanged, provided that escapement from the catch is reasonably consistent from year to year and that there is no change in the adult natural mortality rate. As the density of spermatozoa in the vas deferens of sperm whales appears to be related to body size, and probably age (Best, 1974), the effect of intensive whaling on larger animals may result in younger, less fertile males becoming involved in reproductive activity. Clearly successful management will demand that both an effective sex ratio and a suitable age composition for adult males are maintained.

A situation more analogous to that postulated for the sperm whale has been described for southern elephant seals (Carrick et al., 1962; Laws, 1960). At South Georgia an industry based exclusively on adult males was carried on from the start of the century to the 1960s. As a result there were apparently major changes in the age and size composition of the adult bulls (Laws, 1960). The extent of these changes became apparent when a comparison was made with the situation at Heard and Macquarie Islands, where elephant seals had remained unexploited for many years and the herds were large and stable. At Macquarie Island, a few precocious 6-year-old males appeared in the breeding season, and by 10 or 11 years of age all males were ashore, but the few dominant bulls which mated with the majority of females must have been at least 15 years old (Ingham, 1967). In South Georgia, however, the ages of the bulls killed ranged from 4 to 12 years, with a mean around 7–8 years, and bulls at 6 years of age were holding harems. The onset of spermatogenesis seemed to occur at 4 years of age in the South Georgia population (Laws, 1956), so that males were holding harems within two or three years of attaining sexual maturity. This situation, however, appeared to have had no deleterious effect on the reproductive rate, the pregnancy rate being estimated as about 82% (Laws, 1960).

If a straight comparison could be made with the sperm whale, it would suggest that the mean age of males attending mixed schools could decline to as little as 7.5/4 or 1.875 times the age at sexual maturity (onset of spermatogenesis). This would be equivalent to an age of about 19 years in the sperm whale, if an age of 10 years at the onset of spermatogenesis is accepted (Best, 1974).

Changes in the sex ratio of mature elephant seals on the breeding beaches apparently accompanied the reduction in size of the harem bulls at South Georgia. In the reserve area (where no exploitation took place), the ratio of cows to bulls ashore during the breeding season (13 : 1) was similar to that at Macquarie Island (11 : 1), while in the exploited areas at South Georgia the overall ratio was 30 : 1 (Carrick et al., 1962). At Macquarie Island, where the situation can be said to approximate to that of an unexploited population, the average number of cows in a harem was 48. From this it can be calculated that, on average; 48/11 or 4.36 bulls were available per harem. The mean harem size on the exploited beaches at South Georgia appears to have been similar to that at Macquarie (approx. 50), but as the cow–bull ratio was so much greater, there would only be 48/30 or 1.6 bulls available per harem. On this basis, and assuming only one bull is actually in possession of a harem, the ratio of active to idle bulls can be calculated as 1 : 3.36 in an unexploited population and 1: 0.6 in an exploited population. This is equivalent to an 82% reduction in the idle bull population.

These examples of the northern fur seal and southern elephant seal suggest that a considerable reduction in the idle-to-active bull ratio and a substantial decline in the mean age of mature males are possible in a polygynous species without adverse effects on its reproductive rate. In these pinnipeds, however, the extreme gregariousness of adult females during the breeding season greatly facilitates the rendezvous between the sexes. In view of the far greater dispersion of mixed schools of sperm whales during the breeding season and the relatively small size of the individual schools, it would be unwise to assume that reductions of a similar magnitude in the "harem" reserve or the mean age of schoolmaster bulls are necessarily applicable.

Although sperm whale catches throughout most of this century have been composed predominantly of males, there has been a recent increase in the catch of females, in both relative and absolute terms. This has been at least partly due to a deliberate policy of the International Whaling Commission to bring female populations to the level giving the maximum sustainable catch of both sexes (Anon., 1973). Such an intensification of the exploitation of females, however, could have significant effects on the social structure and (possibly) reproductive success of the stocks concerned.

In the African elephant, an animal whose social organization and life history are remarkably similar in several respects to those of the sperm whale, the selective culling of large individuals from family units caused considerable disturbance within such herds, and a tendency for them to bunch and form larger aggregations was observed (Douglas-Hamilton, 1973; Laws *et al.*, 1975). The older, larger females seemed to exert a strong coordinating and protective influence over the remainder of the herd, so that their removal caused the leaderless survivors to amalgamate with other family units. Vast herds of 100–200 animals occurred, usually in elephant populations in conflict with man, where the older age classes are normally the first to disappear. The selective removal of the largest and oldest females therefore adds to the disturbance factor caused by cropping and tends to produce much larger social aggregations than usual, with a consequent intensification in the localized use of the habitat.

Because of the nature of whaling, where the quota system and minimum size limit combine to produce a strong incentive to catch the largest animals available, the exploitation of female sperm whales might produce a similar effect to that observed in some elephant populations that have been subject to selective cropping. A considerable amount of modern data on school size in sperm whales exists (especially that given by Gambell, 1972; and Ohsumi, 1971), but an equivalent amount of quantitative data for the early years of the open-boat fishery in the eighteenth and nineteenth centuries does not exist (or has not been extracted from logbook records). Based on accounts such as those given by Beale (1839), Bennett (1840), Seabury (in Clark, 1887), and Scammon (1874), however, it seems in general as though school sizes of sperm whales did not differ markedly in the nineteenth century from those now observed, and several of these authors referred to schools of over 100 (and up to 500 or 600) animals. It is possible of course that the operations of the open-boat whalers themselves were selective by nature and that their activities had already affected school size and composition by the time that Beale and others made their observations. However, herds of over 100 sperm whales are now rarely seen on whaling grounds (Gambell, 1972; Ohsumi, 1971), and it seems clear that any "bunching" reaction in sperm whales as the result of the selective removal of older females must be much less than that observed in African elephants. This may confirm the suggestion made in Section VII.B that an upper limit to the school size of sperm whales is set by a compromise between the degree of gregariousness in their prey and the energy demands of the school.

The mathematical models developed so far for sperm whale populations have usually assumed a relationship between population density and certain parameters such as pregnancy rate, age at sexual maturity, and natural mortality rate. While such a relationship has been observed in several other animal species, no density-related changes have yet been detected in these parameters for sperm whales. Furthermore, the complex social organization of the species may mean that a simple direct relationship with over-all population density itself may not apply. In the African elephant, for instance, whose social organization closely resembles that of the sperm whale, Douglas-Hamilton (1973) has proposed that natural regulation of population size is not necessarily directly density dependent. Even at high population densities, the apparent lack of territorial behavior in elephants prevented the development of a correspondingly high level of intraspecific aggression, and there were no indications that social interactions related to overcrowding in any way hindered reproduction or contributed to natural regulation. Douglas-Hamilton considered it more likely that food supply is probably the ultimate controlling factor of elephant populations.

While little is known at present of the social behavior of the sperm whale, observations or inferences of aggressive interactions have been limited almost exclusively to intermale disputes (Caldwell *et al.*, 1966; and Section III.B.2). Territoriality has not been demonstrated and seems unlikely to be very marked, given the large "home ranges" recorded for the species (Section VI). In fact the size of individual schools appears likely to be controlled ultimately by nutritional supply and demand (Section VII.B). Hence it is believed that the ultimate controlling factor for sperm whale populations also may be food supply rather than absolute density of individuals. Under these circumstances it is important to understand the trophic interrelationships between mixed and bachelor schools. If (as seems likely from their migratory patterns and diving behavior) male sperm whales are utilizing different components of the cephalopod fauna from the females and young, then trends in the abundance of large and medium-sized males may not necessarily result in changes in the abundance of females and young, or in their reproductive and mortality parameters, even though the over-all density of the population may change significantly. The design of realistic sperm whale population models should therefore take into account the relative trophic positions of the different social groupings before a decision is reached on the nature of the biological response to a change in the density of any component of the population.

ACKNOWLEDGMENTS

Most of the field work for this paper was carried out while the author was under contract to the Fisheries Development Corporation of South Africa. Laboratory facilities were provided by the Sea Fisheries Branch, Cape Town, and I am indebted to the Director of Sea Fisheries for permission to publish this paper. The cooperation of both Saldanha Whaling Limited and Union Whaling Company Limited is gratefully acknowledged.

I would also like to express my thanks to the following people. Sidney G. Brown, Whale Research Unit, Institute of Oceanographic Sciences, United Kingdom, for providing cyamid specimens; Charles Griffiths, Department of Zoology, University of Cape Town, for identification of cyamids; Einar Vangstein, Bureau of International Whaling Statistics, Norway, for provision of catch statistics; Dennis V. Hansen and Michael A. Meÿer, Sea Fisheries Branch, Cape Town, for practical assistance in the field; and Kathy Ralls, National Zoological Park, Smithsonian Institution, Washington D.C., for constructive criticism of this paper.

REFERENCES

Anon., 1962, North Pacific Fur Seal Commission Report on Investigations from 1958 to 1961, Tokyo, Kenkyusha Co., 183 pp.
Anon., 1973, Sperm whale assessment meeting. Parksville, Vancouver Island, BC, 3-10 May 1972, *Rep. Int. Whal. Comm.* **23**:55–88.
Anon., 1977, Report of the sperm whale meeting. La Jolla, California, 16–25 March 1976. *Rep. Int. Whal. Comm.* **27**:240–252.
Bartholomew, G.A., 1970, A model for the evolution of pinniped polygyny, *Evolution* **24**:546–559.
Bartholomew, G.A., 1974, The relation of the natural history of whales to their management, in: *The Whale Problem. A Status Report* (W. E. Schevill, ed.), pp. 294–302, Harvard University Press, Cambridge, Massachusetts.
Beale, T., 1839, *The Natural History of the Sperm Whale*, 2nd ed., John Van Voorst, London.

Bennett, F.D., 1840, *Narrative of a Whaling Voyage around the Globe from the Year 1833 to 1836*, 2 vols., Richard Bentley, London.

Bertrand, B.C.R., 1975, The social system of lions, *Sci. Am.* **232**:54–65.

Berzin, A. A., 1971, The Sperm Whale (A. V. Yablokov, ed.), Pishchevaya Promyshlennost, Moscow; translation by Israel Program for Scientific Translations, 1972, 394 pp.

Best, P. B., 1967, The sperm whale *(Physeter catodon)* off the west coast of South Africa 1. Ovarian changes and their significance, *S. Afr. Div. Sea Fish. Invest. Rep.* **61**:1–27.

Best, P. B., 1968, The sperm whale *(Physeter catodon)* off the west coast of South Africa 2. Reproduction in the female, *S. Afr. Div. Sea Fish. Invest. Rep.* **66**:1–32.

Best, P. B., 1969a, The sperm whale *(Physeter catodon)* off the west coast of South Africa 3. Reproduction in the male, *S. Afr. Div. Sea Fish. Invest. Rep.* **72**:1–20.

Best, P. B., 1969b, The sperm whale *(Physeter catodon)* off the west coast of South Africa 4. Distribution and movements, *S. Afr. Div. Sea Fish. Invest. Rep.* **78**:1–12.

Best, P. B., 1970, The sperm whale *(Physeter catodon)* off the west coast of South Africa 5. Age, growth and mortality, *S. Afr. Div. Sea Fish. Invest. Rep.* **79**:1–27.

Best, P. B., 1974, The biology of the sperm whale as it relates to stock management, in: *The Whale Problem. A Status Report* (W.E. Schevill, ed.), pp. 257–293, Harvard University Press, Cambridge, Massachusetts.

Betesheva, Ye. I., 1961, Feeding of commercial whales of the Kuriles, *Tr. Soveshch. Ikhtiol. Kom. Akad. Nauk SSSR* **12**:104–111.

Brodie, P. F., 1971, A reconsideration of aspects of growth, reproduction, and behavior of the white whale *(Delphinapterus leucas)*, with reference to the Cumberland Sound, Baffin Island, population, *J. Fish. Res. Board Can.* **28**: 1309–1318.

Budylenko, G. A., Pervushin, A. S., and Naumov, A. G., 1970, Why female sperm whales entered the Antarctic, *Tr. Atl. Nauchno-Issled. Inst. Rybn. Khoz. Okeanogr.* **29**:203–215.

Caldwell, M. C., and Caldwell, D. K., 1966, Epimeletic (care-giving) behavior in cetacea, in: *Whales, Dolphins and Porpoises* (K. S. Norris, ed.), pp. 755–789, University of California Press, Berkeley.

Caldwell, M. C., and Caldwell, D. K., 1972, Behavior of marine mammals, in: *Mammals of the Sea, Biology and Medicine* (S. H. Ridgway, ed.), pp. 419–465, Charles C. Thomas, Springfield, Illinois.

Caldwell, D. K., Caldwell, M. C., and Rice, D. W., 1966, Behavior of the sperm whale, *Physeter catodon* L., in: *Whales, Dolphins and Porpoises* (K. S. Norris, ed.), pp. 677–717, University of California Press, Berkeley.

Carrick, R., Csordas, S. E., and Ingham, S. E., 1962, Studies on the southern elephant seal, *Mirounga leonina* (L.). IV. Breeding and development, *C.S.I.R.O. Wildl. Res.* **7**: 161–197.

Christensen, A. F., 1926, Hvalfetet på Ecuador, *Nor. Hvalfangst-Tid.* **15**:111–112.

Clark, A. H., 1887, The whale fishery 1. History and present condition of the fishery, in: *The Fisheries and Fishery Industries of the United States* (G. B. Goode, ed.), 5. History and Methods of the Fisheries, Vol. 2, pp. 3–218, U.S. Commission on Fish and Fisheries, Washington, D.C.

Clarke, M. R., 1970, Function of the spermaceti organ of the sperm whale, *Nature (London)* **228**:873–874.

Clarke, R., 1956, Sperm whales of the Azores, *Discovery Rep.* **28**:237–298.

Clarke, R., 1962, Whale observation and whale marking off the coast of Chile in 1958 and from Ecuador towards and beyond the Galapagos Islands in 1959, *Nor. Hvalfangst-Tid.* **51**:265–287.

Clarke, R., Aguayo, L. A., and Paliza, O., 1968, Sperm whales of the Southeast Pacific. Part I: Introduction. Part II: Size range, external characters and teeth, *Hvalradets Skr.* **51**:1–80.

Deacon, G.E.R., 1937, The hydrology of the southern ocean, *Discovery Rep.* **15**:1–124.

Douglas-Hamilton, I., 1973, On the ecology and behaviour of the Lake Manyara elephants, *E. Afr. Wildl. J.* **11**:401–403.

Eberhardt, R. L., and Norris, K. S., 1964, Observations of newborn Pacific gray whales on Mexican calving grounds, *J. Mammal.* **45**:88–95.
Filippova, J. A., 1972, New data on the squids (Cephalopoda: Oegopsida) from the Scotia Sea (Antarctic), *Malacologia* **11**:391–406.
Fink, B. D., 1959, Observation of porpoise predation on a school of Pacific sardines, *Calif. Fish Game* **45**:216–217.
Gambell, R., 1967, Seasonal movements of sperm whales, *Symp. Zool. Soc. London* **19**:237–254.
Gambell, R., 1968, Aerial observations of sperm whale behaviour based on observations, notes and comments by K. J. Pinkerton, *Nor. Hvalfangst-Tid.* **57**:126–138.
Gambell, R., 1972, Sperm whales off Durban, *Discovery Rep.* **35**:199–358.
Gambell, R., Lockyer, C., and Ross, G. J. B., 1973, Observations on the birth of a sperm whale calf, *S. Afr. J. Sci.* **69**:147–148.
Gaskin, D. E., 1970, Composition of schools of sperm whales *Physeter catodon* Linn. east of New Zealand, *N.Z. J. Mar. Freshwater Res.* **4**:456–471.
Gaskin, D.E., and Cawthorn, M.W., 1973, Sperm whales *(Physeter catodon* L.) in the Cook Strait region of New Zealand: some data on age, growth and mortality, *Norw. J. Zool.* **21**:45–50.
Gilmore, R.M., 1959, On the mass strandings of sperm whales, *Pac. Nat.* **1**:9–16.
Harrison, R. J., 1969, Reproduction and reproductive organs, in: *The Biology of Marine Mammals* (H. T. Andersen, ed.), pp. 253–348, Academic Press, New York.
Ingham, S. E., 1967, Branding elephant seals for life-history studies, *Polar Rec.* **13(85)**:447–449.
Iukov, V. L., 1971, Some data on the feeding of sperm whales in high latitudes of the Antarctic, *Tr. Atl. Nauchno-Issled. Inst. Rybn. Khoz. Okeanogr.* **39**:54–59.
Ivanov, A. P., 1972, Small sperm whales in the Antarctic, *Tr. Vses. Nauchno-Issled. Inst. Morsk. Rybn. Khoz. Okeanogr.* **90**:182–183.
Ivashin, M. V., 1967, Whale globe-trotter, *Priroda (Moscow)* **1967(8)**:105–107.
Ivashin, M V., and Budylenko, G. A., 1970, Female sperm whales in the Antarctic? *Priroda (Moscow)* **1970(2)**:103–104.
Johnson, A. M., 1968, Annual mortality of territorial male fur seals and its management significance, *J. Wildl. Manage.* **32**:94–99.
Jonsgård, Å., 1960, On the stocks of sperm whales *(Physeter catodon)* in the Antarctic, *Nor. Hvalfangst-Tid.* **49**:289–299.
Kanwisher, J., and Sundnes, G., 1966, Thermal regulation in cetaceans, in: *Whales, Dolphins and Porpoises* (K. S. Norris, ed.), pp. 397–409, University of California Press, Berkeley.
Kenyon, K. W., Scheffer, V. B., and Chapman, D. G., 1954, A population study of the Alaska fur-seal herd, *U.S. Fish Wildl. Serv. Spec. Sci. Rep. Wildl.* **12**:1–77.
Kleinenberg, S. E., Yablokov, A. V., Bel'kovich, B. M., and Tarasevich, M. N., 1964, Beluga *(Delphinapterus leucas)* —Investigation of the species, Akad. Nauk. SSSR; translation by Israel program for Scientific Translations, 1969, 376 pp.
Knox, G. A., 1970, Antarctic marine ecosystems, in: *Antarctic Ecology* (M. W. Holdgate, ed.), Vol. 1, pp. 69–96, Academic Press, London.
Korabelnikov, L. V., 1959, The diet of sperm whales in the Antarctic seas, *Priroda (Moscow)* **1959(3)**:103–104.
Laws, R. M., 1956, The elephant seal *(Mirounga leonina* Linn.). III. The physiology of reproduction, *F.I.D.S. Sci. Rep.* **15**:1–66.
Laws, R. M., 1960, The southern elephant seal *(Mirounga leonina* Linn.) at South Georgia, *Nor. Hvalfangst-Tid.* **49**:466–476, 520–542.
Laws, R. M., Parker, I.S.C., and Johnstone, R. C. B., 1975, *Elephants and their Habitats. The Ecology of Elephants in North Bunyoro, Uganda*, Clarendon Press, Oxford, 376 pp.
Lockyer, C., 1976, Body weights of some species of large whales, *J. Cons. Int. Explor. Mer* **36**:259–273.
Masaki, Y., Wada, S., and Ohsumi, S., (unpublished MS), Preliminary report on investigation of

sperm whale schools off the coast of Japan, Report prepared for a special meeting of the scientific committee of the International Whaling Commission, Parksville, B.C., Canada, May 3–10, 1972.

Matthews, L. H., 1938, The sperm whale, *Physeter catodon, Discovery Rep.* **17**:93–168.

Mead, J. G., 1975, Anatomy of the external nasal passages and facial complex in the Delphinidae (Mammalia: Cetacea), *Smithson. Contrib. Zool.* **207**:1–72.

McNab, B. K., 1971, On the ecological significance of Bergmann's rule, *Ecology* **52**:845–854.

Morozov, D. A., 1970, Dolphins hunting, *Rybn. Khoz.* **46(5)**:16–17.

Murphy, R. C., 1947, *Logbook for Grace*, MacMillan, New York.

Nishiwaki, M., 1962, Aerial photographs showing sperm whales' interesting habits, *Nor. Hvalfangst-Tid.* **51**:395–398.

Nishiwaki, M., Hibiya, T., and Ohsumi (Kimura), S., 1958, Age study of sperm whale based on reading of tooth laminations, *Sci. Rep. Whales Res. Inst. Tokyo* **13**:135–153.

Nishiwaki, M., Ohsumi, S., and Maeda, Y., 1963, Change of form in the sperm whale accompanied with growth, *Sci. Rep. Whales Res. Inst. Tokyo* **17**:1–14.

Norris, K. S., and Harvey, G. W., 1972, A theory for the function of the spermaceti organ of the sperm whale *(Physeter catodon* L.), in: *Animal Orientation and Navigation* (S. R. Galler, K. Schmidt-Koenig, G. J. Jacobs, and R. E. Belleville, eds.), pp. 397–417, National Aeronautics and Space Administration, Washington, D.C.

Ohsumi, S., 1965, Reproduction of the sperm whale in the northwest Pacific, *Sci. Rep. Whales Res. Inst. Tokyo* **19**:1–35.

Ohsumi, S., 1966, Sexual segregation of the sperm whale in the North Pacific, *Sci. Rep. Whales Res. Inst. Tokyo* **20**:1–16.

Ohsumi, S., 1971, Some investigations on the school structure of sperm whale, *Sci. Rep. Whales Res. Inst. Tokyo* **23**:1–25.

Ohsumi, S., and Masaki, Y., 1975, Japanese whale marking in the North Pacific, 1963–1972, *Bull. Far Seas Fish. Res. Lab.* **12**:171–219.

Ohsumi, S., and Nasu, K., (unpublished), Range of habitat of the female sperm whale with reference to the oceanographic structure, Report prepared for a special meeting of the scientific committee of the International Whaling Commission, Honolulu, March 13–24, 1970.

Ohsumi, S., Kasuya, T., and Nishiwaki, M., 1963, Accumulation rate of dentinal growth layers in the maxillary tooth of the sperm whale, *Sci. Rep. Whales Res. Inst. Tokyo* **17**:15–35.

Omura, H., 1950, On the body weight of sperm and sei whales located in the adjacent waters of Japan, *Sci. Rep. Whales Res. Inst. Tokyo* **4**:1–13.

Pervushin, A. S., 1966, Observation on delivery of sperm whales, *Zool. Zh.* **45**:1892–1893.

Peterson, R. S., 1968, Social behavior in pinnipeds with particular reference to the northern fur seal, in: *The Behavior and Physiology of Pinnipeds* (R. J. Harrison, R. C. Hubbard, R. S. Peterson, C. E. Rice, and R. J. Schusterman, eds.), pp. 3–53, Appleton-Century-Crofts, New York.

Ralls, K., 1976, Mammals in which females are larger than males, *Quart. Rev. Biol.* **51**:245–276.

Robson, F. D., and Van Bree, P. J. H., 1971, Some remarks on a mass stranding of sperm whales, *Physeter macrocephalus* Linnaeus 1758, near Gisborne, New Zealand, on March 18, 1970, *Z. Saeugetierkd.* **36**:55–60.

Rosenblum, E. E., 1962, Distribution of sperm whales, *J. Mammal.* **43**:111–112.

Rowe-Rowe, C. T., 1973, Social behaviour in a small blesbok population, *J. S. Afr. Wildl. Manage. Assoc.* **3(2)**:49–52.

Scammon, C. M., 1874, *The Marine Mammals of the Northwestern Coast of North America*, John H. Carmany and Co., San Francisco, 319 pp.

Schenkkan, E. J., and Purves, P. E., 1973, The comparative anatomy of the nasal tract and the function of the spermaceti organ in the Physeteridae (Mammalia, Odontoceti), *Bijdr. Dierkd.* **43(1)**:93–112.

Schubert, K., 1951, Das Pottwalvorkommen an der Peruküste, *Fishereiwelt* **3(8)**:130–131.

Sergeant, D. E., 1962, The biology of the pilot or pothead whale *Globicephala melaena* (Traill) in Newfoundland waters, *Fish Res. Board. Can. Bull.* **132**:1-84.

Sergeant, D. E., 1969, Feeding rates of cetacea, *Fiskeridir. Skr. Ser. Havunders.* **15**:246-258.

Sergeant, D. E., 1973, Environment and reproduction in seals, *J. Reprod. Fertil. Suppl.* **19**:555-561.

Simpson, G. G., Roe, A., and Lewontin, R. C., 1960, *Quantitative Zoology*, Harcourt, Brace and Co., New York.

Stephenson, A. B., 1975, Sperm whales stranded at Muriwai Beach, New Zealand, *N.Z. J. Mar. Freshwater Res.* **9**:299-304.

Tarasevich, M. N., 1967, On the structure of cetacean groupings 1. Structure of the groupings of *Physeter catodon* males, *Zool. Zh.* **46**:124-131.

Tarasevich, M. N., 1968, Dependence of distribution of the sperm whale males upon the character of feeding, *Zool. Zh.* **47**:1683-1688.

Tayler, C. K., and Saayman, G. S., 1972, The social organisation and behaviour of dolphins *(Tursiops aduncus)* and baboons *(Papio ursinus)*: some comparisons and assessments, *Ann. Cape Prov. Mus. Nat. Hist.* **9**:11-49.

Tomilin, A. G., 1957, Whales, in: *Mammals of the USSR and Adjacent Countries*, (V. G. Heptner, ed.), Vol. 9, Akad. Nauk SSSR, Moskva; translation by Israel Program for Scientific Translations, 1967, 717 pp.

Tormosov, D. D., 1970, The geographical distribution of the various physiological and age groups of sperm whales in the southern hemisphere, *Tr. Atl. Nauchno-Issled. Inst. Rybn. Khoz. Okeanogr.* **29**:41-52.

Tormosov, D. D., 1975, Ecologo-physiological basis of different structures of sperm whale concentrations, *Morskie Mlekopitajushie*, Kiev: 127-129.

Tormosov, D. D., and Sazhinov, E. G., 1974, Nuptial behaviour in *Physeter catodon*, *Zool. Zh.* **53**:1105-1106.

von Richter, W., 1971, Observations on the biology and ecology of the black wildebeest *(Connochaetes gnou)*, *J. S. Afr. Wildl. Manage. Assoc.* **1(1)**:3-16.

Yablokov, A. V., (unpublished), Proposal for agenda item 6.8.3., Report prepared for a special meeting of the scientific committee of the International Whaling Commission, La Jolla, December 3-13, 1974.

Zenkovich, B. A., 1962, Sea mammals as observed by the round-the-world expedition of the Academy of Sciences of the USSR in 1957/58, *Nor. Hvalfangst-Tid.* **51**:198-210.

~ Chapter 8

BEHAVIOR AND SIGNIFICANCE OF ENTRAPPED BALEEN WHALES

Peter Beamish

Department of the Environment
Fisheries and Marine Service
Marine Ecology Laboratory
Bedford Institute of Oceanography
Dartmouth, Nova Scotia B2Y 4A2

I. INTRODUCTION

Every year a large number of baleen whales are entrapped in ice and in fishermen's nets. Many escape on their own. Others die.

In this article we examine the causes, experimental methods, animal behavior, and significance of eastern Canadian entrapments, focusing on a 1974 ice entrapment of a blue whale and a 1975 net entrapment of a humpback whale. Acoustic behavior is discussed relative to the hypothesis of baleen whale echolocation. Results of measuring, tagging, and release are presented.

II. ICE ENTRAPMENT

A. Causes

Large whales often feed along ice fronts. On March 18, 1974, whales were seen in a north–south lead of open water off Cape Ray, southwest Newfoundland, Canada. That night a storm from the southwest sealed the open channel on the south with a field of ice, and westerly winds progressively moved the remaining open water and the whales towards the land. On Wednesday three entrapped blue whales *(Balaenoptera musculus)* were observed from the cliffs south of St. Andrews, Newfoundland.

Fig. 1. (a and b) A 22-m live blue whale temporarily entrapped by ice on the southwest coast of Newfoundland, March 21, 1974. The animal escaped before sunrise the following morning.

Other animals have been reported entrapped, often temporarily, in a field of ice away from land. Once again, the cause is considered to be related to unusual meteorological conditions associated with ice movements closing an open lead in the ice.

B. Experimental Methods

On Thursday, March 21, one of the three whales, measuring 22 m in length, rested alive and apparently in a reasonable state of health in 2.5 m of water, entrapped by surface ice 20–50 cm thick (Fig. 1). A large grounded iceberg to the right of the whale offered a stable platform from which to work. During the evening, gale-force offshore winds removed the ice and the whale escaped.

Sounds were recorded using three calibrated hydrophones and a Pemco (model 110) instrumentation tape recorder. Hydrophone 1 (Celesco LC32) was connected to an Applied Cybernetics preamplifier (LA460). Hydrophones 2 and 3 (Clevite CHIA sensors) were connected to RC high-pass filters and preamplifiers (Clevite CE-25) and then to Applied Cybernetics preamplifiers (LA260). The system frequency response was flat (± 3 dB) for hydrophone 1 from 20 Hz to 80 kHz, and for hydrophones 2 and 3 from 200 Hz to 80 kHz. Calibration was accomplished by injecting a known signal at the input of the filters and measuring the recorded output.

We began recording with hydrophones 1 and 2 at 0942 hr. Positions of the hydrophones (± 0.1 m) are shown in Figs. 1a and 2a. The cables were wrapped around pieces of driftwood placed on the ice in order to maintain the hydrophone sensors at mid-depth. Recording from hydrophone 3 commenced at 1145 hr.

The animal was resting along a sandy, gently sloping beach with the anterior tip of its rostrum at mid-depth. The acoustic regime was therefore one of a 2.5-m-thick uniform layer of clear, approximately isothermal sea water, bounded below by sand, above by surface ice, and at one corner by a grounded iceberg.

Between 1226 and 1233 hr a 1-cm-radius lead weight was repeatedly dropped through a hole in the ice. Recording was terminated at 1530 hr because of increased wind velocity.

C. Results

1. Acoustic Behavior

Seven sequences of short repetitive pulses were recorded (Table I). In each case the animal exhaled and inhaled 5–13 sec before the sounds were emitted.

Figures 2a, 3b, and 3c illustrate the marked differences in frequency

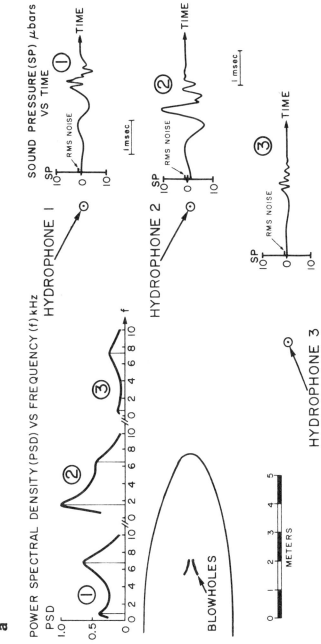

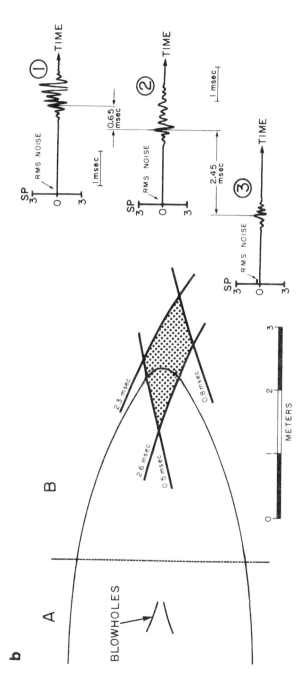

Fig. 2. (a) Positions of the hydrophones relative to the blue whale, unfiltered sound pressure at the receivers vs. time curves for each hydrophone and corresponding power spectral density vs. frequency curves. (b) Relative times of arrival of the sounds, after filtering with a pass-band from 6 to 8 kHz, are useful to locate the dotted area where higher-frequency sounds are being projected into the water.

Table I. Pulse Sequences Recorded from a Blue Whale *(Balaenoptera musculus)*

Sound no.	Time	No. of pulses	Mean pulse interval (msec)	Maximum source level in db at 1 m (re 1 μbar)
1	1023:04	11	19.5	31.2
2	1149:50	27	19.4	30.5
3	1226:47	9	15.0	27.0
4	1228:28	7	19.5	26.5
5	1229:58	18	19.7	32.4
6	1236:20	19	20.9	31.6
7	1318:08	12	18.3	29.7

characteristics of the pulses as observed at different directions of transmission. Lower-frequency sounds were received directly ahead of the animal, higher frequencies to the sides. This may have been caused by multipath interference. Number of pulses per sequence and interval between sequences varied with time (Table I), while the shape of the power spectral density curves (Fig. 2a) varied only with position of the receiver and not with time. This remarkable repetition of the frequency distribution within each sound pulse was measured over a time period of more than 175 min (103 sounds).

Ice noise, which occurs on the recordings, was not the cause of the aforementioned pulse sequences because (1) it was always of relatively lower sound pressure level; (2) it was isotropic, appearing with similar frequency characteristics on each hydrophone; and (3) its sound source location was never near the source location of the pulse sequences. At the times of the recorded pulses neither the animal, the hydrophone cables, nor the two observers were moving in any way that could make mechanical sounds similar to the recorded pulses.

An experiment intended to enhance the possibility of sound production by the animal was performed between 1226 and 1233 hr. A 1-cm-radius lead weight was dropped four times through a hole in the ice approximately 4 m directly ahead of the whale. Three of the seven recorded bursts of pulses, including the loudest sounds (Table I and Fig. 3), were recorded during these 7 min. A fourth burst occurred approximately 3 min later. Sounds on all hydrophones were received each time the weight splashed into the water. It was possible from this experiment to obtain a horizontal acoustic source location of the weight by measuring the pulse arrival-time differences between sets of hydrophones. All such acoustic locations were within 0.7 m of the actual location. This was useful in testing the system for reliability of source location.

The relative times of arrival of the short repetitive pulses, at the three hydrophones, are shown in Figs. 2, 3c, and 3d. By filtering the signals with a pass-band from 6 to 8 kHz (Figs. 2b and 3d), the arrival-time differences measured from all signals were 0.65 ± 0.01 msec between hydrophones 1 and 2 and 2.45 ± 0.01 msec between hydrophones 2 and 3. Filters used were Krohn-Hite (model 3550) with attenuation slopes of 24 dB/octave.

Defining the exact start of a pulse is often a major source of error. However, in the filtered whale sounds, the amplitude of the third cycle rises significantly above the amplitude of the second and, therefore, the maximum pressure at this point was chosen as a reference point for time measurement. To best illustrate this characteristic in an oscilloscope photograph (Fig. 3d), the tape was played backwards so that the oscilloscope could be more consistently triggered and the initial stages of the pulses could be more clearly viewed. Ringing of the filters causes stretching of the signals in time, and this was determined for the 6- to 8-kHz pass band to be less than 0.1 msec for a 4-cycle sine wave calibration pulse. This source of error, which could influence arrival-time differences between pulses, is markedly reduced when two hydrophone signals are passed through

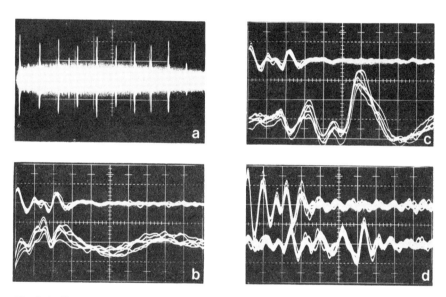

Fig. 3. Oscilloscope photographs of sound sequence number 5 (Table I). For photographs b, c, and d, the original tape recording was played backward so that the oscilloscope could be more consistently triggered (chop-mode triggered by upper trace) allowing the initial stages of the pulses to be more clearly viewed. (a) The horizontal sweep speed is 20 msec/major division. (b) The horizontal sweep speed is 0.2 msec/major division. The upper trace shows the superposition of successive clicks from hydrophone 1 after filtering with a pass-band from 3 to 20 kHz in order to obtain consistent triggering for both traces. The lower trace shows the identical clicks after filtering with a pass-band from 0.5 to 20 kHz. (c) This photograph is identical to Fig. 3(b) except that the lower trace shows clicks recorded from hydrophone 2, after filtering with a pass-band from 0.5 to 20 kHz. (d) The horizontal sweep is 0.2 msec/major division. The upper and lower traces show superposition of successive clicks from hydrophones 1 and 2, respectively, after both signals were passed through identical filters with pass-bands from 6 to 8 kHz.

identical filters (Fig. 3d), so that time stretching is approximately the same for each signal. Playing the tape in the same direction as it was recorded yielded identical estimates of arrival-time differences.

Another possible error in picking the start of a 7-kHz pulse is the time corresponding to a cycle, or approximately 0.14 msec. With this number added to the previously mentioned standard deviations of 0.01 msec, hyperbolae have been drawn to indicate a probable sound source location (Fig. 2b). The curve marked 0.5-msec, for example, represents the locus of positions for which a sound source would cause a 0.5-msec arrival-time difference on hydrophones 1 and 2. Differences of 0.65 ± 0.15 msec and 2.45 ± 0.15 msec would be obtained from a sound source lying within the dotted area. It is significant to note that a sound made in the position of the blowhole (which was submerged when all of the sounds reported in Table I were recorded) would cause an arrival-time difference on hydrophones 1 and 2 of approximately 0.4 msec, a significantly shorter time than measured (Figs. 3c and d).

2. Other Behavior

Observations of eye movement revealed that the animal was continuously scanning forward and aft with an approximate total angular eye rotation of 30° and a period for one complete scan, from forward to aft and return, of approximately 1.5 sec. These estimates were made independently by two observers, during a time period of approximately 10 min, directly following the acoustic recording session.

D. Discussion

These data provide new evidence that (1) a blue whale was the source of short repetitive sounds and (2) that a higher-frequency component of the sounds was apparently projected into the water from the anterior part of the head, some distance from any probable sound source location.

It has been difficult to identify with certainty the source of sounds that might emanate from free-ranging baleen whales; only three such recordings have been reported for blue whales (Poulter, 1968; Beamish and Mitchell, 1971; Cummings and Thompson, 1971). In the first of these, any evidence that the sounds were emitted by a whale, such as a correlation of the position of the animal with recorded acoustic parameters, was lacking. In the latter two reports, signal amplitude increased as a blue whale approached the hydrophone and decreased thereafter, and trained observers reported that no other cetaceans were seen in the vicinity. Other recordings have not been reported in the literature, partly because the identification of the species has been uncertain. In addition, data are sparse because blue whales are rare; few chances occur to study these magnificent animals.

The evidence suggests that a higher-frequency component of the sounds (6–8 kHz) was projected into the water from the forward part of the head (Fig. 2b). The front half of the rostrum (upper jaw) is devoid of air cavities and movable muscles, thus it is unlikely that sounds are produced there (region B in Fig. 2b). On the other hand, a large source of acoustic power exists in the process of moving air through passages associated with the respiratory system (at or posterior to region A in Fig. 2b), and therefore it can be suggested that the animals may have a waveguide to carry the acoustic energy forward before projection into the water. In addition, the repetition of the shape of the unfiltered signals and the power spectral density curves (Fig. 2a) at three different locations, over a time period of more than 175 min, is very atypical of sounds emitted directly from a biological source. One might expect a temporal variation greater than was measured, for sounds from a biological source that did not make use of a passive directional filter. On the other hand, variable sounds produced in region A (Fig. 2b), and then passed along a waveguide of fixed geometry, would more likely emerge with the aforementioned measured frequency–time characteristics.

These separate results give complementary support to the hypothesis that there exists an acoustic waveguide in region B (Fig. 2b). Theoretical work in progress further supports this hypothesis. We therefore suggest that the long, tapered head of some mysticete whales may contain a directional acoustic antenna, perhaps not found in other animals.

These sounds may have been useful for echolocation. They are short and repetitive (Figs. 3b–d), and of amplitude and frequency which may be sufficient for the whale to obtain recognizable echoes from the hydrophones or possibly from the lead weight. Similar sounds have been recorded in the presence of free-ranging blue whales (Poulter, 1968; Beamish and Mitchell, 1971), gray whales (Painter, 1963; Wenz, 1964; Asa-Dorian and Perkins, 1967; Fish et al., 1974), fin whales (Perkins, 1966; Winn and Perkins, personal communication), humpback whales (Winn et al., 1971; Payne, personal communication; this paper), minke whales (Winn and Perkins, 1976; Beamish and Mitchell, 1973) and right whales (Payne, personal communication).

Norris (1969) presented a review of our knowledge of marine mammal echolocation in which he stated "It is highly probable that echolocation in some form is widespread, if not universal, among odontocete cetaceans. . . ." It should not be surprising that this method of detecting food and other targets in the ocean could be employed by the baleen whales as well. There exists, however, no adequate evidence for the hypothesis that baleen whales echolocate, because it is necessary to determine if animal performance depends on the combined processes of transmission, reflection, and reception of the sound energy. Norris presented negative evidence in the case of a gray whale. Certainly the eyes of mysticetes are useful for locating objects underwater.

In support of the hypothesis that some baleen whales echolocate are the recordings of this experiment, particularly the fact that the signals were stronger and more frequent during and directly following the introduction of a lead weight into the water. This change in the acoustic performance of the animal may have been caused by its curiosity about the changing environment. The position, the color, and the size of the lead weight may have caused the animal in this case to use sound energy as a means of possible perception. In further support of the hypothesis that some baleen whales echolocate, there exist the previously mentioned reports of short repetitive underwater sounds received at a time when baleen whales were positively identified. It is somewhat mystifying why more such recordings are not made in the presence of these animals; one explanation may be that their eyes offer sufficient sensory information near the surface of the sea during daylight, the conditions under which most mysticete bioacoustic data are collected.

III. NET ENTRAPMENTS

A. Causes

Large whales often feed among fishermen's nets. On June 23, 1975, a 10.2-m male humpback *(Megaptera novaeangliae)* was found entangled in a fish trap near Branch, Newfoundland (Fig. 4). Events related to the entrapment were as follows. Several days previously, whale movements had been correlated with migrations of capelin *(Mallotus villosis)* northward along a steep shoreline, where a long net had been placed from surface to bottom, at right angles to both shoreline and fish movement. Customarily, the capelin and whales swam through the net, and indeed this particular net had been hauled and repaired frequently for whale damage.

Cod fish *(Gadus morhua)*, presumably chasing the capelin, did not swim through the net but turned, and while swimming along the net entered a door of a large circular trap. The doors of the trap were not wide enough for the flukes of a whale, and presumably they closed around the tailstock of the young humpback after he had made the similar turn. Continuing into the trap, other doors closed around the tail. A subsequent struggle for freedom added further entanglement leading to the observed situation of fourteen lines around the tailstock and a further two restraining the lateral motion of the animal.

Other causes for entrapment have included contact with gill nets which are placed in deeper water away from shorelines, often in long rows. Several humpback whales have been observed entangled and towing the nets at slow speeds. The reduced maneuverability probably leads to an inability to feed and occasionally to exhaustion and drowning.

Fig. 4. A 10.2-m live humpback whale temporarily entrapped by a net near Branch, Newfoundland, June 24–26, 1975. The animal was released on June 26.

B. Experimental Methods

The young humpback was studied from June 23–25, during which time arrangements were made for measuring, tagging, and release. Water depth was approximately 20 m.

Sounds were recorded using seven hydrophones (Clevite CH-17U) and the

Pemco (model 110) recorder. A Clevite seven-channel preamplifier incorporated flashing diodes for overload warning and a 1000-Hz calibration signal oscillator. The system frequency response was flat within 3 dB from 20 Hz to 18 kHz and within 6 dB from 18 to 50 kHz.

The seven hydrophones were placed in a linear array with a horizontal separation of 0.75 m and a depth of 2.0 m. The distance from the array to the whale varied from 1 to 20 m. Objects such as fishing weights and marbles were thrown into the water at ranges of 5–50 m from the whale.

Eight divers observed the whale from beneath the surface of the water. Observations from surface craft using facemask and snorkel were especially useful at the times of measurement and release.

C. Results

1. Acoustic Behavior

Many sounds were recorded, principally when no divers were in the water. A series of clicks (Fig. 5a) is particularly significant in that it was recorded without the presence of other cetaceans and because it is similar to a series of clicks recorded previously in the presence of a free-swimming humpback in the Gulf of St. Lawrence (Fig. 5b). The spectral peaks occur at 2.1 and 2.0 kHz, respectively.

A series of higher frequency clicks (Fig. 5c) with spectral peaks at 8.2 kHz was recorded only once during the three days.

Figures 5d and 6a show a frequently recorded "ratchet" sound which always occurred when the animal was observed to be exhaling underwater (Fig. 7a). Other types of similar pulses were recorded when the animal was not observed exhaling underwater (Figs. 5e, 6b). Airborne noises of exhalation (Watkins, 1967) were barely audible underwater and were recorded only when the hydrophones were a few meters from the animal. The loud recorded sounds associated with exhalation underwater were in fact abruptly terminated (Fig. 6a) as soon as the extremity of the blowholes appeared above the sea surface. The anterior segment of the head always remained submerged. Source levels did not exceed 58 dB at 1 m (ref. 1 dyne/cm^2).

Other pulsed sounds (Figs. 5f, 6c) were uncorrelated with any noticeable behavior. These and additional recorded sounds are similar to components of humpback whale "songs" recorded in more tropical waters (Payne and McVey, 1971; Winn *et al.*, 1971). However, no indication of a "song" was present in any of the data.

2. Other Behavior

The eyes were very active, continuously focusing on the divers. As one moved toward the tail and upon obtaining a position several meters from the tailstock, the animal would rotate its head laterally, presumably to keep the diver within its field of vision. Occasionally, the eyelids would close completely.

Another humpback of approximately the same length was observed on two occasions to circle within 20 m of the entrapped animal (Fig. 7b). During the remainder of the daytime, this "stand-by" whale remained several kilometers away, apparently feeding.

Upon release, the animal (now measured, tagged, and named Nicholas) moved immediately to a position alongside the 12-m fisheries patrol vessel that was assigned to track it. A few moments later he headed south at approximately 3 knots and continued on a straight course for the remaining 2 hr before darkness.

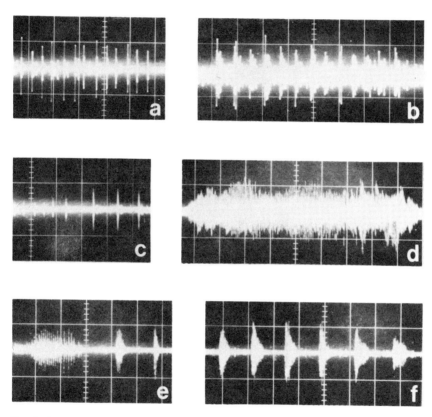

Fig. 5. Oscilloscope photographs of humpback sounds. (a) Clicks of the entrapped animal with horizontal sweep speed of 0.1 sec/major division. (b) Clicks of a free-ranging humpback with horizontal sweep speed of 0.05 sec/major division. (c) Higher-frequency clicks of entrapped animal with horizontal sweep speed of 0.2 sec/major division. (d) Clicks recorded during underwater exhalation (also shown in Fig. 6a). The horizontal sweep speed is 0.5 sec/major division. (e) Clicks recorded without underwater exhalation (also shown in Fig. 6b). The horizontal sweep speed is 0.5 sec/major division. (f) Other typical pulsed sounds (also shown in Fig. 6c). The horizontal sweep speed is 0.5 sec/major division.

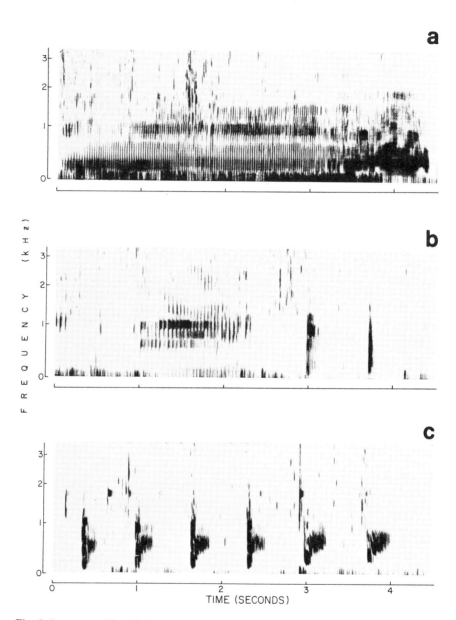

Fig. 6. Sonograms of humpback sounds with an analyzing filter bandwidth of 300 Hz. (a) Clicks recorded during underwater exhalation (also shown in Fig. 5d). (b) Clicks recorded without underwater exhalation (also shown in Fig. 5e). (c) Other typical pulsed sounds (also shown in Fig. 5f).

Fig. 7. (a) Entrapped humpback exhaling underwater which always correlated with sounds represented by Figs. 5d and 6a. (b) The "stand-by" whale.

Fig. 8. (a) Appearance of the tagged whale. (b) The rope and ribbon tag. Four numbered spaghetti fish tags can also be seen.

D. Discussion

These data provide additional evidence that a baleen whale can produce short sharp repetitive sounds suitable for echolocation. In fact, most of the recorded sonic emissions of the whale were pulsed, such that substantial echoes would occur from the surface and sea floor before the emission of the next pulse. The pulse repetition rate shown in Fig. 6a can be seen to have increased as the animal approached the sea surface (perhaps this is related to the decreasing time needed for a surface echo to return to the animal's ears). Once again, this is not proof of echolocation, but it is additional evidence to support the contention that baleen whales might echolocate.

The correlation of an abrupt decrease in underwater sound, and the emergence of the blowholes above the sea surface, indicates that at least one underwater sound source lies very near the dorsal extremity of the blowholes and not near the anterior segment of the head which remained submerged (see frontispiece). These data support the waveguide hypothesis as developed in relation to regions A and B in Fig. 2b.

The pulsed sound produced during underwater exhalation (Figs. 5d, 6a) and just prior to exhalation in air is similar to a part of a humpback song (sung in tropical waters) called the "surface ratchet" by Winn *et al.* (1971). Therefore, our observations are additionally useful in supporting their contention that this sound defines the termination of one song and the beginning of the next.

Another interesting observation is that the number of pulses recorded during the time when the animal was exhaling underwater counted in the hundreds (Fig. 6a), yet the number of air bubbles extruding from the blowholes at this time counted only in the tens. Therefore, if a pulse is produced by air movement, air storage must take place before exhalation.

Very little is known about the usefulness of the eyes of baleen whales. The eyes of this animal, however, were extremely useful in keeping close watch on a diver. Head and body lateral motion enhanced the ability of the eyes to follow a diver moving towards the tail, presumably when reaching the edge of the field of vision. This lateral head motion occurred when the diver was approximately 2 m from the tailstock.

IV. TECHNIQUES OF MEASURING, TAGGING, AND RELEASE

Measuring entrapped whales is important for determination of growth rate, presuming the animal can be identified and again measured at a later date. The technique employed for the entrapped humpback was to attach a line as near as possible to the notch in the flukes (see frontispiece) and then to mark the line with an inserted wire opposite the anterior portion of the upper jaw. The back profile of the animal is not flat, therefore the line was passed along the dorsal ridge leading to a length measurement approximately 5 cm longer than a horizontal distance from fluke notch to the anterior rostral edge. This must be taken into account if the animal is again measured by aerial photography adjacent to an object of known length.

The tag was designed to be visually apparent from a surface vessel or an aircraft and to be of minimum drag and distraction to the animal. Two pieces of yellow polypropylene line were spliced and interwoven with international orange ribbons and white spaghetti numbered tags (Fig. 8). They were fastened on either side of the dorsal fin using a single ¼-in. stainless-steel bolt.

A fiberglass radio tag that had been designed to fit neatly onto the dorsal fin of a blue, fin, sei, or minke whale could not be modified for the irregular shape of the humpback dorsal fin in the time available. A mold for such a tag should really be made after the entrapment in order to ensure a streamlined configuration. A radio tag must fit tightly against the dorsal ridge of the animal in order to prevent any vertical drag component. The force of horizontal drag must be entirely borne by the leading edge of the dorsal fin and as evenly distributed as possible.

Release of an entrapped whale is far from simple. The most critical aspect is that when strain is placed on a number of lines by a whale, all such lines must be cut simultaneously. Failure to cut a final line before the full power of the animal is exerted may result in the whale towing quantities of fishing gear which could well surround the vessels and personnel involved in the release operation. If the net fails to break cleanly away, then the animal would likely not survive with its added burden in tow.

A solution to this problem involves moving the largest vessel behind the whale and tying a sufficiently strong rope from this vessel around the tailstock in the form of a slip knot. A method of quickly releasing or cutting this rope should then be arranged. The next step is to cut away lines and net which are not restraining the animal. In the case of the humpback, only four of 16 lines were necessary to hold the whale; these were all of new ⅝-in. dacron. Next, one should carefully observe the periodicity at which strain comes on the main holding lines; this may change before and after a blow. The animal should then be tired by any frightening action, such as emission of loud high-frequency sonic pulses underwater. The objective is to cause, following the activity, a momentary release of tension on the restraining lines. At this time, all lines should be cut simultaneously, freeing the animal.

V. SIGNIFICANCE AND SUMMARY

Some entrapped baleen whales die, but with renewed effort and understanding many could be studied, tagged, and released. The rare accessibility of these animals leads to tagging opportunities not paralleled in any of the noncaptive baleen whale research programs. The significance of tagging is enormous. A radio tag permits study at night, prolonged tracking and behavioral studies, and transmission of oceanographic data (Evans, 1974). A visual tag could give repeated daylight sightings indicating migratory patterns, as well as allowing for repeated measurements of the animal, for growth-rate studies.

Entrapped baleen whales give an opportunity for controlled acoustic experiments; previously a major problem has been identifying which animal made which sound. Directivity of sonic emission can be studied by attaching acoustic receivers onto the whale's head as well as using arrays of hydrophones at a known orientation to the animal. Most important of all are experiments to determine if the animals can echolocate objects underwater and, if so, at what limits of target size and range. If baleen whales have a biological sonar system, then the acoustic properties of the signals used for food-finding could be useful for man to study zoogeography in the oceans.

In summary, this paper presents new evidence that baleen whales can produce short, sharp, repetitive sounds suitable for echolocation. The blue whale recordings indicate that higher-frequency sounds are apparently projected into the water from the anterior part of the animal. The humpback recordings associate movement of air with an underwater source of sound, which source is apparently located very near the dorsal extremities of the blowholes, several meters from the anterior part of the animal. Together, these results support the hypothesis that the long tapered head of baleen whales acts as an acoustic waveguide.

Entrapped baleen whales offer opportunities to further test this hypothesis, to test for echolocation ability, and to track tagged whales in the natural environment. We must then learn to retrieve the quantities of information available from more continuous observations of the natural animal behavior.

ACKNOWLEDGMENTS

Thanks are due to David Prentiss, Duncan Findleson, Scott Krauss, Harry Mangalam, Colin Craig, and Ernest O'Rourke for valuable assistance in the field experiments, and to Dr. R.G. Busnel and Albin Dziedzic, Laboratoire de Physiologie Acoustique, France, and David Heffler, Atlantic Geoscience Center, for aid in signal processing.

REFERENCES

Asa-Dorian, P. V., and Perkins, P. J., 1967, The controversial production of sound by the California gray whale, *Eschrichtius gibbosus, Nor. Hvalfangst-Tid.* **56**:74–77.
Beamish, P., and Mitchell, E., 1971, Ultrasonic sounds recorded in the presence of a blue whale *Balaenoptera musculus, Deep-Sea Res.* **18**:803–809.
Beamish, P., and Mitchell, E., 1973, Short pulse length audio frequency sounds recorded in the presence of a minke whale (*Balaenoptera acutorostrata*), *Deep-Sea Res.* **20**:375–386.
Cummings, W. C., and Thompson, P. O., 1971, Underwater sounds from the blue whale, *Balaenoptera musculus, J. Acoust. Soc. Am.* **50**:1193–1198.
Evans, W. E., 1974, Telemetering of temperature and depth data from a free ranging yearling California gray whale, *Eschrichtius robustus, Mar. Fish. Rev.* **36(4)**:52–58.
Fish, J. F., Sumich, J. L., and Lingle, G. L., 1974, Sounds produced by the gray whale, *Eschrichtius robustus, Mar. Fish. Rev.* **36(4)**:38–45.
Norris, K. S., 1969, The echolocation of marine mammals, in: *The Biology of Marine Mammals* (S. Anderson, ed.), pp. 391–423, Academic Press, New York.
Painter, D. W. II, 1963, Ambient noise in a coastal lagoon, *J. Acoust. Soc. Am.* **35**:1458–1459 (L).
Payne, R. S., and McVey, S., 1971, Songs of humpback whales, *Science* **173**:587–597.
Perkins, P. J., 1966, Communication sounds of finback whales, *Nor. Hvalfangst-Tid.* **55**:199–200.
Poulter, T. C., 1968, Marine mammals, in: *Animal Communication, Techniques of Study and Results of Research* (T. A. Sebeok, ed.), pp. 405–465, Indiana Univ. Press, Bloomington.
Watkins, W. A., 1967, Air-borne sounds of the humpback whale, *Megaptera novaeangliae, J. Mammal.* **48(4)**:573–578.
Wenz, G. M., 1964, Curious noises and the sonic environment in the ocean, in: *Marine Bio-acoustics* (W. N. Tavolga, ed.), pp. 101–123, Pergamon Press, New York.
Winn, H. E., and Perkins, P. J., 1976, Distribution and sounds of the minke whale, with a review of mysticete sounds, *Cetology* **19**:1–12.
Winn, H. E., Perkins, P. J., and Poulter, T. C., 1971, Sounds of the humpback whale, Proc. 7th Ann. Conf. Bio. Sonar and Diving Mammals, pp. 39–52.

Chapter 9

THE VOCAL AND BEHAVIORAL REACTIONS OF THE BELUGA, *Delphinapterus leucas*, TO PLAYBACK OF ITS SOUNDS

David W. Morgan

Aquatic Behavior Laboratory
Department of Biology
University of Notre Dame
Notre Dame, Indiana 46556

I. INTRODUCTION

The playback of natural sounds has been used as an experimental technique with diverse groups of animals and with varied degrees of success to elucidate the biological significance of the sounds. Additionally, the playback of pure tones has been used to investigate which parameters of an acoustic signal are directly concerned with communication. Such parameters have included frequency limits of hearing, effectiveness of auditory masking, direction-finding and frequency discrimination capabilities, and intensity limits of frequency detection (audiograms).

Playback experiments were first used with cetaceans in 1952 to determine the upper limits of hearing of the bottlenose porpoise, *Tursiops truncatus*. Kellogg and Kohler (1952) and Kellogg (1953) determined that the upper limit of hearing in this species reached at least to 80 kHz, and this limit was further extended to 120 kHz by Schevill and Lawrence (1953) and to 150 kHz by Johnson (1967), all of whom used playback of pure tones as their experimental technique.

Pure-tone playbacks were also used by Dudok van Heel (1959) to determine auditory direction finding in *Phocoena phocoena* and by Johnson (1968) to measure masked frequency thresholds in *T. truncatus*. Other uses of this technique have included determinations of audiograms for several species, for example, *Inia geoffrensis* (Jacobs and Hall, 1972) and *Orcinus orca* (Hall and Johnson, 1972), and of auditory frequency discrimination limens in *T. truncatus* (Jacobs, 1972).

Playback of the sounds of a conspecific animal has been used since 1961 as a means of demonstrating acoustical exchanges between dolphins. In that year, Lilly and Miller (1961) showed that acoustic stimuli from one *T. truncatus* immediately elicited whistles and click trains from an isolated conspecific. Lang and Smith (1965) showed that an isolated *T. truncatus* would respond to the sounds of a conspecific with response varying according to the sound type being played back, suggesting that different sounds had differing significance. Dreher (1966) also found varied vocal and behavioral reactions to playback of six different whistle contours of *T. truncatus*. Caldwell *et al.* (1972b) found that a *T. truncatus* could discriminate between eight conspecifics on the basis of a wide assortment of their whistle emissions, and concluded that certain whistles were specific to the individual producing them. The existence of such "signature whistles" had also been suggested by Lang and Smith (1965).

A third type of playback experiment has been the playback of sounds of one cetacean species to another. Fish and Vania (1971) used playbacks of the sounds of the killer whale *(Orcinus orca)* to keep belugas from entering the Kvichak River in Alaska during the salmon spawning run, and Cummings and Thompson (1971) used these sounds to alter the behavior of the gray whale *(Eschrichtius robustus)* during its southward migration past California. In both cases the animals responded with a flight reaction. Caldwell *et al.* (1972a) found that *T. truncatus* was able to discriminate between individuals of *Delphinus delphis* solely on the basis of the whistles played back. Davies (1962), carrying this type of work even further, played back killer whale sounds to the Zambezi River shark *(Carcharhinus zambezensis)* and found that the largest of five sharks was disturbed by the sounds, swimming around the tank at greatly increased speed.

Thus, sound playbacks are seen to be a powerful tool in the investigation of many aspects of cetacean sound. However, the playback of conspecific sounds to animals in captivity and in the field for the purposes of correlating behavior and vocal emissions, and determining the significance of sounds, has hardly been utilized. As mentioned above, Lilly and Miller (1961), Lang and Smith (1965), Dreher (1966), and Caldwell *et al.* (1972b) have started work in this direction with captive animals, but no results of investigations of this nature with cetaceans in the field have been published. In fact, the only published field work of this type done with any of the marine mammals is that carried out by Watkins and Schevill (1968) with Weddell seals (*Leptonychotes weddelli*) in Antarctica. They found that the seals seemed to respond to playbacks of good fidelity, whereas a playback of poor quality elicited only silent interest or annoyance.

The research described in this report was designed to determine whether or not the conspecific playback technique could be used as an experimental tool for the investigation of acoustic communication by modification of sound emissions and/or behavioral patterns in the beluga, *Delphinapterus leucas* Pallas. The beluga was chosen as the experimental subject because of its availability in captivity and in the field and because of its known ability to produce a wide range of sounds (Schevill and Lawrence, 1949, 1950; Fish and Mowbray, 1962).

II. MATERIALS AND METHODS

The experiments were carried out with both captive and free-ranging belugas. The captive animals were held at the New York Aquarium and included four adult animals: two females from Hudson's Bay (Frances and Ethel), one male from the St. Lawrence River (Blanchon), and one male from Kvichak Bay, Alaska (Alex). Blanchon was the dominant male, and Frances the dominant female.

The field experiments were carried out in the Saguenay River, Quebec Province, Canada, during the summers of 1970 and 1971. At least some of the animals in the herd of 40–60 whales were recognized both years by scars or pits on their dorsal surface. The age composition of the herd was mixed, including very small, dark calves; medium-sized, gray, young animals; and large pure-white adults. Playback experiments were performed at various positions on the river, both when the whales were swimming up or down the river and when they were milling in the quiet bays and estuaries.

At the Aquarium, the whales' sounds were recorded by either a Hydro-Products R-130 or a Chesapeake LF-310 hydrophone, fed through a preamplifier box, and recorded on an Ampex 1260 stereophonic tape recorder. Visual observations were recorded on the second track of the same tape, allowing real-time correlation of behavior and sounds. Playbacks were effected from a Uher 4000-L monaural tape recorder, amplified by a Kudelski-Paudex amplifier and emitted through an LTV University MM-2PPS underwater loudspeaker or a Chesapeake J-9 omidirectional sound transducer. The University loudspeaker was used only for the first series of playbacks at the aquarium. Field recordings were made from a 16-ft Boston whaler using a Uher 4400-S Report Stereo tape recorder, and the playback amplifier used was a Realistic PA-25. Flat frequency response of the system was 50–10,000 Hz, with the possible exception of the University loudspeaker. All recordings were made at 7½ in./sec tape speed.

The captive group was recorded over a period of three days in July 1968. From these recordings, 12 sounds and a control (background tank noise) were chosen as the primary playback sounds. These sounds are designated throughout this report as PBS 1–13 (playback sounds 1–13), and are shown in Fig. 1.

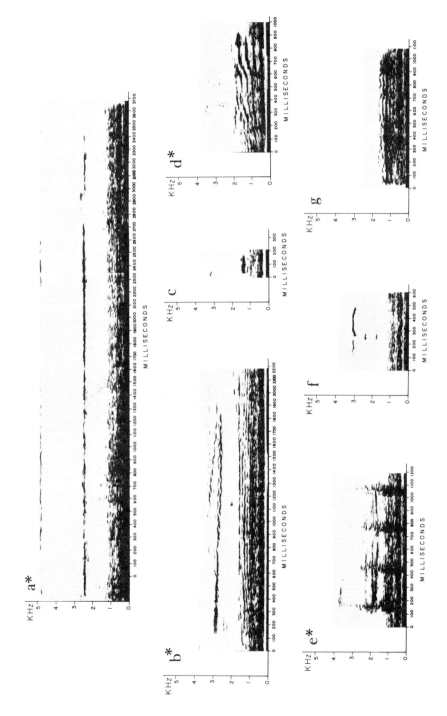

The Vocal and Behavioral Reactions of the Beluga

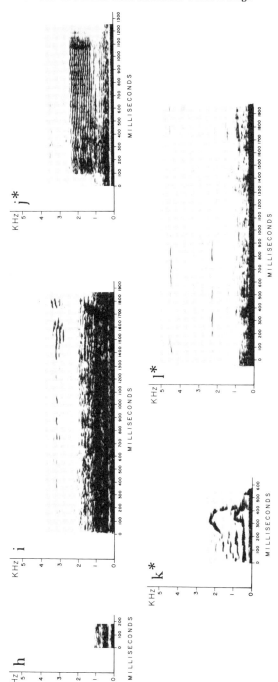

Fig. 1. Typical examples of the sound-types used in the PBS and FldPBS series of playbacks. Analyzer effective bandwidth 60 Hz. (a) harmonic long, loud whistle (harmonic LLW); (b) buzz; (c) bark; (d) squawk (type 1); (e) jaw clap; (f) whistle; (g) squawl; (h) chirp; (i) whinny; (j) blare; (k) squawk (type 2); (l) pure long, loud whistle (pure LLW). Note the high level of background tank noise below 1.5 kHz.

Complete descriptions of the sounds can be found in Morgan (1973). Each sound was placed on a tape loop in combination with a 10-sec piece of blank tape and re-recorded for 3 min, resulting in alternation of the sound and 10 sec of silence.

Each 9-min playback experiment was made up of 3-minute preplayback (PrPb), playback (Pb), and postplayback (PtPb) periods. A time lapse of at least 30 min was allowed between experiments at the aquarium, and the sounds were presented in random order. In the field, playback experiments were performed whenever the opportunity arose.

Six series of playback experiments were performed:

1. Playback of sounds from the captive population to the captive animals. The first experiments involved the PBS collection of sounds to the four captive whales, all in the same tank.

2. Playback of sounds from the captive population to the Saguenay herd. In July and August of 1970, eight of the PBS sounds, plus a 4.8-kHz pure tone, were played back to the Saguenay herd. This collection is designated as FldPBS 1–4; 6–10 (field playback sounds 1–4; 6–10) and is marked with asterisks in Fig. 1.

3. Playback of sounds from the Saguenay animals to the Saguenay herd. Seven of the sounds recorded from the Saguenay herd in July of 1970 were re-recorded in the manner described above. These seven sounds, designated as SagPBS 1–7 (Saguenay playback sounds 1–7) and shown in Fig. 2, were then played back to the Saguenay herd in August of 1970 and July–August of 1971.

4. Playback of sounds from the Saguenay herd to three captive animals. In 1970 Alex was placed in a separate tank leaving Blanchon, Frances, and Ethel in the large beluga pool. In 1971, the SagPBS series was played back to these three animals.

5. Playback of sounds from the Saguenay herd to a single captive animal. The SagPBS series was also played back to Alex alone.

6. Playback of synthesized sounds to a captive animal. Nine synthesized sounds (SynPBS 1–9), based on the characteristics of one sound that uniformly elicited a vocal response from Alex, were played back to him in his isolated pool. The results of this series of playback experiments will be reported in a later paper.

Each sound emission recorded from the animals during the playback experiments was identified and counted. All behavioral observations were correlated with each type of sound emission. Activity in the tank was measured by counting instances of investigation of the hydrophone and the speaker and number of circuits of the tank (only when working with one animal). In the field, dive times and approach toward or withdrawal from the boat were noted.

Fig. 2. Typical examples of the sound types used in the SagPBS series of playbacks. Analyzer effective bandwidth 60 Hz. (a) moan; (b) ping; (c) scream (upper band) and wail (lower band); (d) roar; (e) blat; (f) jaw clap; (g) squeal; (h) cry; (i) Saguenay long, loud whistle (Saguenay LLW).

The Vocal and Behavioral Reactions of the Beluga

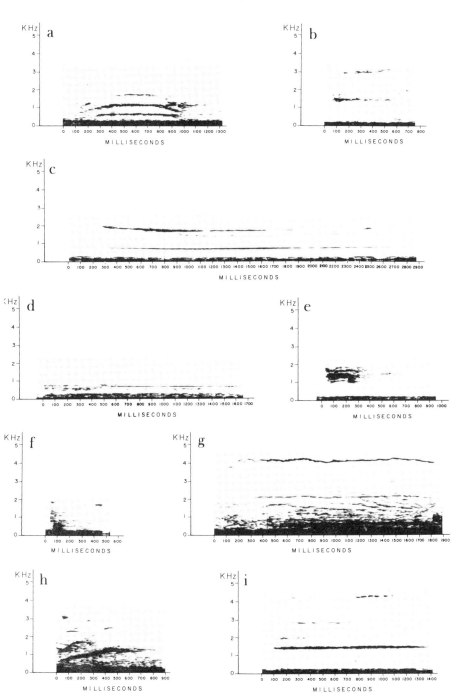

Each sound was graphed on a Kay Electric Company Vibralyzer sound spectrograph and analyzed for frequency, duration, waveform, and harmonic structure. Duration of harmonic long, loud whistles (LLWs) was measured on a B + K Level Recorder (model 2305) at a paper drive speed of 1 cm/sec.

III. RESULTS

A. Playback of Sounds Recorded from the Captive Animals

1. Playback to Captive Animals at the New York Aquarium

Vocal Reaction. The harmonic LLW, the contact sound series, and the whinny elicited significant vocal responses (Table I), while the other ten sounds did not. Since it was usually impossible to determine which whale was making the sounds, total numbers of each sound type were counted. However, the harmonic LLW was generally accompanied by a stream of bubbles from the blowhole of the emitting animal.

The harmonic LLW was originally recorded from Alex in 1968, was emitted only by Alex, and was seldom heard during later recording sessions except in response to the playback. Altogether, it was played back to Alex 27 times: 11 when Frances, Ethel, and Blanchon were also present, and 16 when isolated in his separate tank. During these 27 playbacks Alex increased the number of harmonic LLWs emitted during Pb 23 times, never decreased the number emitted, and made no changes four times ($P < 0.01$; sign test). With regard to PtPb as compared with Pb, there were three further increases, 20 decreases, and four no-changes ($P < 0.01$; sign test). Finally, in comparing PrPb and PtPb, the positive response was significantly carried over into PtPb from Pb ($P < 0.05$; sign test) so that the number of harmonic LLWs remained higher than in PrPb in 15 of the 27 cases, with three repetitions showing decreases and nine showing no change.

Playback of the contact sound series caused an increase in this vocalization series during Pb, but this effect did not carry over into PtPb. Jaw claps and total sounds also decreased in response to the contact sound series between Pb and PtPb. Playback of the whinny elicited a significant decrease in the number of whistles emitted in PtPb as compared with Pb, and the harmonic LLW elicited a decrease in number of whistles emitted in Pb.

Changes in Interest Directed toward Hydrophone and Speaker. Table II shows significant changes of interest directed toward the equipment placed in the tank for all four whales considered together. Interest was expressed in two ways: by orienting the head toward the hydrophone or speaker while remaining in position, or by approaching one of the two and often nudging or biting it.

Table I. Significant Increases and Decreases in Frequency of Emission of Five Sound Types during and after Playback of PBS 1, 3, and 6 to the Four Captive Belugas[a]

PBS	N	Period change	Harmonic LLW			Jaw clap			Contact sound series			Whistle			Totals		
			+	−	0	+	−	0	+	−	0	+	−	0	+	−	0
1 harmonic LLW	11	PrPb-Pb	9	0													
		Pb-PtPb			2[b]												
		PrPb-PtPb	8	0	3[b]												
3 contact sound series	10	PrPb-Pb							6	0							
		Pb-PtPb				0	7	3[c]	0	6	4[c]						
		PrPb-PtPb									4[c]						
6 Whinny	11	PrPb-Pb										0	6	5[c]			
		Pb-PtPb													1	9	0[c]
		PrPb-PtPb										0	6	5[c]			

[a] Key to the symbols used: PBS = number and descriptive name of the sounds played back; N = number of repetitions of each sound played back; + = increase in frequency of emission; − = decrease in frequency of emission; 0 = no change in frequency of emission; PrPb-Pb = changes occurring between preplayback and playback; Pb-PtPb = changes occurring between playback and postplayback; PrPb-PtPb = changes occurring between preplayback and postplayback.
[b] Change is significant at the 0.05 level (sign test: Siegel, 1956).
[c] Change is significant at the 0.01 level (sign test: Siegel, 1956).

Table II. Significant Increases and Decreases of Interest Directed toward Hydrophone and Speaker during and after Playback of Nine PBS Playback Sounds to the Four Captive Belugas[a]

PBS	N	Period change	Orient toward hydrophone			Orient toward speaker			Approach hydrophone			Approach speaker		
			+	−	0	+	−	0	+	−	0	+	−	0
1 Harmonic LLW	11	PrPb-Pb				6	0	5[c]	0	6	5[c]			
		Pb-PtPb				0	6	5[c]						
		PrPb-PtPb												
2 Buzzes	11	PrPb-Pb				7	0	4[c]						
		Pb-PtPb												
		PrPb-PtPb												
4 Buzz and whinny	11	PrPb-Pb				6	0	5[c]				7	0	4[c]
		Pb-Ptb				0	6	5[c]						
		PrPb-PtPb												
6 Whinny	11	PrPb-Pb				6	0	5[c]				8	1	2[c]
		Pb-PtPb				0	5	5[c]				0	8	3[b]
		PrPb-PtPb												

The Vocal and Behavioral Reactions of the Beluga

7	Whistles	10	PrPb-Pb	7	0	3^c
			Pb-PtPb	0	7	3^c
			PrPb-PtPb			
9	Type 1 squawk	11	PrPb-Pb	7	0	4^c
			Pb-PtPb			
			PrPb-PtPb			
10	Jaw clap, buzz and whinny	10	PrPb-Pb	0	6	4^c
			Pb-PtPb			
			PrPb-PtPb			
11	Type 2 squawk	10	PrPb-Pb	6	0	4^c
			Pb-PtPb			
			PrPb-PtPb			
12	Control	9	PrPb-Pb	6	0	3^c
			Pb-PtPb			
			PrPb-PtPb			

^{a} An increase (+) means that more of the four whales evidenced interest in the equipment in Pb or PtPb than had done so in the preceding period. Conversely, a decrease (−) means that fewer whales showed interest. A no-change designation (0) indicates that the same number of whales showed interest in both periods, although different combinations of animals may have been involved. Refer to Table I for explanation of other symbols.
^{b} Change is significant at the 0.05 level.
^{c} Change is significant at the 0.01 level.

There were no significant changes between periods in orientation toward the hydrophone, but orientation toward the speaker increased significantly during playback of seven of the thirteen sounds: buzzes, whinny, buzz-and-whinny, whistles, harmonic LLW, type-1 squawk, and control. Significant decreases from Pb to PtPb were found with harmonic LLW, whinny, buzz-and-whinny, and whistles. Smaller decreases occurred in response to all other playback sounds except the jaw claps and the jaw clap–buzz-and-whinny combination.

Significant decreases of approach toward the hydrophone in Pb occurred only in response to the harmonic LLW and the jaw clap–buzz-and-whinny combination. Approach toward the speaker increased significantly in reaction to only three of the playback sounds (buzz-and-whinny, whinny, and type-2 squawk), although smaller increases were noted to occur in association with all other sounds except the contact sound series, the jaw clap–buzz-and-whinny, and the pure LLW, which were associated with decreases of interest, and the harmonic LLW, which caused no changes. The whinny led to a distinct loss of interest in PtPb.

Table III presents the number of playback sounds, of the entire PBS collection, that elicited a change of interest directed toward either the hydrophone or the speaker. There was no significant change in head orientation toward the hydrophone, but head orientation toward the speaker increased for every playback sound in Pb, and also decreased for all 13 sounds in PtPb to near-PrPb levels. The whales approached the hydrophone significantly less in Pb, but approached the speaker significantly more in Pb, with a decrease in approach once more in PtPb. This demonstrates that an initial orientation toward the speaker by at least some of the whales during Pb was the usual reaction and that interest fell off soon afterward.

Considering the whales individually, the only animal that showed a significant decrease of interest in the hydrophone during Pb was Alex. However, all four showed highly significant increases in interest directed toward the speaker during Pb and equally significant decreases in interest during PtPb. Complete results for the individual whales can be found in Morgan (1973).

In Table IV is shown the total number of times interest in the equipment (hydrophone and speaker) was shown by all four whales considered together, summed over all repetitions of each PBS. The largest increases in approaching the speaker in Pb occurred with five playback sounds which can be grouped into two series: the buzz, the whinny, and the buzz-and-whinny; and the type-1 and type-2 squawks. In all of these cases, interest fell off rapidly in PtPb. Only two playback sounds, the contact sound series and the jaw clap–buzz-and-whinny combination, were associated with a decrease in approaching the sound source in Pb. Both of these playback sounds were made up of a natural grouping of sounds, rather than being one discrete sound. The other six playback sounds elicited only small increases in approaching the speaker.

A comparison of the over-all effect of each of the PBS sounds in eliciting

Table III. Number of the 13 PBS Playback Sounds that Elicited Increases and/or Decreases of Interest Directed toward Hydrophone and Speaker from All Four Whales[a]

Period change	Orient toward hydrophone			Orient toward speaker			Approach hydrophone			Approach speaker		
	+	−	0	+	−	0	+	−	0	+	−	0
PrPb-Pb	2	3	8	13	0	0[b]	2	10	1[c]	11	2	0[c]
Pb-PtPb	3	2	8	0	13	0[b]	7	3	3	2	10	1[c]
PrPb-PtPb	4	2	7	6	2	5	4	7	2	7	4	2

[a] Refer to Table II for a key to the symbols used.
[b] Change is significant at the 0.05 level.
[c] Change is significant at the 0.01 level.

Table IV. Total Number of Times Interest Was Shown by All Four Whales to the Hydrophone and Speaker, Summed over All Repetitions of Each Playback Sound[a]

PBS	N	Orient toward hydrophone			Orient toward speaker			Approach hydrophone			Approach speaker		
		PrPb	Pb	PtPb	PrPb	Pb	PtPb	PrPb	Pb	PtPb	PrPb	Pb	PtPb
Harmonic LLW	11	0	0	0	0	13	1	11	5	10	4	6	5
Buzzes	11	0	0	2	2	19	4	11	4	9	6	17	9
Contact sound series	10	0	0	0	1	14	2	5	3	3	9	7	5
Buzz-and-whinny	11	0	2	1	1	13	1	11	7	12	3	13	7
Jaw claps	8	0	0	0	2	9	2	5	8	6	7	8	5
Whinny	11	1	0	1	2	14	2	8	6	6	1	14	1
Whistles	10	1	0	0	1	15	1	11	9	6	8	9	8
Blare	9	0	0	0	2	13	1	9	7	8	5	6	8
Type-1 squawk	11	1	4	2	0	12	3	8	6	9	6	10	7
Jaw clap–buzz-and-whinny	10	0	0	0	0	5	0	10	7	3	17	9	9
Type-2 squawk	10	0	0	1	2	9	1	6	7	11	6	15	8
Control	9	0	0	0	0	6	3	6	2	6	4	5	7
Pure LLW	8	0	0	0	0	14	1	7	7	7	8	9	6

[a] Refer to Table II for a key to the symbols used.

Table V. Number of Times that Interest in Speaker Was Increased (+) or Decreased (−) from PrPb to Pb by Each Whale for Each Playback Sound

PBS	N	Alex	Blanchon	Ethel	Frances	Total
Buzzes	11	+8	+7	+7	+6	+28
Whinny	11	+6	+5	+8	+6	+25
Buzz-and-whinny	11	+8	+7	+5	+2	+22
Type-1 squawk	11	+6	+6	+2	+2	+16
Harmonic LLW	11	+7	+3	+4	+2	+16
Type-2 squawk	10	+7	+4	+2	+2	+15
Pure LLW	8	+2	+5	+5	+3	+15
Whistles	10	+5	+3	+3	+4	+15
Blare	9	+4	+2	+4	+2	+12
Contact sound series	10	0	+2	+5	+4	+11
Jaw claps	8	+2	+2	+1	+3	+8
Control	9	+3	+2	0	+2	+7
Jaw clap–buzz-and-whinny	10	0	−3	−1	1	−3

nonvocal behavioral changes is shown in Table V, which compares the relative strengths of the changes of interest from PrPb to Pb toward the speaker by each whale for each playback sound. The buzzes, the whinny, and the buzz-and-whinny combination elicited much interest from all four whales. However, when the jaw clap was added to the buzz-and-whinny (jaw clap–buzz-and-whinny combination), the responses elicited were the lowest of all playback sounds for all whales. The jaw clap itself also elicited a low response. The contact sound series, which elicited several vocal responses, elicited relatively low interest toward the speaker.

2. Playback to the Saguenay River Herd

Vocal Reaction. The significant changes in numbers of emissions of four Saguenay-herd sound types during and after playback of the FldPBS playback series to the Saguenay herd are presented in Table VI. The sounds are illustrated in Fig. 2, except for the ring which was very similar to the ping but had a longer reverberation time, and the click train which was a series of rapid-repetition-rate clicks. Numbers of sounds decreased during Pb and increased during PtPb. Since the speaker was relatively close to the hydrophone, the playback sounds may have masked some of the sounds emitted by the herd during playback. However, all significant changes between PrPb and PtPb, with the exception of one sound, were also decreases and suggest that the overall decrease of sound production observed between PrPb and Pb was real. No strong specific response was elicited by any of the playback sounds recorded at the New York Aquarium such as was observed from Alex in response to playback of the harmonic LLW.

Table VI. Significant Increases and Decreases in Frequency of Emission of Four Sound Types and Total Sound to Playback of FldPBS 2, 3, 4, 6, 9, and 10 to Saguenay herd[a]

FldPBS	N	Period change	Ping +	Ping −	Ping 0	Ring +	Ring −	Ring 0	Squeal +	Squeal −	Squeal 0	Click train +	Click train −	Click train 0	Total sounds +	Total sounds −	Total sounds 0
2 Buzzes	10	PrPb-Pb				6	0	4[c]	0	7	3[c]	0	6	4[c]	0	9	1[b]
		Pb-PtPb															
		PrPb-PtPb															
3 Blare	12	PrPb-Pb				7	0	5[c]									
		Pb-PtPb				8	1	3[c]									
		PrPb-PtPb															
4 Type-1 squawk	10	PrPb-Pb													0	9	1[b]
		Pb-PtPb	0	9	1[b]										1	9	0[c]
		PrPb-PtPb															
6 Contact sound series	10	PrPb-Pb										0	7	3[c]	1	9	0[c]
		Pb-PtPb													10	0	0[b]
		PrPb-PtPb										0	6	4[c]	1	8	1[c]
9 Pure LLW	10	PrPb-Pb															
		Pb-PtPb													8	0	2[b]
		PrPb-PtPb															
10 4.8-kHz pure tone	10	PrPb-Pb	0	7	3[c]												
		Pb-PtPb															
		PrPb-PtPb															

[a] Refer to Table I for a key to the symbols used.
[b] Change is significant at the 0.05 level.
[c] Change is significant at the 0.01 level.

Behavioral Reactions. With one exception the whales did not appear to react to the playback of the captive-animal sounds. However, a gray animal, indicating that it was a young whale, was sometimes observed close to the boat during or shortly after PtPb. It could not be determined whether this was always the same individual. This happened most frequently after playback of the jaw claps, being observed in five of the ten instances that this sound was used. With the other playback sounds, the most often this event was noted was one out of ten or two out of twelve repetitions.

Diving Times. Measurement of diving times during each playback experiment showed no discernible effect of the captive-animal sounds (data and statistical test in Morgan, 1973). The jaw claps were associated with the greatest change in the mean dive, showing a trend toward increase in length of dive in Pb (+18.12 sec), followed by a decrease in PtPb (−17.70 sec). Neither of these trends was statistically significant. The mean dive time for 761 dives was 25.7 sec.

B. Playback of Sounds Recorded from the Saguenay Herd

1. Playback to the Saguenay River Herd

Vocal Reaction. Table VIIa shows the significant vocal changes that occurred during playback of the SagPBS playback series to the wild herd in 1970. Table VIIb shows the same for the playbacks of 1971. None of the significant changes of 1970 were repeated in 1971, although it is believed that the same herd was involved both years. However, all changes of significance involved decreases in sound production, whether in Pb or in PtPb. Once again, the playback sounds may have masked sounds being made by the whales during Pb, but the overall decreases shown in PtPb (to below the levels of PrPb or Pb) suggest that the overall reaction in most cases was a decrease in sound production.

Behavioral Reactions. Observations were recorded for whales coming toward or retreating from the boat, any deviation in the path of the animals in the cases when they were traveling up or down river, and any pause in such a transit (Table VIII). Reactions to the playbacks of certain sounds were much more common when the animals were milling than when making their daily passage up and down the Saguenay River.

Three playback sounds often drew the whales toward the boat when milling: the screams-and-wails, the blats-and-ping, and the jaw claps (see Fig. 2 for description of sounds). The screams-and-wails playback was an extended series of sounds, and the blats-and-ping playback included three 'blats' and one 'ping.' The response to the jaw claps recorded from the Saguenay animals was elicited from both young and adults. The only one of the seven sounds that elicited a strong positive reaction when in transit was the squeals.

Table VIIa. Significant Increases and Decreases in Frequency of Emission of Three Sound Types during and after Playback of SagPBS 1, 3, and 6 to Saguenay herd in 1970[a]

SagPBS	N	Period change	Ping +	Ping −	Ping 0	Ring +	Ring −	Ring 0	Total sounds +	Total sounds −	Total sounds 0
1 Moans	10	PrPb-Pb Pb-PtPb PrPb-PtPb							0	9	1[b]
3 Screams-and-wails	12	PrPb-Pb Pb-PtPb PrPb-PtPb				0	6	6[c]			
6 Squeals	9	PrPb-Pb Pb-PtPb PrPb-PtPb	0	6	3[c]						

[a] Refer to Table I for a key to the symbols used.
[b] Change is significant at the 0.05 level.
[c] Change is significant at the 0.01 level.

Table VIIb. Significant Increases and Decreases in Frequency of Emission of Two Sound Types during and after Playback of SagPBS 1, 3, 6, and 7 to Saguenay Herd in 1971[a]

SagPBS	N	Period change	Click train −	Click train +	Click train 0	Total sounds −	Total sounds +	Total sounds 0
1 Moans	10	PrPb-Pb Pb-PtPb PrPb-PtPb	0 0	6 6	4[c] 4[c]	0	8	2[b]
3 Screams-and-wails	12	PrPb-Pb Pb-PtPb PrPb-PtPb				1	10	1[c]
6 Squeals	10	PrPb-Pb Pb-PtPb PrPb-PtPb				1	8	1[c]
7 Saguenay LLW	9	PrPb-Pb Pb-PtPb PrPb-PtPb	1	8	0[c]			

[a] Refer to Table I for a key to the symbols used.
[b] Change is significant at the 0.05 level.
[c] Change is significant at the 0.01 level.

Table VIII. Approaches toward (+) and Withdrawals from (−) Research Boat during Playback of SagPBS Playback Series to Saguenay Herd in 1970 and 1971[a]

SagPBS	N_m	Milling			N_t	Transiting		
		+	−	0		+	−	0
Moans	6	3	1	2	15	1	0	14
Pings	9	1	2	6	11	0	1	10
Screams-and-wails	22	15	1	6[b]	13	0	0	13
Blats-and-ping	16	10	2	4[c]	14	2	0	12
Jaw claps	19	13	0	6[b]	9	1	1	7
Squeals	9	2	1	6	15	7	0	8[c]
Saguenay LLW	4	1	0	3	8	0	0	8

[a] N_m: number of repetitions of each SagPBS while the whales were milling.
 N_t: number of repetitions of each SagPBS while the whales were in transit up or down river.
[b] Change is significant at the 0.05 level.
[c] Change is significant at the 0.01 level.

Diving Times. Playbacks had no effect on diving times (data and analyses in Morgan, 1973).

2. Playback to Three Captive Animals

The sounds from the Saguenay herd were played back to the three animals remaining in the large beluga tank, and to Alex alone in the separate pool. Over-all numbers of sounds were much lower in the group tank after Alex's removal. Alex had been either the most vocal of the four whales when they were all together or his presence had caused more vocalizations from the other three. In fact, he was found to be more vocal alone than were the other three belugas together.

Vocal Reaction. There were no significant changes in the vocal emissions of Blanchon, Frances, or Ethel in response to any of the Saguenay playback sounds.

Changes of Interest Directed toward Hydrophone and Speaker. A summary of the significant changes of interest directed toward the hydrophone and speaker by all three whales considered together is presented in Table IX. As shown in columns one and three, there were no significant changes of interest shown toward the hydrophone in reaction to any of the Saguenay playback sounds. Orientation toward the speaker increased significantly in Pb for five of the Saguenay sounds, the exceptions being the moans and the screams-and-wails. These increases were all followed by decreases in PtPb. No significant changes in approaching or touching the speaker were elicited.

Table X shows that the whales oriented toward the speaker more often in Pb than in PrPb in response to all seven playback sounds, and approached it more often in Pb in response to all sounds except the squeals. There were no significant changes associated with either orientation or approach toward the hydrophone.

Table IX. Significant Increases and Decreases of Interest Directed toward Hydrophone and Speaker during and after Playback of SagPBS 1-7 to Three Captive Belugas[a]

SagPBS	N	Period change	Orient toward hydrophone			Orient toward speaker			Approach hydrophone			Approach speaker		
			+	−	0	+	−	0	+	−	0	+	−	0
1 Moans	10	PrPb-Pb												
		Pb-PtPb												
		PrPb-PtPb												
2 Pings	10	PrPb-Pb				8	0	2[b]						
		Pb-PtPb				0	7	3[c]						
		PrPb-PtPb												
3 Screams-and-wails	10	PrPb-Pb												
		Pb-PtPb												
		PrPb-PtPb												
4 Blats-and-ping	10	PrPb-Pb				7	0	3[c]						
		Pb-PtPb												
		PrPb-PtPb												
5 Jaw claps	11	PrPb-Pb				8	0	3[b]						
		Pb-PtPb				0	5	5[c]						
		PrPb-PtPb												
6 Squeals	10	PrPb-Pb				8	0	2[b]						
		Pb-PtPb				0	8	2[b]						
		PrPb-PtPb												
7 Saguenay LLW	10	PrPb-Pb				7	0	3[c]						
		Pb-PtPb												
		PrPb-PtPb												

[a] Refer to Table II for a key to the symbols used.
[b] Change is significant at the 0.05 level.

Table X. Summary of Number of Playback Sounds that Elicited Increases and/or Decreases of Interest Directed toward Hydrophone and Speaker from All Three Whales[a]

Period change	Orient toward hydrophone			Orient toward speaker			Approach hydrophone			Approach speaker		
	+	−	0	+	−	0	+	−	0	+	−	0
PrPb-Pb	0	0	7	7	0	0[c]	2	2	3	6	0	1[c]
Pb-PtPb	0	0	7	0	7	0[c]	1	2	4	0	6	1[c]
PrPb-PtPb	0	0	7	4	1	2	0	2	5	4	2	1

[a] Refer to Table II for a key to the symbols used.
[b] Change is significant at the 0.05 level.
[c] Change is significant at the 0.01 level.

Considering each beluga separately, no whale showed any significant changes of interest toward the hydrophone, but all three increased interest in the speaker in Pb and decreased interest again in PtPb in response to all seven sounds. None of the sounds caused a consistently greater increase of interest than any of the others (data in Morgan, 1973).

Behavioral Reactions. Although there were no consistent behavioral reactions observed to any single Saguenay sound, a series of sexual encounters that may have been induced by the playbacks was observed on February 26, 1971. On that date, successive presentation of five different Saguenay sounds, 25 or more minutes apart, was associated with sexual behavior directed by Blanchon toward Frances. Blanchon initiated the interaction and in all cases Frances seemed unreceptive. At 1110 hr, during the PtPb period of the playback of the pings, Blanchon attempted intromission with Frances. This caused emission of a contact sound series as Frances drew away and Blanchon followed, swimming upside-down under her. At 1145 hr, the blats-and-ping was played back and Blanchon's penis erected fully in late PtPb.

The screams-and-wails, the squeals, and the moans were presented at 1235, 1324, and 1425 hr, respectively, and attempted intromissions occurred in late Pb or in PtPb in all three cases. The attempt elicited by the moans seemed less intense and was not followed by the contact sound series as was the case after the screams-and-wails and the squeals. Following this less intense attempt, Blanchon emitted a 'hissing' sound, rapidly turned over, and emitted bubbles from his blowhole. As these events occurred, jerking movements were observed around the penis, and ejaculation may have occurred. The sixth sound played back, at 1520 hr, was the harmonic LLW originally recorded from Alex, the adult male that had previously been in the tank with Blanchon, Frances, and Ethel. There was no indication of sexual behavior in reaction to this playback, but Blanchon became very excited, swimming rapidly around the tank in Pb and early PtPb. During PrPb he had been lying quietly on the bottom of the tank, and in late PtPb he seemed to calm down, swimming more slowly around the tank.

These sounds were played back repeatedly to those same whales in July and November of 1971, but no sexual behavior was observed during any of these playbacks.

3. Playback to One Captive Animal (Alex)

Vocal Reaction. Two of the sounds recorded from the Saguenay herd, the screams-and-wails and the squeals, elicited one response each at the 0.05 level of significance. When the screams-and-wails was played back, the number of harmonic LLWs emitted by Alex increased in Pb. Upon playback of the squeals, the total number of sounds emitted decreased in PtPb to less than in Pb.

Changes of Interest Directed toward Hydrophone and Speaker. The only significant change of interest in the equipment elicited from Alex by any one sound during this playback series was an increase in orientation toward the speaker during Pb of the jaw claps, although orientation toward the speaker did increase during Pb to a lesser extent in response to all seven SagPBS sounds (data in Morgan, 1973).

Activity as Measured by Number of Tank Circuits. Alex's activity increased during playback for all seven Saguenay sounds. With six of the seven, activity remained higher in PtPb than in PrPb (Table XI).

IV. DISCUSSION

A. Playback of Sounds Recorded from the Captive Animals

1. Playback to Captive Animals at the New York Aquarium

With the exceptions noted below, the playback of these sounds did not cause any response in vocalizations. However, in general, the belugas in

Table XI. Total Number of Circuits of Tank Completed by Alex during SagPBS Playback Experiments

SagPBS	Number of tank circuits			
	N	PrPb	Pb	PtPb
Moans	11	13.50	24.50	30.25
Pings	11	19.50	30.00	25.75
Screams-and-wails	11	22.50	29.25	30.75
Blats-and-ping	11	25.50	32.50	27.50
Jaw claps	11	20.33	24.00	32.50
Squeals	10	17.00	25.50	19.00
Saguenay LLW	11	26.50	34.25	25.25

captivity exhibited increased interest toward the speaker during Pb. With some sounds, this increased interest remained high in PtPb. It thus seems that new sounds are investigated by a beluga, but it is impossible to say if the interest is in the "meaning" of the sound itself or in the sound as an indication of the presence of another animal. Depending on the communicatory significance, if any, of the individual sounds, increased interest for either of the above reasons could be of advantage to the animal. However, the evidence presented in this paper suggests that beluga sounds do have varying effects individually and in different combinations (for example, see the discussions of the buzz, whinny, jaw-clap complex below).

The most significant vocal response elicited by any of the playback sounds was that elicited from Alex upon playback of the harmonic LLW, whereby he increased LLW production during Pb, with some carryover into PtPb.

The biological significance or "meaning" of the harmonic LLW is not known, but it seemed to occur in situations that might have been described in humans as productive of "impatience" or "expectation," such as the expected time of feeding or before times of training sessions or public exhibitions. The very fact that it occurred so seldomly, but was so uniformly elicited by playback of the same sound, suggested that it occurred only in a very specific context. This was in sharp contrast to such commonly occurring sounds as the shorter, less strident whistle which was heard at any time of day and under almost any circumstance. Also, since the effect of the harmonic LLW playback carried over into PtPb, it can be assumed that the effect on Alex was not merely that of a stimulus–response reflex action, but was instead an increase in his overall level of arousal that continued after cessation of the initiating stimulus.

The harmonic LLW was never heard from the free-swimming herd in the Saguenay River, nor did it elicit any reaction from the Saguenay animals. It also was not recorded from any of the other animals held captive at the New York Aquarium and did not elicit a vocal response from any of them. On the other hand, Alex's vocal reaction was strong and stereotyped. Why should such a dichotomy of response be shown by animals that had been in captivity together for several years? One possible explanation is the different localities of origin of the captive belugas. Blanchon, Frances, and Ethel were all captured on the east coast of North America, whereas Alex was obtained from a west coast population. It is suggested that different populations of belugas may possess different dialects, "in which (the sounds are) similar among most or all individuals living in a particular locality, but are different from one locality to another" (Lemon, 1967). As Lemon states in reference to birds, it seems reasonable that animals exhibiting dialects should respond more to sounds of their own particular dialect. If indeed dialects do occur in cetaceans, one would expect to find them in populations as widely separated as those of the belugas of the east and west coasts of North America.

It has been shown by numerous investigators that sounds differ between cetacean species and also between individuals of one species with enough regularity that individual recognition is possible (Caldwell et al., 1972b), even to the point of recognition of individuals of a different species (Caldwell et al., 1972a). It would seem reasonable, then, to expect differences between sounds of geographically isolated beluga populations. Such dialects might even be expected to occur between herds along one coast, in which context they would serve to maintain individual herd integrity during the times of congregation which are reported to occur for fattening and migration (Kleinenberg et al., 1964).

The contact sound series was an extended combination of various sound types (barks, squawks, jaw claps, whistles, squawls, buzzes, whinnys, and chirps) emitted at times of physical contact or close proximity between two or more whales, or at times of major disturbance. Apparently the contact sound series was indicative of a high state of arousal. Further, the specific sounds responsible for conveying this state may have been the squawk and the jaw clap, two of the more prominent sounds occurring at the time of maximum disturbance during the emission of the contact sound series (Morgan, 1970). Playback of a contact sound series to the captive animals resulted in increased emission of the contact sound series, with the usual interindividual contact, during the Pb period. This response could have been elicited in two ways. First, the playback might have merely increased the level of activity in the tank by conveying a disturbance context to the animals, thereby increasing the chances for contact. Secondly, the playback might have directly initiated contact between individuals. It is not known which of these two mechanisms produced the observed reaction, but the data on interest directed toward the hydrophone and sound source support the latter. Approach toward both pieces of equipment decreased in Pb indicating that this interest was directed elsewhere in the tank, possibly at the tankmates.

The other vocal changes noted during playback of the contact sound series were decreases in the number of jaw claps not involved in a contact sound series, and total sounds emitted during playback. The latter change might have been expected as a secondary effect of the increase of contact sound series since the contact sound series was a relatively long sound emission, thus leaving less time for the production of other sounds. The jaw clap was one of the sounds included in the contact sound series, and thus its emission was included in the count of contact sound series rather than as a separate sound emission.

However, the three other sounds nearly always found associated with the contact sound series, the squawk, the whistle, and the chirp, neither increased nor decreased significantly as separate sounds during playback of the contact sound series. Because of the context of emission of these three sound types, they would not have been expected to show such a decrease. The squawk was not a common sound out of the context of the contact sound series. It was apparently a sound associated with a high state of arousal (Morgan, 1970), brought on by either fright or interindividual contact or proximity, and was thus of uniform low occurrence as a separate sound in all playback experiments. The situation with regard to the whistle and the chirp was exactly the reverse. These sounds were associated with any disturbance, however slight, inside or outside the tank, and were thus taken as being indicative of a very low state of arousal when occurring alone (Morgan, 1970). They were of uniform high occurrence throughout most of the playback experiments and would not have been expected to decrease in occurrence during playback. They were also the "finishing sounds" of nearly all contact sound series, seeming to occur for some time after the cessation of the contact sound series, much as a small bird will occasionally "peep" while calming down after being frightened.

Closer inspection of the reactions to the buzz, the whinny, and the jaw clap show some interesting relations between these sounds, suggesting that syntax is important in the conveyance of the significance of cetacean sounds (presumably by placing the sounds in a meaningful context).

The whinny was never observed to occur alone; it was always emitted in combination with the buzz, and may have been incidental to the production of the buzz, although the buzz was often heard without the accompanying whinny. The PBS designated the whinny was actually such a buzz-and-whinny combination, with the whinny being louder relative to the buzz than was generally the case.

During Pb, the buzz elicited only an increase in orientation toward the speaker, while the whinny elicited increase of both orientation and approach toward the speaker. None of these increases carried over in PtPb, although total sounds decreased in PtPb of the whinny playback. The buzz-and-whinny combination again elicited the increases in orientation and approach toward the speaker, but in this case approach remained higher in PtPb.

When the jaw clap was played back alone or in combination with the buzz-and-whinny, there were no significant changes of interest toward the speaker. Thus the jaw clap inhibited the effect of the buzz-and-whinny noted above. In fact, the jaw clap–buzz-and-whinny combination was one of only two sounds of the 13 in the PBS series that showed a trend toward eliciting a decrease of interest in the speaker during Pb. The other sound showing this tendency was the contact sound series which, like the jaw clap–buzz-and-whinny combination, was a combination of various sound types, was of relatively longer duration, and included the jaw clap. As discussed earlier, the contact sound series may have affected the whales by redirection of interest from speaker to tankmates. Perhaps a similar response to the jaw clap–buzz-and-whinny combination was responsible for the decrease of interest shown toward the equipment during playback of that sound. There was, however, no vocal response associated with playback of the jaw clap–buzz-and-whinny combination such as occurred in the case of the contact sound series playbacks.

The major differences between the two types of sound series just mentioned is that the squawk occurred in the contact sound series but did not occur in the jaw clap–buzz-and-whinny combination. When two types of squawks were played back to the four belugas, interest toward the speaker increased significantly during Pb in reaction to both sounds, and there was no vocal reaction to either sound. The two dominant sounds of the contact sound series are the jaw clap and the squawk. The jaw clap caused no changes of interest in the equipment and no vocal changes. When these two sounds were combined in the context of the contact sound series and played back to the animals, approach toward the sound source decreased, and number of emissions of contact sound series during Pb increased significantly. It is suggested that the decrease in approach is due to the inhibitory effect of the jaw clap on interest in the speaker, perhaps by redirection of interest toward tankmates, and that the increase in contact sound series emissions is due to the squawk, which, when added to an extended series of sounds, increases levels of arousal to the point where contact or near-contact is made between animals leading to the emission of a contact sound series.

Syntax may be defined as the ways in which individual sounds or words are combined to form a code or message. The results presented here support the conclusion that syntax was important in the transfer of information from one beluga to another by sound, with series of sounds carrying additional or different meaning to the receiving animal than did the component sounds individually. This necessarily leads one to the conclusion that at least some of the sounds of the beluga are communicative in function. Syntax is meaningful only in a communicatory context.

2. Playback to the Saguenay River Herd

The only significant changes in vocalizations of the Saguenay animals in response to playback of the aquarium-recorded sounds were a decrease of total sounds during Pb followed by a partial increase during PtPb. It is possible that some sounds were masked during Pb, but this seems unlikely due to the decrease of several sounds observed from Pb to PtPb. In general, no changes in behavior were noted. This was perhaps due to tank noise or distortion by the enclosed recording situation. Conversely, the sounds may have been changed by the whales during their period of captivity or they may have been unrecognizable by virtue of originating from animals possessing a dialect different from that of the Saguenay herd. Other possibilities that would account for the lack of a specific reaction include the lack of a proper context or syntax. Previously we have shown that syntax is important, and later we shall show that context is also fundamental.

In several instances, a young whale came close to the boat after playback of the jaw clap. The significance of this, if any, is unknown. Perhaps, in as much as this sound does not rely on subtle frequency or amplitude modulations, it was least modified when recorded in the tank. Also, possibly the young animals had not yet learned to differentiate between the "reliability" or "unreliability" of slightly modified sounds and responded as to a normal jaw clap. Adult animals would have learned to make this differentiation through longer experience and accordingly would not have responded.

B. Playback of Sounds Recorded from the Saguenay Herd

1. Playback to the Saguenay River Herd

The only vocal reactions to the seven Saguenay-recorded sounds were some decreases in numbers of sounds emitted in Pb and PtPb. This seems to be a general reaction of the beluga to new sounds, as the same effect was noted in the presence of noise from passing boats. Behavioral reactions depended on context. Three sounds attracted whales to the boat when they were milling but had no effect when they were moving up and down the river; one sound produced the opposite effect.

Two of the sounds that attracted the whales while milling were in reality combinations of sounds. The blats-and-ping was a combination of three blats with one ping at the end, and the screams-and-wails was an extended series of several sounds (see Fig. 2). Again, combinations of sounds were more effective in eliciting a reaction from the belugas, and thus presumably were more effective in conveying information to the animals.

The third sound found effective during milling was the jaw clap, which in this case attracted both young and adult belugas. The jaw clap, crack, or bang has been described as either an alarm or threat call for several cetacean species (Wood, 1953; Caldwell *et al.,* 1962; Fish and Mowbray, 1962), including the beluga. In the light of the present experiments, it is felt that a better description of the significance of this sound might be as an "attention" or "alerting" call, produced in reaction to either an alarm or antagonistic context. A sound that had evolved for this purpose would be expected to be startling, loud, and definitive. The jaw clap, with its abrupt onset, wide frequency spectrum, and high intensity fulfills these requirements. The reaction to such an "alerting" sound would be expected to be an approach toward the animal producing the sound in order to gather more information concerning the cause of the disturbance or for mutual protection. The jaw clap was the only single sound used as a playback that produced a significant response in the milling context. In captivity, when combined with other sounds, it inhibited reaction to the other sounds, suggesting that it dominated the significance of the combination. Regarding this abrupt, intense sound as an attention or alerting call allows it to serve as either a threat or an alarm call, with its particular meaning being determined by the context in which it is emitted and/or perceived. Further reaction would then be dependent upon other information (visual, acoustical, or tactile) perceived after the receiving animals had been alerted to a situation of immediate and overriding significance.

Only one of the seven Saguenay sounds used as playbacks, the squeals, elicited an approach from at least a part of the herd while it was moving up or down the river. When the whales were making such a transit the calves were found concentrated toward the rear of the herd, nearly always accompanied by an adult beluga, presumably the mother. This was the portion of the herd from which the most squeals were recorded, and it is suggested that these sounds were associated with the calves. They could have been produced by the young themselves, by the females accompanying the calves, or by both. Although the squeals were heard while the whales were either milling or transiting, the playback was effective in attracting them only while they were transiting. Possibly it was more likely that a young beluga would have become separated from its attendant female while the herd was moving than while it was milling in quiet water. If the squeal functions for maintenance of contact between calf and mother, between calf and entire herd, or as a general distress call of the young, it would have been most effective as a playback in the situation where loss of contact or distress was most likely. The response was usually elicited from more than one adult beluga, often accompanied by calves.

The fact that context is important in determining reactions of whales to at least some sounds (see also Caldwell and Caldwell, 1967) leads one to wonder whether cetacean sounds and behavior patterns are produced in reaction to a particular context (are responsive) or whether the sounds themselves are the stimulus for the production of other sounds or behavior patterns (are causal). Sounds such as the harmonic LLW and the jaw clap seem to be both responsive and causal. However, not all sounds produced by a species would necessarily be expected to act in this manner. Some would be expected to occur only in response to a particular context and not elicit further reactions from other animals, while others might be produced for the purpose of eliciting a reaction from another animal. This latter type of sound is the type used for communication in the human species, assuming purposive thought and action on the part of the emitting individual, and so far not definitively shown to be used by any nonhuman species, although such use has been suggested in at least two cetacean species, *T. truncatus* (Lilly, 1963) and *T. gilli* (Evans and Dreher, 1962). A third type of sound would be that uttered in response to a particular stimulus context which, when received by another animal, transmits information regarding that context to the receiving animal, causing it to react vocally and/or behaviorally as it would to the original context. If one looks at the playback of natural sounds to animals in the light of these divisions, it is easy to realize that not all sounds would be expected to elicit a reaction regardless of the context in which they were played back. Only causal sounds would be expected to elicit a reaction.

The causal sound is normally emitted in reaction to a certain set of conditions, or context, which serves as the stimulus for the production of that sound. It would seem reasonable, then, to assume that some of these sounds would be effective in eliciting a normal reaction only if played back in a context similar to that under which the sounds are normally produced. The results presented herein support this contention. Of the four Saguenay sounds found to be effective in eliciting a response from the belugas, three were significantly effective only while the herd was milling, and one was significantly effective only while the herd was moving. In addition, none of the sounds recorded from the belugas at the New York Aquarium elicited a significant response from the Saguenay animals, and none of the Saguenay sounds elicited a specific response from the captive animals.

In contrast to the context-dependent causal sounds, there are many types of sounds of general meaning that would be expected to convey their meaning in any context to the animal receiving the sound. Examples of such sounds would be those associated with danger, distress, threat, and other states of high emotional arousal, and the signature whistle, which serves for individual recognition (Caldwell et al., 1972b). Since the harmonic LLW elicited responses from Alex at all times of day or year and when alone or with other belugas, it is concluded that it was one of these types of sounds, being produced in response to some particular context, but having meaning to the receiving animal regardless of the context in which it was received. Thus, causal sounds can be further subdivided as being context-dependent or context-independent regarding reception. Responsive sounds, by definition, are context-dependent only as regards their production. They are context-independent regarding reception since they are uniformly ignored by the receiving animal in any context.

Usually, when a positive reaction was elicited by one of these playback sounds, only a few of the animals were involved. However, on one occasion during the 1970 season, the response was shown by the entire herd. This was the strongest response observed during both years of field work and was elicited by two playback sounds, the screams-and-wails and the blats-and-ping. This is regarded as further evidence that syntax is important in the transfer of information and thus in eliciting responses.

On the morning of August 8, 1970, the beluga herd was slowly working its way across Baie Ste. Catherine toward the mouth of the Saguenay River. The research boat was positioned in their path, and during the 3-min PrPb period the herd was passing by the boat and heading toward the river. The experiment began with the playback of the screams-and-wails, and the animals turned from their previous course and came toward the boat. The screams-and-wails was played back four times in succession, with short intervals between while the playback tape was being rewound. During two of these intervals the belugas began to move away, but turned and reapproached when playback began again. The blats-and-ping was then played back and the animals continued to approach, to within approximately 8 ft of the boat, swimming on top of the water without submerging.

The blats-and-ping was immediately followed by the playback of the squeals. During the first 2 min of this 3-min playback, no whales were observed on the surface. Three groups then surfaced about 30 yards from the boat and remained at that distance until the playback ended. After the squeals ended the animals moved on past the boat and started to swim away. The screams-and-wails was then played back again, and the herd turned once more to approach the boat. At this time an adult white animal passed beneath the speaker at an estimated depth of 6–7 ft. Once again, when this playback ended the belugas started to move off. When the screams-and-wails started again, the herd turned and approached to within 25 yards, where they stopped and began milling about. At this time two large adult belugas separated from the herd and swam directly and steadily toward the boat. One of these two was not seen again, but the other came to a position directly under the speaker, stopped, turned onto its side, then onto its back, and inclined its head upward toward the sound source. This animal then swam from view, and although the screams-and-wails playback continued for more than 5 min afterward, the entire herd moved away and proceeded on toward the mouth of the Saguenay River. Further playback was not effective in drawing the herd back toward the boat.

This reaction, elicited by the playback of natural sound combinations, was a clear demonstration of scouting behavior in the beluga. Other instances of scouting behavior in this species were observed during the field work and will be fully described in a separate report. Scouting behavior has also been reported in other cetacean species (Evans and Dreher, 1962; Caldwell and Caldwell, 1964; Caldwell et al., 1965).

2. Playback to the Captive Animals

For the most part, playback of the Saguenay River sounds to the three animals together and to Alex alone elicited no recordable responses except for an over-all increase of interest in the speaker during Pb. In contrast to the playback of the aquarium-recorded sounds, there was no gradation of response to these sounds from a different population. As discussed previously, the lack of a specific response may have been related to improper context or syntax or to different dialects.

As has been seen throughout this paper when working with both captive and free-swimming animals, a general reaction was observed to all sounds played back which, although different between the captive and free contexts, would serve the purpose of putting the animals in a better condition for receiving further information about the sound stimulus. With captive animals this general reaction was an orientation or actual approach toward the speaker (except in the cases of the jaw clap and the two sound series, as discussed previously). With the free-swimming animals the general response was a decrease in number of sound emissions, a reaction which was also noted upon the approach of a motor boat toward the herd. Whether these reactions were merely those prompted by curiosity or were a conscious attempt to discover further information about the stimulus cannot be said at this time. The end result, however, would have been the same in either case; i.e., the animals would have been in a more appropriate state for the reception of more information about the stimulus, its source, and its reason for occurrence. The single instance of scouting behavior described would suggest a purposeful attempt to learn more about the sound and its source, but such a teleological explanation can be advanced only as a suggestion at present.

It is impossible to know if there was any meaning to the once-observed correlation between playback and mating other than to say that the reaction to the sound recorded from a previous tankmate was very different from that to the sounds from the Saguenay herd.

In the playbacks to Alex, the screams-and-wails elicited more harmonic LLWs during Pb and the squeals caused a decrease in total numbers of sounds in PtPb. The significance of these reactions is unknown, but both playbacks also elicited responses from the wild animals. Also, Alex's activity increased in response to all seven Saguenay-recorded sounds, although no increase in sexual arousal was noted at any time.

In summary, it does appear that playback of conspecific sounds to cetaceans can be a useful tool in examining communication capabilities. In particular in this paper it is shown that both syntax and context are important in the communication of the beluga. Also, although there appears to be a general reaction to all new sounds within the hearing range of the beluga, either by decrease of sound emissions or approach toward the source of the sound, certain specific sounds (such as the harmonic long, loud whistle; the jaw clap; the squeal; etc.) and certain combinations (such as the contact sound series or the screams-and-wails) have specific meaning if played back in a proper context.

ACKNOWLEDGMENTS

Thanks are due to many people who aided by giving advice and help during the course of this study. Primary among these is Dr. Howard E. Winn, who gave advice and guidance throughout and who supplied the equipment. The work at the New York Aquarium was aided by the entire staff, in particular Dr. Ross Nigrelli, Dr. James Oliver, Robert A. Morris, and Douglas Kemper. The field work on the Saguenay River could not have succeeded without the cooperation of many of the people of Tadoussac, Quebec Province, especially Mr. and Mrs. Lewis Evans, Mr. and Mrs. Harold Price, and the Rev. and Mrs. Russell Dewart. Alan Evans and Martin Hyman ably served as boatmen on the Saguenay River in 1970 and 1971, respectively. I also profited greatly from the advice of Dr. David E. Gaskin of Guelph University, Dr. David E. Sergeant of the Fisheries Research Board of Canada, and Mr. William E. Schevill and Mr. William A. Watkins of the Woods Hole Oceanographic Institution. Financial support for the work came from the Office of Naval Research (Grant No. N00014-68-A-0215-0003 to H.E. Winn).

REFERENCES

Caldwell, M. C., and Caldwell, D. K., 1964, Experimental studies on factors involved in care-giving behavior in three species of the cetacean family Delphinidae, *Bull. South. Calif. Acad. Sci.* **63**:1–20.

Caldwell, M. C., and Caldwell, D. K., 1967, Intraspecific transfer of information via the pulsed sound in captive Odontocete cetaceans, in: *Animal Sonar Systems: Biology and Bionics* (R. G. Busnel, ed.), pp. 879–936, Vol. II, N.A.T.O. Advanced Study Institute, Jouy-en-Josas, France.

Caldwell, M. C., Haugen, R. M., and Caldwell, D. K., 1962, High-energy sound associated with fright in the dolphin, *Science* **138**:907–908.

Caldwell, M. C., Caldwell, D. K., and Siebenaler, J. B., 1965, Observations on captive and wild Atlantic bottlenosed dolphins, *Tursiops truncatus*, in the northeastern Gulf of Mexico, *Los Angeles County Mus. Contrib. Sci.* **91**:1–10.

Caldwell, M. C., Caldwell, D. K., and Hall, N. R., 1972a, Ability of an Atlantic bottlenosed dolphin *(Tursiops truncatus)* to discriminate between, and potentially identify to individual, the whistles of another species, the common dolphin *(Delphinus delphis)*, *Marineland Res. Lab. Tech. Rep.* **9**:1–7.

Caldwell, M. C., Hall, N. R., and Caldwell, D. K., 1972b, Ability of an Atlantic bottlenosed dolphin to discriminate between, and respond differentially to, whistles of eight conspecifics, *Marineland Res. Lab. Tech. Rep.* **10**:1–5.

Cummings, W. C., and Thompson, P. O., 1971, Gray whales, *Eschrichtius robustus*, avoid the underwater sounds of killer whales, *Orcinus orca*, *U.S. Fish. Bull.* **69**:525–530.

Davies, D. H., 1962, Note on the use of killer whale sounds as a shark repellent, *S. Afr. Assoc. Mar. Biol. Res. Bull.* **3**:32–33.

Dreher, J. J., 1966, Cetacean communication: Small-group experiment, in: *Whales, Dolphins, and Porpoises* (K. S. Norris, ed.), pp. 597–602, University of California Press, Berkeley.

Dudok van Heel, W. H., 1959, Audio-direction finding in the porpoise *(Phocoena phocoena)*, *Nature, London* **183**:1063.

Evans, W. E., and Dreher, J. J., 1962, Observations on scouting behavior and associated sound production on the Pacific bottlenose porpoise *(Tursiops gilli,* Dall), *Bull. South. Calif. Acad. Sci.* **61**:217–226.

Fish, J. F., and Vania, J. S., 1971, Killer whale, *Orcinus orca*, sounds repel white whales, *Delphinapterus leucas*, *U.S. Fish. Bull.* **69**:531–535.

Fish, M. P., and Mowbray, W. H., 1962, Production of underwater sound by the white whale or beluga, *Delphinapterus leucas* (Pallas), *J. Mar. Res.* **20**:149–162.

Hall, J. D., and Johnson, C. S., 1972, Auditory thresholds of a killer whale, *Orcinus orca* Linnaceus, *J. Acoust. Soc. Am.* **51**:515–517.

Jacobs, D. W., 1972, Auditory frequency discrimination in the Atlantic bottlenose dolphin, *Tursiops truncatus* Montague: A preliminary report, *J. Acoust. Soc. Am.* **52**:696–698.

Jacobs, D. W., and Hall, J. D., 1972, Auditory thresholds of a fresh water dolphin, *Inia geoffrensis* Blainville, *J. Acoust. Soc. Am.* **51**:530–533.

Johnson, C. S., 1967, Sound detection thresholds in marine mammals, in: *Marine Bio-Acoustics* (W. N. Tavolga, ed.), Vol. 2, pp. 247–260, Pergamon Press, New York.

Johnson, C. S., 1968, Masked tonal thresholds in the bottlenosed porpoise, *J. Acoust. Soc. Am.* **44**:965–967.

Kellogg, W. N., 1953, Ultrasonic hearing in the porpoise, *Tursiops truncatus*, *J. Comp. Physiol. Psychol.* **46**:446–450.

Kellogg, W. N., and Kohler, R., 1952, Reactions of the porpoise to ultrasonic frequencies, *Science* **116**:250–252.

Kleinenberg, S. E., Yablokov, A. V., Bel'kovich, B. M., and Tarasevich, M. N., 1964, Beluga *(Delphinapterus leucas),* Investigation of the Species, translated by Israel Program for Scientific Translations, 376 pp.

Lang, T. G., and Smith, H. A. P., 1965, Communication between dolphins in separate tanks by way of an electronic acoustic link, *Science* **150**:1839–1844.

Lemon, R. E., 1967, The response of cardinals to songs of different dialects, *Anim. Behav.* **15**:538–545.

Lilly, J. C., 1963, Distress calls of the bottlenose dolphin: Stimuli and evoked behavioral response, *Science* **139**:116–118.

Lilly, J. C., and Miller, A. M., 1961, Vocal exchanges between dolphins, *Science* **134**:1873–1876.

Morgan, D. W., 1970, The reactions of belugas to natural sound playbacks, *Proc. 7th Ann. Conf. Biol. Sonar and Diving Mammals,* pp. 61–66, Stanford Research Institute, Menlo Park, California.

Morgan, D. W., 1973, The vocal and behavior reactions of the white whale, or beluga, *(Delphinapterus leucas)* to underwater playback of natural and synthetic beluga sounds, PhD thesis, University of Rhode Island.

Schevill, W. E., and Lawrence, B., 1949, Underwater listening to the white porpoise *(Delphinapterus leucas), Science* **109**:143–144.

Schevill, W. E., and Lawrence, B., 1950, A phonograph record of the underwater calls of *Delphinapterus leucas, Woods Hole Oceanogr. Inst. Ref. No.* 50-1.

Schevill, W. E., and Lawrence, B., 1953, Auditory response of a bottlenosed porpoise, *Tursiops truncatus,* to frequencies above 100 kc, *J. Exp. Zool.* **124**:147–165.

Siegel, S., 1956, *Nonparametric Statistics for the Behavioral Sciences,* McGraw-Hill, New York. 312 pp.

Watkins, W. A., and Schevill, W. E., 1968, Underwater playback of their own sounds to *Leptonychotes* (Weddell seals), *J. Mammal.* **49**:287–296.

Chapter 10

THE WHISTLE REPERTOIRE OF THE NORTH ATLANTIC PILOT WHALE (*Globicephala melaena*) AND ITS RELATIONSHIP TO BEHAVIOR AND ENVIRONMENT

Algis G. Taruski

Graduate School of Oceanography
University of Rhode Island
Kingston, Rhode Island 02881

I. INTRODUCTION

Very little is known about the normal sounds and behavior of the North Atlantic pilot whale, *Globicephala melaena,* or of other pilot whales. Schevill (1964) stated that the whistles of *Globicephala melaena* range from 0.5 to 5.0 kHz in sound frequency, while those of *G. macrorhynchus* range from 2 to 12 kHz. Busnel *et al.* (1971) described double clicks produced by *Globicephala*. Busnel and Dziedzic (1966) presented a detailed report on a single hour-long encounter with a school of 11 pilot whales. During the encounter one animal was harpooned. The authors described and figured five signal types and several variants. In addition, they noted abrupt frequency shifts and amplitude variations.

Most studies have dealt with captive odontocetes. Caldwell *et al.* (1969, 1970), working with *Tursiops truncatus,* reported that this species' vocalizations are highly stereotyped signature whistles, which vary slightly depending on arousal and emotional state of the emitter. Other investigators (Dreher and Evans, 1964; Dreher, 1966; Poulter, 1968) regard dolphin vocalizations as a quasi-language, i.e., particular sounds have precise meanings. Lilly (1961, 1967) has argued that dolphin sounds and human language are equivalent.

Most of the above arguments are based on studies of captive animals. There are, however, disadvantages in working with captives. Stereotyped behavior in captive individuals of a social species should not necessarily be considered representative of the species' behavior in the wild. Additionally, close work with these seemingly friendly and intelligent animals can lead to anthropomorphism. Studies of wild cetaceans have different disadvantages, such as restricted vision, rough weather, unstable platform, noisy environment, and difficulty in locating animals, to name a few. However, in the field, one is dealing with the animals in their normal acoustic, social, and spatial environment.

This study will concern itself with sounds recorded and behavior observed at sea. The organization as well as the behavioral and environmental correlates of pilot whale whistles will be examined.

II. MATERIALS AND METHODS

Field recordings of wild North Atlantic pilot whales were made on three cruises of the *R/V Trident*. The total duration of recordings sampled was greater than 30 hr. Sounds were recorded from Newfoundland south to Hudson Canyon. In most cases simultaneous behavioral notes were recorded on a second channel of the tape recorder. In addition, written notes on behavior and other data were recorded. These written and spoken notes were the basis for selection of behavioral/environmental categories. In most cases only those behavioral/environmental categories were studied for which at least four or five events could be found. In three categories, only one encounter each was recorded, but for these events the recordings were of long duration and were sampled at multiple locations in order to obtain a representative sample.

Whistles were randomly selected in each event and a sample size of about 50 whistles was selected since this number of whistles gives 95% confidence limits that the error in the mean frequency will not exceed 0.8 kHz and the error in mean duration will not exceed 0.08 sec. In all there were 14 behavioral/environmental categories, and 822 whistles were sampled. This gives a mean sample size of 58.7 whistles. The range of sample sizes was 50–77. The largest number of encounters sampled for each category was six, with a mean of four.

When an appropriate behavioral or environmental encounter was located on a field recording from the accompanying notes, I listened to the entire event, counted whistles, and timed the event to determine a calling rate. The recording was then rewound and a random one-digit number was used to select whistles to be recorded onto another audio tape. When the appropriate number of whistles had been selected, the tape of selected whistles was processed by playing the sounds into a Saicor SA1 53B spectrum analyzer, with a Honeywell 1856 Visicorder printout. Tapes were played at ¼ speed (1⅞ in./sec) and the 0- to

The Whistle Repertoire of the North Atlantic Pilot Whale

5-kHz and 0.5-cm/sec scales of the analyzer were used. This gave a continuous real-time printout of 0–20 kHz at 2 cm/sec time scale.

Data were recorded by transcribing the whistle shape onto a table where frequency and duration were also recorded. This was done for all 822 whistles. After all whistles were transcribed, an inspection of the shapes of the sound frequency versus time traces confirmed an earlier suspicion that clear categories of whistle types would be difficult to define. It appears that the whistle shapes in pilot whales can be arranged in a continuum such that any particular whistle can be derived from any other whistle through a series of intermediates, each only slightly different from its antecedent. This matter will be discussed more fully in a later section.

It was decided that seven somewhat arbitrary categories of whistles could be defined as follows (Fig. 1): (1) level frequency—essentially no change in frequency throughout the entire duration of the whistle; (2) falling frequency—a noticeable decrease in frequency throughout all or most of the duration of the whistle; (3) rising frequency—a noticeable increase in frequency throughout all or most of the whistle; (4) up–down—a whistle in which frequency first rises and then falls: a hump-shaped display, or inverted U or V; (5) down–up—a whistle in which frequency first falls and then rises: a U or V-shaped display; (6)

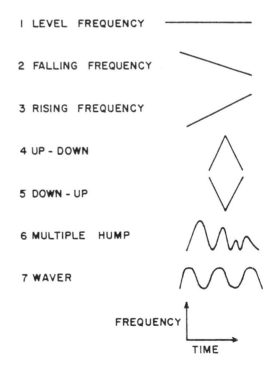

Fig. 1. The seven most common whistle types emitted by the North Atlantic pilot whale.

multiple humps—a whistle in which there are at least two frequency inflections; and (7) waver—a whistle in which there are at least two frequency inflections, but the frequency excursions are symmetrical about some mean; frequency inflections occur repeatedly at the same maximum and minimum.

Sound frequency was measured as a duration-weighted mean of the frequency excursion. For level or only slightly inclined whistles, this presents no problem. For whistles with large frequency changes, this method tends to level the extent of the frequency excursion but does provide a weighted representative index of the frequency of the whistle. If each increment of frequency change were measured and a mean derived for many whistles, the result would be the same. This method seems better than simply ignoring those whistles with significant frequency changes.

Duration was measured as the distance in centimeters from the onset to cessation of a whistle trace. This was then converted to seconds, and measured to the nearest 0.05 sec.

The 14 behavioral/environmental categories were:

1. *Lolling-Milling*. During such behavior the animals are relatively inactive, staying near the surface of the water, swimming slowly, if at all, and not in any particular direction. Nothing but blowing and slow swimming are observed.

2. *Transiting*. During this behavior, animals are moving as a group in a particular compass direction for at least several minutes, and often longer. Speed is generally between 2 and 5 knots.

3. *Excitement*. During this behavior the animals are relatively stationary, i.e., not transiting, but very active at the surface, spy-hopping, tail slapping, lobtailing, and in general thrashing and splashing at the surface.

4. *Fear-Kill*. During this single encounter, a herd of about 150 pilot whales was driven ashore in Newfoundland and shot by their captors.

5. *One Hour Pre-kill*. During this single encounter the pilot whales mentioned above were being herded by small boats toward a shoal. Loud sounds were used to drive the animals. This was presumed to be a slightly less stressful situation than number 4.

6. *Pre-kill*. During this single event recording, lasting overnight, the above-mentioned whales were being held in a relatively shallow bay by a barrier of nets and small boats; later the whales were driven ashore and slaughtered. They had not been exposed to intensive herding and were thus presumed to be only slightly stressed. Calling rate during these last three categories was highly variable and therefore not measured.

Four time-of-day environmental categories were studied, with all recordings made during late summer and early fall.

7. *Dawn*. One hour before to one hour after sunrise, from 0530 to 0730 hr.

8. *Midday*. Generally from 1200 to 1400 hr.

9. *Dusk*. One hour before to one hour after sunset, from 1830 to 2030 hr.

10. *Midnight.* From 0000 to 0200 hr.
11. *Large Herd.* There were estimated to be 50 or more individuals in the herd.
12. *Small Herd.* There were estimated to be 25 or fewer individuals, but never less than five in the herd.
13. *Ship Near.* Sounds were recorded while the animals were within 250 yards of the ship.
14. *Ship Far.* Sounds were recorded while the animals were between 0.25 and 1 mile away from the ship.

Seventeen pairwise comparisons between behavioral/environmental categories were made.

All comparisons were made on the basis of: whistle category, using the Spearman rank order correlation coefficient; mean frequency; mean duration; and mean calling rate, using either the t test or t' test where variances were unequal.

Several larger groupings of behavioral/environmental categories were compared on the basis of all four parameters using the Kendall coefficient of concordance W for whistle categories and the analysis of variance parametric test or an approximate test where the analysis of variance could not be used because of large differences in variance (Sokal and Rohlf, 1969). As a rule, parametric tests were used whenever assumptions for such tests were not seriously violated.

The following pairwise comparisons were made because the contexts were considered to represent opposite states or behaviors: lolling vs. transiting; lolling vs. excitement; lolling vs. fear; large herd vs. small herd; ship near vs. ship far; and ship near vs. lolling.

The following comparisons were made because the contexts were suspected of being similar or identical: pre-drive vs. one hour pre-drive; pre-drive vs. fear; fear vs. one hour pre-drive; all six possible pairs of time of day; and ship near vs. fear.

Multiple comparisons were made because some similarity among the contexts was suspected, or it was desired to test for such similarity. The comparison among the first three behaviors was made to determine how well the excitement context could be differentiated from the lolling and transiting contexts.

III. RESULTS

It has become evident after examining over 1500 pilot whale whistle sonograms that there are no clear mutually exclusive whistle types. Their frequency vs. time traces can be arranged in a continuum or matrix from simple to complex through a series of intermediates. Each whistle is only very slightly different from its two adjacent whistles in the continuum (Fig. 2). Bastian (1967) noted a similar property among the whistles of two captive *Tursiops truncatus*.

There are several basic components of whistles which exist either alone or in combination to form more complex whistles. These are (Fig. 3):

1. *Level Frequency Whistle* — various frequencies and durations: (a) combination of two level frequency components with abrupt frequency shift.

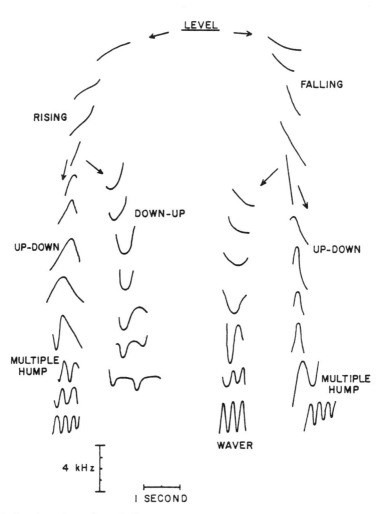

Fig. 2. Sample tracings of actual pilot whale whistles which show a continuum of intermediate contour shapes between seven basic whistle types in Fig. 1. Arrows indicate alternative paths. All tracings are drawn to the same frequency and time scales, but each whistle does not have the identical position relative to the 0-kHz baseline on the original sonogram.

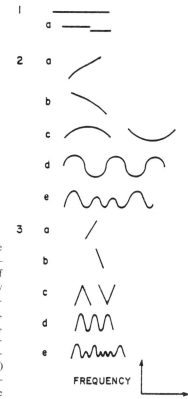

Fig. 3. Schematic diagram of pilot whale whistle continuum elements. (1) Level frequency whistle—various frequencies and durations: (a) combinations of two level frequency components with abrupt frequency shift. (2) Gradual frequency change: (a) rising; (b) falling; (c) both rising and falling; (d) multiples of (c), i.e. a sinusoidal, wavering whistle; (e) multiples of (c), i.e. asymmetrical—multiple hump. (3) Abrupt frequency changes: (a) chirp—short duration, rising frequency; (b) chirp—short duration, falling frequency; (c) V-shape—combining 3a and b; (d) multiples of 3c—waver, symmetrical; (e) multiples of 3c—multiple hump, asymmetrical.

2. *Gradual Frequency Change:* (a) rising; (b) falling; (c) both rising and falling; (d) multiples of (c), i.e., a sinusoidal, wavering whistle; and (e) multiples of (c), i.e., asymmetrical, multiple hump.

3. *Abrupt Frequency Changes:* (a) chirp, short duration, rising frequency; (b) chirp, short duration, falling frequency; (c) V-shape, combining a and b; (d) multiples of c), waver, symmetrical; and (e) multiples of c), multiple hump, asymmetrical.

Obviously whistles in categories 2 and 3 above are nearly identical, except for the sharpness of the frequency inflection or steepness of the frequency change.

The basic components can also be combined in different ways to produce all the whistles found in nature for these animals. Many combinations exist, but in nature the most common are the seven types studied in the behavioral/environmental categories (Fig. 1).

The basic and combined types differ in the following ways (see Figs. 1–3):

(1) sound frequency; (2) extent of frequency excursion; (3) slope of frequency excursion; (4) amplitude emphasis or actual omission of parts of whistles; (5) slight quaver of basic sound trace; (6) rasp or tuning sound at onset of whistle; (7) duration; and (8) number and abruptness of frequency inflections.

The abruptly rising or falling frequency whistles (3a and b in Fig. 2) are produced less often than the more gradual frequency changes (2a and b in Fig. 3), while the abrupt-inflection humped types such as multiple-humped, V-shaped, and wavering whistles (3c, d, and e) are more common than these same whistles with more gradual frequency inflections (2c, d, and e in Fig. 3).

Table I presents data on the frequency of occurrence of seven whistle types in each of the 14 contexts studied. Several categories are identified as high and low arousal states. The level of arousal or alertness is believed to reflect the level of fear or excitement, with high arousal meaning either severe stress or great excitement and alertness resulting from other causes. The significance of these states will be discussed later.

Seventeen pairwise comparisons of whistle category distribution data (Table I) were made (Table II). Of these only five had correlation coefficients high enough to justify rejection of the null hypothesis (unrelated distributions) at the 0.05 level of significance. There were no significant negative correlations.

Only lolling vs. transiting, and pre-drive vs. fear-kill pairs showed a high degree of correlation, among the behavior pairs. Two other sets of comparisons, lolling vs. excitement and lolling vs. fear-kill, showed correlation coefficients very close to 0. The other three categories showed r's between 0.5 and 0.7, indicating a moderate level of association. It seems from Table I and the correlation coefficients in Table II that the lolling and transiting categories show quite different distributions from the other four behavior categories, which are reasonably similar among themselves. This may indicate that the level of arousal has an effect on the types of whistles produced, since lolling and transiting probably represent low arousal states, while the other four behavior categories are representative of a high level of arousal or alertness (stress and excitement). Lolling and transiting are represented by a very high proportion of level frequency, falling, and rising whistles, 83% and 78%, respectively, while the other four behavior categories show 48%, 40%, 54%, and 50% respectively.

If the seven whistle categories are examined, it is possible to divide them into two larger categories: simple (level, falling, rising) and complex, (up–down, down–up, multiple-hump, and waver).

If the 14 behavioral/environmental categories are examined on this basis, there appear to be three somewhat natural groupings: (1) lolling and transiting, which contain 17% and 22% complex whistles respectively; (2) excitement (52%), fear (60%), 1-hr pre-drive (46%), pre-drive (50%), small herd (64%), and ship near (55%); and (3) the remaining six categories, large herd, ship far, and the four time-of-day periods, which have intermediate percentages, from 32% to 41%. The above groupings may be somewhat related to state of arousal,

The Whistle Repertoire of the North Atlantic Pilot Whale

Table I. Frequencies of Occurrence of 7 Whistle Types in 14 Behavioral and Environmental Categories (First Six Categories Are Considered Behaviors)

	N	Level	Falling	Rising	Up–down	Down–up	Multihump	Waver
Lolling[a]	77	26	21	17	4	0	4	5
Transiting[a]	65	15	26	10	9	0	4	1
Excitement[b]	58	4	8	16	11	2	17	0
Fear-Kill[b]	50	10	6	4	11	2	13	4
1 hr pre-drive[b]	50	8	5	14	6	1	12	4
Pre-drive[b]	50	12	4	9	12	0	11	2
Dawn[c]	50	17	5	11	8	6	5	0
Midday	64	21	12	7	10	3	11	0
Dusk	54	18	8	6	10	0	9	3
Midnight	62	22	14	6	2	6	9	3
Large herd	62	11	19	12	9	3	6	2
Small herd[b]	50	4	9	5	19	1	12	0
Ship near[b]	71	4	16	12	27	1	8	3
Ship far	59	9	16	11	14	2	7	0
Totals	822	181	169	140	150	27	128	27

[a] 2 presumed low arousal states.
[b] 6 presumed high arousal states.
[c] Remaining 6 presumed moderate arousal states.

Table II. Pairwise Whistle Category Comparisons, Using Spearman Rank Order Correlation Coefficient, rs[a]

Comparison	Correlation coefficient (rs)	rs significant at 0.05 level
Lolling vs. transiting	0.848	Yes
Lolling vs. excitement	−0.027	No
Lolling vs. fear-kill	0.156	No
Fear-kill vs. excitement	0.634	No
Pre-drive vs. 1 hr pre-drive	0.670	No
Pre-drive vs. fear-kill	0.821	Yes
1 hr pre-drive vs. fear-kill	0.527	No
Day vs. night	0.777	Yes
Dawn vs. dusk	0.482	No
Dawn vs. day	0.390	No
Day vs. dusk	0.821	Yes
Dusk vs. night	0.438	No
Dawn vs. night	0.304	No
Large herd vs. small herd	0.464	No
Ship near vs. ship far	0.893	Yes
Ship near vs. fear-kill	0.563	No
Ship near vs. lolling	0.241	No

[a] H_0 (Null hypothesis): The rankings of whistle categories by frequency of occurrence are unrelated (Siegel, 1956).

in which lolling and transiting are low arousal states. Again, excitement, fear, herding, and capture; presence of a nearby ship; and relatively few conspecifics nearby may constitute high arousal or alertness states. The remaining six categories are intermediate, or may contain whistles recorded during events of both high and low arousal.

Caldwell *et al.* (1970) suggest that high arousal levels or emotion often are associated with relatively more simple whistle types in *Tursiops truncatus*. This is directly opposite to what was found in this study of the North Atlantic pilot whale.

The four time-of-day categories showed two significantly high ($P < 0.05$) correlations among the six pairwise comparisons made (Table II). These four categories had similar rankings of level, multiple-hump, and waver whistles (Table II), but rankings of the other four whistle types were dissimilar. The whistle category distributions for small herds and large herds were not significantly correlated, while whistle types were very similar whether the ship was near or far from the animals.

Two other comparisons were made, ship near vs. fear, since both situations may represent stress situations. Correlation was moderate, but not significant. Ship near vs. lolling were also compared since the first may be a stressful situation, while the second is probably a low stress situation, and correlation here was low.

Clearly, there are some significant differences in the frequency of occurrence of difference whistles related to the behavior of the emitting animals. There are also a number of comparisons in which the frequencies of occurrence of whistles are nearly identical, even though the two behaviors or environments compared are seemingly very different. Since a large number of pairwise statistical comparisons can sometimes produce significant results by chance alone, multiple comparisons of whistle rankings in behavioral and environmental categories are presented in Table III. The Kendall coefficient of concordance, W, is used. This measures the degree of association among the ranks given each whistle type in each behavioral/environmental category. The null hypothesis tested is that there is no association among the rankings, i.e., the rankings are unrelated. In all comparisons listed in Table III, concordance is significantly different from 0, except for the comparison among lolling, transiting, and excitement. All other comparisons show concordance significant at the 0.05 level, including the postulated groups with different arousal states, although the W for the six moderate arousal states is not much higher than W for all 14 categories or all six behavior categories.

The statistical comparisons presented in Table III support the conclusion that there is, in nearly all cases, similarity in the types of whistles produced, regardless of the context situation.

Table IV presents mean whistle frequencies, mean whistle durations, and mean calling rates in the same 14 behavioral and environmental categories of Table I. The same sample of whistles was used to obtain data in both tables. Subsequent tables present statistical comparisons among the parameters presented in Table IV.

Table III. Kendall Coefficient of Concordance, W, to Measure Degree of Association among Whistle Category Rankings by Several Behavioral/Environmental Categories[a]

Comparisons among	W	SS	Significant concordance or association at 0.05 level
All 14 categories	0.514	2821	Yes
6 behavioral categories	0.564	569	Yes
4 times of day	0.621	278	Yes
8 environmental categories	0.536	961	Yes
First 3 behaviors	0.581	146.5	No
Second 3 behaviors	0.77	194	Yes
6 high arousal states	0.709	715	Yes
6 moderate arousal states	0.582	587	Yes
2 low arousal states (rs) from Table II	0.848		Yes

[a] H_0: Rankings of whistle types by frequency of occurrence in the contexts listed are not associated (Siegel, 1956). See Table I for identification of contexts listed.

Table IV. Frequency, Duration, and Calling Rate Measured in 14 Behavioral/Environmental Categories (First 6 Categories Are Considered Behaviors)

	Frequency (kHz)			Duration (sec)			Calling rate (calls/min)		
	Whistles (n)	Mean	Standard deviation	Whistles (n)	Mean	Standard deviation	Encounters (n)	Mean	Standard deviation
Lolling[a]	77	3.4	0.95	77	0.65	0.21	6	23.8	8.4
Transiting[a]	65	3.5	1.4	65	0.68	0.27	4	35.9	2.7
Excitement[b]	58	4.5	1.5	58	0.83	0.33	5	34.9	12.6
Fear-kill[b]	50	4.4	1.8	50	0.65	0.22			
1 hr pre-drive[b]	50	3.7	0.91	50	0.72	0.35			
Pre-drive[b]	50	3.4	1.6	50	0.76	0.35			
Dawn[c]	50	4.3	1.8	50	0.66	0.35	5	14.6	7.8
Midday	64	4.0	1.8	64	0.71	0.29	5	18.1	9.0
Dusk	54	4.0	1.5	54	0.79	0.32	4	24.6	6.3
Midnight	62	4.2	1.8	61	0.65	0.29	5	16.0	9.0
Large herd	62	3.5	1.5	62	0.81	0.33	5	41.4	6.5
Small herd[b]	50	3.9	1.7	50	1.0	0.42	4	15.9	1.8
Ship near[b]	71	4.7	2.0	71	0.74	0.29	6	26.7	2.3
Ship far	59	4.5	2.4	59	0.67	0.27	5	14.7	3.4

[a] Low arousal states.
[b] High arousal states.
[c] Remaining—moderate arousal states.

Table V presents results of the same 17 pairwise comparisons of Table II but, in this case, on the basis of mean whistle frequency. Only five comparisons exhibit a significant difference between mean whistle frequencies. In all others, it can be assumed that the means are equal. The lolling category is involved in three of the pairs with significantly different mean whistle frequencies. Lolling had the lowest mean frequency of any context category. The other two comparisons in which significant differences were found involved the fear category. Apparently the extreme stress of this context caused production of higher-frequency whistles, as compared to the pre-drive and 1-hr pre-drive contexts.

Table VI presents the same comparisons on the basis of mean whistle duration. Eight pairs differed significantly in mean duration. Among the behavior pairs, lolling and excitement differ significantly in mean duration, just as in mean frequency. Fear and excitement (both presumed high-arousal states) differ in

Table V. Pairwise Comparisons of Mean Sound Frequency (t or t' Test)[a]

Comparison	Mean frequencies (kHz)		t or t'	Means unequal at 0.05 significance level
Lolling vs. transiting	3.4	3.5	0.6842[b]	No
Lolling vs. excitement	3.4	4.5	5.389	Yes
Lolling vs. fear	3.4	4.4	3.7956[b]	Yes
Fear vs. excitement	4.4	4.5	0.3149	No
Pre-drive vs. 1 hr pre-drive	3.4	3.7	1.104	No
Pre-drive vs. fear	3.4	4.4	2.9199	Yes
1 hr pre-drive vs. fear	3.7	4.4	2.6639[b]	Yes
Day vs. night	4.0	4.2	0.8303	No
Dawn vs. dusk	4.3	4.0	0.9460	No
Dawn vs. day	4.3	4.0	0.8888	No
Day vs. dusk	4.0	4.0	0	No
Dusk vs. night	4.0	4.2	0.7683	No
Dawn vs. night	4.3	4.2	0.1749	No
Large herd vs. small herd	3.5	3.9	1.254	No
Ship near vs. ship far	4.7	4.5	0.4204	No
Ship near vs. fear	4.7	4.4	0.8	No
Ship near vs. lolling	4.7	3.4	5.1448[b]	Yes

[a] H_0: means are equal.
[b] t' used because of unequal variances.

mean duration. The pre-drive and fear contexts differ in mean duration as they did in mean frequency.

Only two pairs of the four time-of-day categories differ significantly in mean duration. These differences seem due to the relatively long mean duration for the dusk context. The large and small herd contexts have the two highest mean durations of all contexts but also differ significantly. The ship-near context mean duration is not equal to the mean durations of either the lolling or fear contexts.

Table VII presents 11 pairwise comparisons of mean calling rate. Six of the 17 comparisons in previous tables involved encounters for which calling rate could not be determined, and thus these six comparisons were not included in Table VII.

As can be seen, three pairs differ significantly. The first pair, lolling and transiting, differ. Transiting behavior is accompanied by a high calling rate, while lolling is closer to the mean for all contexts, and the transiting calling rate would be significantly different from nearly all other calling rates measured. It is

Table VI. Pairwise Comparisons of Mean Duration (t or t' Test)[a]

Comparison	Mean durations (sec)		t or t'	Means unequal at 0.05 significance level
Lolling vs. transiting	0.65	0.68	0.6842[b]	No
Lolling vs. excitement	0.65	0.83	3.636[b]	Yes
Lolling vs. fear	0.65	0.65	0	No
Fear vs. excitement	0.65	0.83	3.3639[b]	Yes
Pre-drive vs. 1 hr pre-drive	0.76	0.72	0.5275	No
Pre-drive vs. fear	0.76	0.65	1.9619[b]	Yes
1 hr pre-drive vs. fear	0.72	0.65	1.326	No
Day vs. night	0.71	0.65	1.156	No
Dawn vs. dusk	0.66	0.79	2.0	Yes
Dawn vs. day	0.66	0.71	0.8339	No
Day vs. dusk	0.71	0.79	1.424	No
Dusk vs. night	0.79	0.65	2.461	Yes
Dawn vs. night	0.66	0.65	0.1646	No
Large herd vs. small herd	0.81	1.0	2.823	Yes
Ship near vs. ship far	0.74	0.67	1.429	No
Ship near vs. far	0.74	0.65	1.9732	Yes
Ship near vs. lolling	0.74	0.65	2.172[b]	Yes

[a] H_0: means are equal.
[b] t' used because of unequal variances.

Table VII. Pairwise Comparisons of Mean Calling Rate (t Test)[a]

Comparison	Mean calling rates (calls/min)		t	Means unequal at 0.05 significance level
Lolling vs. transiting	23.8	35.9	2.739	Yes
Lolling vs. excitement	23.8	34.9	1.7497	No
Day vs. night	18.1	16.0	0.3689	No
Dawn vs. dusk	14.6	24.6	2.0717	No
Dawn vs. day	14.6	18.1	0.6571	No
Day vs. dusk	18.1	24.6	1.218	No
Dusk vs. night	24.6	16.0	1.611	No
Dawn vs. night	14.6	16.0	0.2629	No
Large herd vs. small herd	41.4	15.9	7.547	Yes
Ship near vs. ship far	26.7	14.7	6.936	Yes
Ship near vs. lolling	26.7	23.8	0.8146	No

[a] H_0: means are equal.

not too surprising that a large herd, with many more potential whistlers, would exhibit the highest calling rate, significantly higher than the rate of a small herd. It is possible that the difference in calling rates between the two ship-proximity categories can be ascribed to an inability to detect all the whistles of distant animals.

Table VIII summarizes all pairwise comparisons. While lolling vs. transiting and pre-drive vs. 1-hr pre-drive contexts are similar on nearly all measures, the other five behavior comparisons indicate that the respective pairs differ on most measures. No pair differs or agrees on all measures, and the pre-drive and 1-hr pre-drive pair are similar on two measures, differing only slightly (Table II) on whistle category rankings. It is probably safe to conclude that these two contexts are essentially equivalent. Lolling and excitement show the greatest degree of difference on all measures, and although the calling rates are not different to a statistically significant degree, the difference (see Table IV) is large. Lolling and transiting are very different behaviors to an observer, yet whistle parameters are surprisingly similar. The whistle parameters measured indicate that real differences exist in sound production, and that these differences may relate to behavior, but that alternatively seemingly different behaviors may be very similar on the basis of whistle parameters.

The four times-of-day categories differ primarily on the basis of whistle type used. Few differences can be detected in the other parameters measured. Apparently time of day may affect, to a degree, the whistle types emitted, but there is overall close concordance in whistle type use when all four categories

Table VIII. Summary of Pairwise Comparison Results[a]

Comparison on basis of	Whistle type (r_s)	Mean sound frequencies (t)	Mean durations (t)	Mean calling rates (t)
Lolling vs. transiting	+	+	+	−
Lolling vs. excitement	−	−	−	+
Lolling vs. fear-kill	−	−	+	NA
Fear vs. excitement	−	+	−	NA
Pre-drive vs. 1 hr pre-drive	−	+	+	NA
Pre-drive vs. fear-kill	+	−	−	NA
1 hr pre-drive vs. fear	−	−	+	NA
Day vs. night	+	+	+	+
Dawn vs. dusk	−	+	−	−
Dawn vs. day	−	+	+	+
Day vs. dusk	+	+	+	+
Dusk vs. night	−	+	−	+
Dawn vs. night	−	+	+	+
Large herd vs. small herd	−	+	−	−
Ship near vs. ship far	+	+	+	−
Ship near vs. fear-kill	−	+	−	NA
Ship near vs. lolling	−	−	−	+

[a] +: distribution or means equal or significantly correlated. −: distribution or means unequal or not significantly correlated. NA: not applicable—calling rate not determined.

are compared together in Table III. It seems safe to conclude that differences in whistle production related to time of day are probably negligible. The few significant differences may be the results of chance alone, since so many pairwise comparisons were made, and this increases the likelihood of a significant difference being found.

Size of herd seems to significantly affect the types of whistles produced, while a ship's presence appears to have little effect on whistle production. Ship-near and ship-far contexts were identical on most measures, while neither the extreme fear nor lolling context whistles were very similar to the ship-near context whistles. This may indicate that whatever effect a ship's presence may have, distance is not critical, but that the animals are neither very fearful nor very relaxed, at least on the basis of their whistling.

Tables IX–XI present analysis of variance results among several means of frequency, duration, and calling rate, respectively. The comparisons are the same ones made in Table III for whistle category use.

Due to unequal variances, the analysis of variance test could not be applied

Table IX. Analysis of Variance—Mean Frequency Comparisons[a]

Comparisons among	F' or F	Means unequal at 0.05 significance level
All 14 categories	5.1487	Yes
6 behavioral categories	6.953	Yes
4 times of day	0.444[b]	No
8 environmental categories	2.791	Yes
First 3 behaviors	11.637	Yes
Second 3 behaviors	4.456	Yes
6 high arousal states	5.913	Yes
6 moderate arousal states	2.274	Yes
2 low arousal states (t')	0.6842	No

[a] H_0: all means equal. See Table IV for identification of contexts listed.
[b] F.

to most comparisons of mean frequencies and durations. In those cases where variances were unequal, an approximate test (F') was used to compare means (Sokal and Rohlf, 1969). This tests whether the samples were drawn from populations with equal means.

Table XII summarizes all multiple comparisons among different contexts. In general, as seen in Table III, there is association among the rankings of whistle types by nearly all context situations. It would be surprising to find that all rankings were independent and not associated in any way, since the ranked whistles were all produced by the same species. The most interesting result is that while the two low arousal states, lolling and transiting, are very similar in

Table X. Analysis of Variance—Mean Duration Comparisons[a]

Means compared among	F' or F	Means unequal at 0.05 significance level
All 14 categories	3.9895	Yes
6 behavioral categories	4.873	Yes
4 times of day	2.396[b]	No[c]
8 environmental categories	7.664[b]	Yes
First 3 behaviors	2.703	No
Second 3 behaviors	1.986	No
6 high arousal states	4.752	Yes
6 moderate arousal states	2.749	Yes
2 low arousal states (t')	0.6842	No

[a] H_0: all means equal. See Table IV for identification of contexts listed.
[b] F.
[c] Marginal $0.05 < P < 0.01$.

Table XI. Analysis of Variance—Mean Calling Rate Comparisons[a]

Comparisons	F	Means unequal at 0.05 significance level
All 11 categories	9.5142	Yes
First 3 behaviors	4.2133	Yes
4 times of day	1.247	No
8 environmental categories	10.456	Yes
6 moderate arousal states	10.116	Yes
2 low arousal states (t)	2.739	No

[a] H_0: all means equal. See Table IV for identification of contexts listed.

whistle use on all measures, when these contexts are compared with the excitement context (first three behaviors) large differences are evident, indicating that whistle production differs significantly when the animals are highly excited. Other results in Table XII do not contradict conclusions drawn from Table VIII (summary of pairwise comparisons). Real differences in whistle production by pilot whales exist, and these differences may relate to behavioral and environmental context.

The earlier conclusion that time of day does not significantly affect whistling, however, is also supported by the multiple comparisons of Table XII. However, when all eight environmental contexts are compared, the resultant differences in means suggest that herd size and ship proximity do affect whistling.

Table XII. Summary of Multiple Comparisons among Behavioral/Environmental Categories[a]

	Whistle categories (W)	Frequency means (F)	Duration means (F)	Calling rate means (F)
All 14 categories	+	−	−	−[b]
6 behaviors	+	−	−	NA
4 times of day	+	+	+	+
8 environmental categories	+			−
First 3 behaviors	−	−	+	−
Second 3 behaviors	+	−	+	NA
6 high arousal states	+	−	−	NA
6 moderate arousal states	+	−	−	−
2 low arousal states (r_s and t')	+	+	+	−

[a] See Table IV for identification of contexts listed. +: distributions or means equal or significantly correlated. −: distributions or means unequal or not significantly correlated. NA: not applicable, calling rates not all determined.
[b] Calling rate measured in 11 categories.

Table XIII. Correlation Coefficients, r^a

Comparison	Between mean frequency and mean duration	Between mean frequency and mean calling rate	Between mean duration and mean calling rate
All 14 or 11 categories, (rate)	−0.08	−0.368	0.168
6 behaviors	−0.334	NA	NA
4 times of day	−0.8	−0.796	0.9754[b]
8 environmental categories	−0.504	−0.509	0.229
First 3 behaviors	0.9978[b]	0.51	0.569
Second 3 behaviors	−0.995[b]	NA	NA
6 high arousal states	−0.159	NA	NA
6 moderate arousal states	−0.837[b]	−0.927[b]	−0.881[b]

[a] H_0: correlation coefficient is 0 (Sokol and Rohlf, 1969). NA: not applicable—calling rates not determined. See Table IV for identification of contexts listed.
[b] Significant correlation at 0.05 significance level.

The second three behavior contexts were all drawn from a single long series of recordings of a large (150 animal) herd in a Newfoundland bay. These animals were subjected to various degrees of stress at various times, but only the mean frequency of whistles varied. Calling rate was highly variable over time and could not be correlated to the degree of harassment.

The three previously postulated groups of contexts with different arousal levels are similar on the basis of whistle type use, but not on other measures, except for lolling and transiting. These two behaviors appear to the observer to be very different, but the whistling behavior during these activities is very similar, except for calling rate. Perhaps the higher calling rate during transiting (Table IV) is useful in maintaining herd cohesion while underway.

While fear and excitement are probably both high arousal states, they are certainly not equivalent states, and this is reflected in the comparisons of six high arousal states in Table XII. Similarly, "moderate" arousal may also be too simplistic a grouping of behaviors in the pilot whale. Likely, many other factors affect whistling, and these cannot all be observed from shipboard.

Table XIII presents correlation coefficients between mean frequency and mean duration, mean frequency and mean calling rate, and mean duration and mean calling rate for the same multiple category comparisons presented in previous tables. Several comparisons show significantly high correlations. The six moderate arousal states show significant negative correlations for all three pairs of parameters. However, since these six contexts are six of the eight environmental contexts compared in Table XIII, and there is no significant correlation for those eight contexts, perhaps there is something atypical about whistle production in the small herd and ship-far context. The ship-far context has the second lowest calling rate and second highest mean frequency among all contexts, while the large herd context exhibits the highest calling rate (Table IV).

Table XIV. Kendall *(W)* Coefficients of Concordance among Ranks of Mean Frequency, Mean Duration, and Mean Calling Rates[a]

Comparisons	SS	W	Significant concordance or association at 0.05 level
All 11 categories	358	0.3616	No
4 times of day	9	0.2	No
8 environmental categories	88.5	0.234	No
First 3 behaviors	14	0.777	No
6 moderate arousal states	23	0.146	No

[a] H_0: rankings of means in contexts listed are not associated (Siegel, 1956). See Table IV for identification of contexts listed.

These extreme values distinguish these contexts from the other environmental categories. Among the behavior categories there is a strong positive correlation between frequency and duration in the first three behaviors (lolling, transiting, and excitement) and a strong negative correlation between frequency and duration for the remaining three categories (fear, 1-hr pre-drive, and pre-drive).

This relationship may be the result of population differences, since the first three behavioral contexts were recorded from animals off Nova Scotia, while the second three were recorded near Newfoundland. Alternatively, this may reflect a difference in whistling depending on whether or not the animals are under severe stress or harassment. However, since the six moderate arousal states also show strong negative mean frequency and duration correlation, and are Nova Scotia animals, neither explanation seems fully satisfactory.

Duration and calling rate are positively correlated for the four times of day only. I see no obvious explanation for this relationship.

The Kendall coefficient of concordance *W* was calculated comparing ranks of mean frequency, duration, and calling rate for all appropriate multiple comparisons (Table XIV). There is no significant association among ranks of means in each behavior or environmental category.

IV. DISCUSSION

An examination of the frequency of occurrence of each whistle type (totals in Table I) reveals that there are two classes of whistles. The five high-frequency-of-occurrence whistles (level, falling, rising, up–down, and multiple hump) do not differ significantly among themselves, but are significantly different from the two categories with the lowest counts (down–up and waver).

Moles (1963) has suggested that such rare sounds may convey specific information, beyond simply signature or individual identification. There seems to be no other obvious explanation for the great difference in frequency of occurrence. The down–up or V-shaped whistle is simply the mirror image of the far more usual up–down or hump-shaped whistle, yet the up–down occurs 5.5 times as often. It has been established that individual signature whistles do occur in porpoises (Caldwell et al., 1969, 1970, 1973; Caldwell and Caldwell, 1971). The frequent repetition of an identical signal in pilot whales, presumably by the same individual, makes it reasonable to conclude that such individual signatures exist for this species also. It also seems reasonable to conclude that the level, falling, rising, up–down, and multiple-hump whistle types represent these signature sounds, while the two rare vocalizations, down–up and waver, may convey other than individual signature information. The down–up whistle was found with greatest frequency in the dawn and night contexts (six each; see Table I). Perhaps this whistle is used for herd organization or cohesion. No clear association between context and the waver whistle is evident.

Pilot whale whistling clearly varies with behavioral and environmental context. In certain cases, striking differences in whistle usage appear to be associated with striking differences in nonvocal behavior. The presumed level of arousal may somewhat account for differences in whistle usage, but the concept of arousal level is an inadequate explanation of variations in mean whistle frequency, duration, and calling rate. It is not at all certain that a larger sample size would produce clearer results since the lolling context, with the largest number of whistles sampled (77; see Table IV) from the largest number of encounters (six), showed a very low standard deviation of both mean frequency and mean duration.

Perhaps there are context-specific sounds produced by pilot whales, beyond what may be their normal individual signature or localization sounds, but it is quite possible that the situations I have designated as different contexts are not recognized as being different by the animals themselves.

Duration is an easily measurable parameter of pilot whale whistles, unlike most others, but perhaps it is of little significance to pilot whales themselves. There are situations where long signal duration would be helpful in conveying information, but perhaps these situations are relatively rare in the environment of the pilot whale. Their vocalizations may primarily be intended to reach nearby animals, and noise or other jamming may be countered by increased amplitude, repetition of signal, or frequency sweeping. No systematic variation in mean whistle duration is evident in the data of Table IV.

The mean whistle frequencies listed in Table IV give some index of the frequency band used in different contexts. The numerical mean reflects the relative proportion of whistles with high and low frequencies. Only a few categories contained enough higher frequency whistles to raise the mean far above 4 kHz. Again there seems to be no clear relationship between context and whistle frequency. There seems to be no clear evidence that this whistle parameter is of great biological significance to the pilot whale or that it clearly reflects the behavioral or environmental context.

Variation in herd size may be the simplest explanation for observed differences in calling rate in different contexts. There was a very great difference between the calling rates of large and small herds. Herd size was not always recorded in field notes, and frequently it is very hard to determine accurately. Also for certain contexts, only a few encounters could be located and, therefore, herd size could not be used as a criterion in selection of encounters sampled.

In conclusion, there are significant variations in the parameters of pilot whale whistles. In some cases these variations can be associated with behavioral or environmental context in a logical manner, while in other cases, the causes of variation, if other than chance, are not clear.

This is the first attempt to specify the entire whistle repertoire of an odontocete. This study sampled whistles from an extensive series of recordings representing a large variety of behavioral and environmental contexts. The recordings were obtained over a period of several years, from a broad geographical area, at roughly the same season of the year. This is the most extensive sample ever used in a study of this type.

While it is impossible to state that this paper describes the *complete* repertoire of the pilot whale, it can be stated that the continuum or matrix concept outlined earlier can fully describe all whistles in the recordings studied. Likely, this concept can be used to describe all whistles of *Globicephala melaena*. From a study of dialects in pilot whales (Taruski, 1976), it also appears that this concept may adequately describe the whistles of pilot whales from other geographic locations such as the West Indies, or the Pacific, whether these pilot whales are recognized as *Globicephala melaena* or *Globicephala macrorhynchus*. Whistle vocalizations of other odontocetes may show a similar organization, if an adequate number of whistles is studied.

Dreher and Evans (1964) presented a figure (Chart 1) depicting 32 whistle contours variously used by four odontocetes. The Pacific pilot whale is listed by them as using only six of the 32 contours. However, their brief recordings were from small groups of pilot whales and may not be representative of the species. The same may be true for the recordings of the other species they studied.

My examination of North Atlantic pilot whale whistles has shown that all but one or two of the 32 contours listed by Dreher and Evans (1964) are used by the North Atlantic pilot whale. In addition, many other somewhat similar contours, not listed by Dreher and Evans (1964), are used by the North Atlantic pilot whale.

Similarly, the whistles depicted by Busnel and Dziedzic (1966) are only a few of the many different contours encountered in this study. Seven archetypal whistle categories were earlier defined, but if a complete chart of every different contour used by the North Atlantic pilot whale were prepared, it would involve several tens of whistles more than Dreher and Evans (1964) figured, and even then slight variants could be found. Therefore, I believe that the continuum or matrix concept outlined earlier is the best method to describe the whistle repertoire of the North Atlantic pilot whale.

V. SUMMARY

The whistle repertoire of the North Atlantic pilot whale *(Globicephala melaena)* is shown to be a continuum or matrix in which no mutually exclusive contour categories can be defined. Each whistle is related to every other whistle through a variable series of naturally occurring intermediates. Whistle parameters other than shape of contour also vary. These parameters are frequency, duration, and calling rate. For this study, seven whistle contour categories were defined. Fourteen behavioral and environmental contexts are compared on the bases of whistle contour use, mean whistle frequency, mean whistle duration, and mean calling rate. Differences in these parameters exist and can in some cases be related to context and arousal state. High arousal is associated with a high proportion of complex whistles. Few differences in whistle parameters were found at different times of day. Two behaviors, lolling and transiting, which appear very different to the observer show few differences in whistling.

The existence of individual signature whistles in pilot whales was discussed. Five of the seven whistle contour categories were suggested as signature whistles.

ACKNOWLEDGMENTS

The author was supported through an Office of Naval Research contract, number N00014-68-0215-003 to H.E. Winn. Funds for equipment and supplies were provided by Professor Howard E. Winn and the Graduate School of Oceanography, University of Rhode Island. Howard E. Winn, Melba C. Caldwell, David K. Caldwell, and René-Guy Busnel kindly criticized the manuscript.

This report was submitted in partial fulfillment of the requirements for the author's PhD at the University of Rhode Island.

REFERENCES

Bastian, J., 1967, The transmission of arbitrary environmental information between bottlenose dolphins, in: *Les Systems Sonars Animaux, Biologie et Bionique* (R.-G. Busnel, ed.), Vol. 2, pp. 803–873, Laboratoire de Physiologie Acoustique, Jouy-en-Josas, France.

Busnel, R.-G., and Dziedzic, A., 1966, Acoustic signals of the pilot whale *Globicephala melaena* and of the porpoise *Delphinus delphis* and *Phocoena phocoena,* in: *Whales, Dolphins, and Porpoises* (K. S. Norris, ed.), pp. 608–648, University of California Press, Berkeley.

Busnel, R.-G., Escudie, B., Dziedzic, A., and Hellion, A., 1971, Structure des clics doubles d'echolocation du globicephale (Cetace odontocete), *C. R. Acad. Sci.* **272**:2459–2461.

Caldwell, M. C., and Caldwell, D. K., 1971, Statistical evidence for individual signature whistles in Pacific whitesided dolphins, *Cetology* **3**:1–9.

Caldwell, M. C., Caldwell, D. K., and Hall, N. R., 1969, An Experimental Demonstration of the Ability of an Atlantic Bottlenosed Dolphin to Discriminate between Whistles of other Individuals of the same Species, Los Angeles County Museum of Natural History Foundation, Tech. Rept. 6, LACMNHF/MRL ONR contract N00014-67-C-0358, 35 pp. (processed).

Caldwell, M. C., Caldwell, D. K., and Turner, R. H., 1970. Statistical Analysis of the Signature Whistle of an Atlantic Bottlenosed Dolphin with Correlations between Vocal Changes and Level of Arousal, Los Angeles County Museum of Natural History Foundation, Tech. Rept. 8, LACMNHF/MRL ONR contract N00014-67-C-0358, 40 pp. (processed).

Caldwell, M. C., Caldwell, D. K., and Hall, N. R., 1973, Ability of an Atlantic bottlenosed dolphin *(Tursiops truncatus)* to discriminate between and potentially identify to individual, the whistles of another species, the common dolphin *(Delphinus delphis)*, *Cetology* **14**:1–7.

Dreher, J. J., 1966, Cetacean communications: A small-group experiment, in: *Whales, Dolphins and Porpoises* (K. S. Norris, ed.), pp. 529–541, University of California Press, Berkeley.

Dreher, J. J., and Evans, W. E., 1964, Cetacean communication, in: *Marine Bio-Acoustics* (W. N. Tavolga, ed.), Vol. 1, pp. 373–393, Academic Press, New York.

Lilly, J. C., 1961, *Man and Dolphin,* Doubleday, Garden City, New York.

Lilly, J. C., 1967, *The Mind of the Dolphin,* Doubleday, Garden City, New York.

Moles, A., 1963, Animal Language and information theory, in: *Acoustic Behavior of Animals* (R.-G. Busnel, ed.), pp. 112 –131, Elsevier, Amsterdam.

Poulter, T. C., 1968, Marine mammals, in: *Animal Communication* (T. A. Sebeok, ed.), pp. 405–465, Indiana University Press, Bloomington.

Schevill, W. E., 1964, Underwater sounds of cetaceans, in: *Marine Bio-Acoustics* (W. N. Tavolga, ed.). Vol. 1, pp. 307–316, Academic Press, New York.

Siegel, S., 1956, *Nonparametric Statistics,* McGraw-Hill, New York.

Sokal, R. R., and Rohlf, F. J., 1969, *Biometry,* W. H. Freeman Co., San Francisco.

Taruski, A. G., 1976, Possible dialects in the pilot whale *(Globicephala* spp.*),* Sounds and Behavior of the Pilot Whale, PhD thesis, University of Rhode Island.

Chapter 11

THE WHISTLE OF THE ATLANTIC BOTTLENOSED DOLPHIN *(Tursiops truncatus)*— ONTOGENY

Melba C. Caldwell and David K. Caldwell

Biocommunication and Marine Mammal Research Facility
University of Florida
Route 1, Box 121
*St. Augustine, Florida 32084**

I. INTRODUCTION

Data gathered on whistles of 126 Atlantic bottlenosed dolphins *(Tursiops truncatus)* of assorted sizes and both sexes indicate that the whistle varies with age in several parameters. Most, but not all, of the changes occur within the first two years of life.

The Atlantic bottlenosed dolphin is a long-lived species, only reaching sexual maturity in northeastern Florida at about 12 years in the female and 13 years in the male (Sergeant *et al.*, 1973). The learning period is therefore prolonged, and there appears to be considerable variability between individuals as to when the fully-developed and generalized adult whistle makes its appearance. This report deals with those changes which we have observed as they occur in time.

We are presenting longitudinal data from a limited number of newborn infants. However, in conjunction with data obtained from animals captured from the wild while still infants and from those animals taken captive at larger sizes, it would appear that sufficient information on the ontogeny of the dolphin whistle has been assembled to warrant an examination of the subject.

* In association with the State University System of Florida Institute for Advanced Study of the Communication Processes and with the Cornelius Vanderbilt Whitney Laboratory for Experimental Marine Biology and Medicine of the University of Florida.

All data presented on newborns were gathered from animals born in commercial oceanaria. Included also are a few larger infants which were taken from the wild when this practice was legal (i.e., prior to the passage of the Marine Mammal Protection Act of 1972 which now forbids the taking of such animals still nursing). Some of the infants were followed closely and recorded frequently, while others were recorded only occasionally.

The results we are presenting here emerged from unstructured rearing conditions. Aronson (1966) suggested the structuring of the acoustic environment of infant dolphins, but for several reasons this is difficult or perhaps impossible.

II. LITERATURE REVIEW

Whistling by infant Atlantic bottlenosed dolphins was noted first by McBride and Kritzler (1951), who stated: "It is vocal from the moment of birth; the high-pitched whistling which all dolphins observed by us use for ordinary communication is very much in evidence. The infant and its mother seem to be in constant vocal communication as shown by the bubbles escaping from their blowholes when whistling is heard." Essapian (1953) referred to "calls" by baby dolphins of this species, and from his context the word "call" refers to the whistle. Wood (1954) also noted whistles by young *Tursiops truncatus* — one a nursing female and the other a young male. In that same paper, Wood also noted the use of bubbles emitted from the dolphin's blowhole as an aid to identification of vocalizing individuals in a colony.

It has been observed that young dolphins rarely emit whistles of more than one sound loop (D.K. Caldwell and M.C. Caldwell, 1968, 1972). The contours drawn by Dreher and Evans (1964, Tables 1 and 2) for three juveniles also indicate a low incidence of multilooped whistles in this group. Again, the young female estimated as a juvenile whose whistle was analyzed and figured by Powell (1966, Fig. 4) apparently emitted only a single-looped whistle.

In spite of the limited references to whistles of infant and juvenile dolphins in the literature, then, certain important points have already been made. The whistle is present at birth; it is not confined to a single sex; there is no indication of seasonality; and multilooped whistles are not commonly reported for young animals. These points receive additional confirmation from the following material.

III. METHODS

Comparative data on the postinfantile groups are summarized from material presently in manuscript (M.C. Caldwell and D.K. Caldwell, manuscript).

Briefly, these data are grouped according to individual and based upon 1–11 recording sessions of each animal. Recording sessions spanned time periods of one day to 4½ years. No differences were found between sexes in the acoustic parameters under consideration here, and the two sexes are therefore grouped in the age-group subsamples.

Age classes were assigned by standard length of the animal (tip of upper jaw to deepest part of fluke notch in a straight line) as follows:

Infancy (age class I) is designated as equivalent to a yearling (i.e., birth through 12 months). There are few published data available for this class, but a regression line based on limited data places this figure at approximately 170 cm in maximum standard length.

Juveniles (age class II) are arbitrarily designated as those animals 170–210 cm in standard length (or about six years of age).

Subadults (age class III) are designated as animals 210–235 cm in standard length in females (about 12 years old) and 210–247 cm in standard length in males (about 13 years old).

Adults (age class IV) are animals above subadult size; but this category may include shorter individuals if they are known to have bred, as the figures for sexual maturity of 235 and 247 cm for females and males, respectively, are only average sizes of sexual maturity for animals in northeastern Florida (see Sergeant *et al.*, 1973).

For infants born in captivity, the true age is known, and this information is included.

The infants are described individually. If born in captivity, brief summaries of rearing conditions are included, as it is conceivable that the acoustic environment may affect the development of the whistle. If captured in the wild, locality of capture is given, as the question of the existence of dialects has yet to be resolved.

Details of age and recording conditions accompany each sound spectrogram. Whistles of newborn infants were found to be too tremulous and quavery to qualify as stereotyped or "signature whistles." The absence or presence of stereotypy during a particular recording session is indicated for each sound spectrogram, and the animals' ages are given. For those recording sessions on infants made prior to the appearance of a stereotyped whistle, whistles which appeared representative of that particular session were selected for sound spectrograms.

Average durations were obtained by reducing the recorded speed eightfold and timing by stopwatch. Number of loops were counted by ear at ⅛ recorded speed and checked at intervals on a Listening, Inc., MSA-1 sound spectrum analyzer. Frequencies were spot checked on a Kay sonograph in addition to the MSA-1 analyzer. The latter is somewhat limited in resolution, but valuable for examining large quantities of data rapidly as an intermediate adjunct between the ear and the Kay sound spectrograms. Many of the infantile whistles were barely audible above background noise, and the Kay sonograms reflect this problem.

IV. EQUIPMENT

Underwater whistles of the animals were usually recorded using a system of an Atlantic Research Corporation model LC-57 hydrophone with a special preamplifier designed and built for the system by William W. Sutherland, then of the Lockheed-California Company. Tape recorders, used interchangeably, were all Uhers (models 4000 Report-S, 4000 Report-L, or 4400 Report-Stereo). On some occasions a hydrophone and preamplifier designed and built by Dr. Thomas C. Poulter of the Stanford Research Institute were used in conjunction with the same Uher recorders.

Recordings were made in air using the same series of Uher recorders with compatible microphones.

All tape recordings were made at a tape speed of 19 cm (7.5 in.) per second.

Sound spectrograms (Sonagrams) were prepared on a Kay Sona-Graph model 662A sound spectrum spectrograph analyzer calibrated in two sections from 85 Hz to 12 kHz (when the recorded tape speed is reduced by half and fed into the analyzer the response of the latter is increased to 24 kHz).

All sound spectrograms illustrated herein were made with an effective filter bandwidth of 600 Hz.

V. DIFFERENCES IN WHISTLES BY AGE CLASS

The parameters examined for differences between age classes are based upon averages for individuals. These parameters include whistle durations, numbers of sound loops, and degree of stereotypy and frequency modulation, in other words, those areas which were quantifiable with some degree of accuracy. Tremulousness, overtones, and breaking whistles as a function of age are discussed, but no attempt has been made to quantify these.

For various reasons, one or more of the 126 animals on which data were analyzed has been eliminated from a particular analysis. The number of animals within each table or subsample may therefore vary.

Table I. Average Number of Loops per Whistle per Individual (Based on a Minimum of 25 Whistles per Individual) by Age Class

	Age class			
	I	II	III	IV
Number	14	37	44	24
Mean	1.1	1.9	2.3	2.3
Mode	1.0	2.0	3.0	2.0

Table II. Maximum Number of Loops per Whistle per Individual (Based on a Minimum of 25 Whistles per Individual) by Age Class

	Age class			
	I	II	III	IV
Number	13	37	43	24
Mean	2.3	3.4	4.7	5.1
Mode	1.0	1.0	5.0	4.0

A. Sound Loops

There is variability between individuals in both the average number of loops per whistle and the maximum number per whistle. This appears to be a size-dependent variable to some extent, particularly as regards the infants. No newborn infant in the sample, for instance, ever recycled the sound loop, and only one adult did not. Only slightly over 20% of all yearlings recycled the sound loops, but almost 80% of the animals were recycling by the end of the juvenile period.

Although even individual dolphins of the same size, sex, and locality of capture showed considerable variability in the average number of loops per whistle, grouped data by age class do indicate a general increase with size through the subadults in average number of loops (Table I) and through the adults in maximum number of loops (Table II). Modal values, however, decrease in the adults, a fact which we attribute to the over-all reduction in activity level which is characteristic of most mature dolphins.

B. Durations

Whistle durations are to a large extent a function of the number of sound loops which they contain. Although considerable variability exists between individual dolphins in the average duration of their normal sound loops, whistle durations in general follow the same pattern of increase with age through the subadults as do the numbers of loops. The adults show no additional increase in average durations over the subadults, and the modal value for the adults is again reduced (Table III). The means of the maximum durations within each age class indicate an increase through the subadult stage, and the modal value again decreases in the adults (Table IV).

C. Frequency Modulations

Newborn infants do not show the degree of frequency modulation in the whistle which most Atlantic bottlenosed dolphins develop later. However, the

Table III. Average Whistle Durations (sec) per Individual (Based on a Minimum of 25 Whistles per Individual) by Age Class

	Age class			
	I	II	III	IV
Number	14	37	44	24
Mean	0.7	0.9	1.0	1.0
Mode	0.5	0.9	1.0	0.9

whistles of the infants in this sample did show a rapid development of frequency sweep. There was some slight additional increase in the range with age class beyond the juveniles, but this varied too widely between individuals and too randomly with sex for us as yet to assign any particular significance to this factor (Table V). The development of the frequency modulation in the infants can be seen more clearly in the section on individuals of age class I.

A postinfantile animal that had a tendency to flatten its whistle when under extreme duress was encountered occasionally. This we attribute to a momentary loss of muscular control. During a given recording session, these flattened whistles gradually phased into the normal frequency-modulated whistles characteristic to the animal being recorded.

D. Stereotypy

Infants are not born with a stereotyped whistle. Although the newborn infants in this sample emitted sounds that were classified as whistles, the sounds themselves were tremulous, quavery, and had little frequency modulation. These qualities were sufficiently characteristic of all the newborns to lead us to believe that we can differentiate infantile whistles from those emitted by older animals and that members of a dolphin school could easily discriminate these whistles from others.

There are also individual differences in "vocal quality" between the whistles of the various infants in this sample, but we have not attempted to

Table IV. Maximum Durations (sec) of Whistles per Individual (Based on a Minimum of 25 Whistles) by Age Class

	Age class			
	I	II	III	IV
Number	14	37	44	24
Mean	1.4	1.6	2.0	2.0
Mode	1.0	1.5	2.0	1.5–2.0

Table V. Frequency Modulation (kHz) in Sound Loops by Age Class

	Age class			
	I	II	III	IV
Number	10	29	41	23
Mean	9.0	10.4	10.0	11.4
Mode		(Too scattered to establish)		

analyze this more complex acoustic problem. One would suspect, however, that a mother could recognize her own infant based only on the vocal quality of the whistle, but this needs experimental investigation. Stereotypy, with its clearly defined frequency and amplitude modulation patterning, is much more easily heard and demonstrated.

The development toward the generalized or typical whistle by the various infants is treated by individual in the next section because of the considerable variability between the subject animals and the ages at which they were recorded.

Due to this individual variability, the 83% figure in age class I in Table VI is a gross oversimplification, i.e., it represents many more recordings of large infants, which were handled or isolated more frequently, than the smaller ones, which were handled only in emergency situations. Most infants had attained stereotypy prior to the end of their first year.

Table VI indicates that, although the modal value for stereotypy remains at 100% across all age classes, the mature animals exhibit a decrease in stereotypy by any criterion. We do not presently attribute this to any maturational process by which the adults are "adding to their vocal repertoire." Rather, it appears that the loss of stereotypy that appears in some individuals, particularly the mature ones, correlates more frequently with lower levels of arousal in which the

Table VI. Percentage of Stereotypy of Whistles of Individuals by Age Class

	Age class			
	I	II	III	IV
Number	13	37	43	25
Mean	83	96	96	91
Mode	100	100	100	100
Percentage of individuals exhibiting less than 95% stereotypy	23	16	23	33
Percentage of individuals exhibiting 100% stereotypy	46	46	51	33

animals were minimally motivated toward vocalizing. We should also note that dolphins, like dogs, vocalize even when they appear to be asleep. The sounds which they emit in this state include abnormal whistles. As the younger animals rest infrequently and rouse more quickly, their whistles do not reflect the same degree of loss of stereotypy from this factor as do those of the adults.

There are other occasions when whistles of postinfantile animals may exhibit a loss of stereotypy. These are mentioned briefly in the next sections, as there are similarities between these and the infantile whistles.

VI. ANALYSES OF WHISTLES OF INDIVIDUAL INFANTS

A. Infant Number 1

Male, born in captivity at Marineland of the Pacific near Los Angeles, California, of parents captured in northeastern Florida. Recording sessions began 19 hr following birth and continued for 17¼ months. Housed in a community of nonperforming dolphins that included up to seven *Tursiops truncatus* of various sizes and both sexes, as well as two female eastern Pacific white-sided dolphins *(Lagenorhynchus obliquidens)*. Recorded aproximately every 2–3 weeks with intermittent sessions interspersed.

The whistles emitted by this animal during the first few days following birth were tremulous and flattened, with only a faint suggestion of a loop structure (Figs. 1 and 2). Greater frequency modulation was apparent after a month (Fig. 3), and even more apparent by the age of three months (Fig. 4).

This dolphin's whistles were never classified as stereotyped, thus making him at 17¼ months the oldest animal in the sample without stereotypy. Although considerable similarity does exist between many of the whistles after three months (Figs. 4 through 10), they were for the most part either too variable or lacking in sufficient acoustic distinctiveness to the human ear to qualify as signature whistles. No whistles of more than a single loop were ever noted.

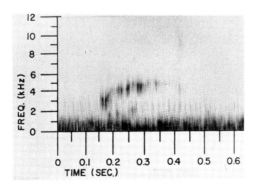

Fig. 1. Whistle emitted by infant number 1 when he was one day old. Recorded by hydrophone in water. Whistles during recording session were not categorized as stereotyped.

The Whistle of the Atlantic Bottlenosed Dolphin   377

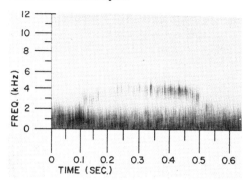

Fig. 2. Whistle emitted by infant number 1 when he was three days old. Recorded by hydrophone in water. Whistles during recording session were not categorized as stereotyped.

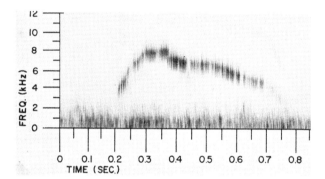

Fig. 3. Whistle emitted by infant number 1 when he was one month old. Recorded by hydrophone in water. Whistles during recording session were not categoriezed as stereotyped.

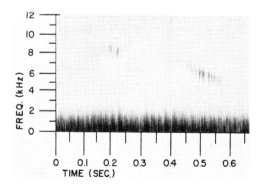

Fig. 4. Whistle emitted by infant number 1 when he was three months old. Recorded by hydrophone in water. Whistles during recording session were not categorized as stereotyped.

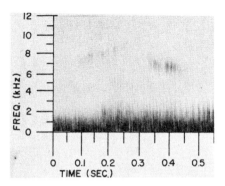

Fig. 5. Whistle emitted by infant number 1 when he was 7½ months old. Recorded by hydrophone in water. Whistles during recording session were not categorized as stereotyped.

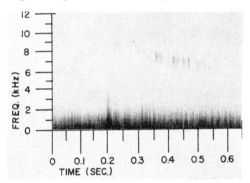

Fig. 6. Whistle emitted by infant number 1 when he was 8 months old. Recorded by hydrophone in water. Whistles during recording session were not categorized as stereotyped.

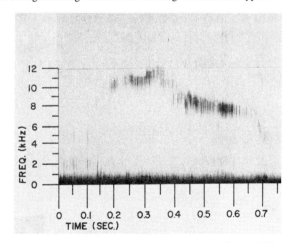

Fig. 7. Whistle emitted by infant number 1 when he was 9½ months old (155 cm standard length). Recorded by microphone in air. Whistles during recording session were not categorized as stereotyped, but noted as being "roughly the same."

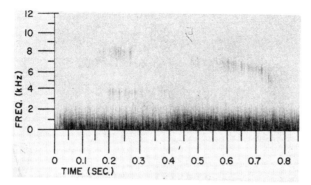

Fig. 8. Whistle emitted by infant number 1 when he was 14 months old. Recorded by hydrophone in water. Whistles during recording session were not categorized as stereotyped.

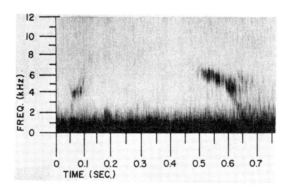

Fig. 9. Whistle emitted by infant number 1 when he was 14½ months old. Recorded by hydrophone in water. Whistles during recording session were not categorized as stereotyped.

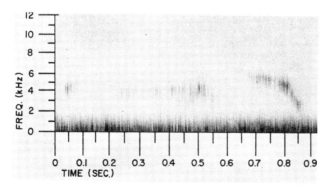

Fig. 10. Whistle emitted by infant number 1 when he was 17½ months old. Recorded by hydrophone in water. Whistles during recording session were not categorized as stereotyped.

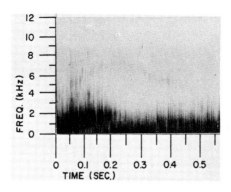

Fig. 11. Whistle emitted by infant number 2 on the day that he was born. Recorded by hydrophone in water. Whistles during recording session were not categorized as stereotyped.

It is not known whether the presence of the two eastern Pacific white-sided dolphins in the tank contributed to this absence of development of stereotypy. There is some slight degree of similarity between the whistles of this infant and those of the more vocal white-sided dolphin of the two. This latter animal also had a more variable whistle than most dolphins (M.C. Caldwell and D.K. Caldwell, 1971, animal number 1, Figs. 2–12). The possibility that these whistles may have affected the whistle development of the infant bottlenosed dolphin cannot be eliminated.

B. Infant Number 2

Male, born in captivity at Marineland of the Pacific of parents captured in northeastern Florida. Recording sessions began during birth and continued for 5½ months. Same community and conditions as infant number 1.

Whistles emitted by this individual on the day he was born were tremulous with limited frequency modulation (Fig. 11). The whistle became less tremulous

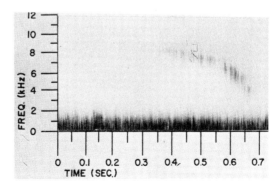

Fig. 12. Whistle emitted by infant number 2 when he was 6 weeks old. Recorded by hydrophone in water. Whistles during recording session were not categorized as stereotyped.

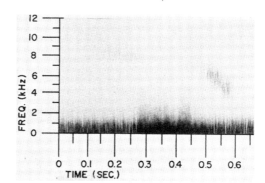

Fig. 13. Whistle emitted by infant number 2 when he was 7 weeks old. Recorded by hydrophone in water. Whistles during recording session were not categorized as stereotyped.

in the succeeding weeks with some stability indicated by the age of 2½ months (Figs. 12–15). The whistle of this infant did not sweep a wider frequency at 5½ months than it did at 2½ months, but was more stable (Fig. 16). The whistle was never classified as stereotyped, although there was considerable similarity between most of his whistles after 2½ months. No recycling of the sound was noted.

As with infant number 1, the effects of exposure to the sounds of the eastern Pacific white-sided dolphins are not known.

C. Infant Number 3

Male, captured off Fernandina Beach, Florida, with his mother and three other animals of the same herd. Estimated age when captured and first recorded was 9–12 months (length not measured). Recorded for three consecutive days only at Marineland of the Pacific.

Recording conditions of analyzed whistles included air recordings during transport, hydrophone recordings when first placed in tank, hydrophone recordings the following day under normative conditions, and hydrophone and air

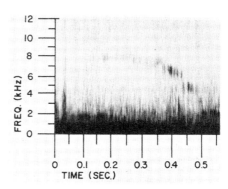

Fig. 14. Whistle emitted by infant number 2 when he was 9 weeks old. Recorded by hydrophone in water. Whistles during recording session were not categorized as stereotyped.

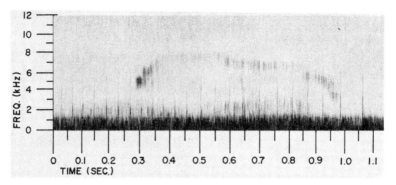

Fig. 15. Whistle emitted by infant number 2 when he was 11 weeks old. Recorded by hydrophone in water. Whistles during recording session were not categorized as stereotyped.

recordings the third day when the animal was stranded in the tank for medical examination. All of the 355 whistles analyzed were classified as stereotyped. They were still infantile in that they were all single-looped and not widely frequency-modulated (Fig. 17).

D. Infant Number 4

Male, born in captivity at the Aquatarium at St. Petersburg Beach, Florida, of parents captured in that area. Tankmates consisted of several *Tursiops truncatus*. Recorded first when 6½ weeks of age and again when four months old.

The 26 whistles recorded by hydrophone at 6½ weeks of age were not classified as stereotyped, although they were noted as being "very distinctive" in

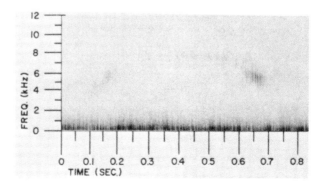

Fig. 16. Whistle emitted by infant number 2 when he was 5½ months old. Recorded by hydrophone in water. Whistles during recording session were not categorized as stereotyped.

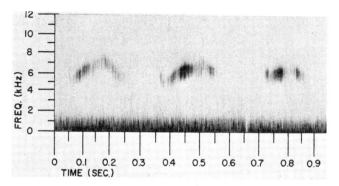

Fig. 17. Three single-looped whistles emitted in rapid succession by infant number 3. His age was estimated as being between 9 and 12 months (length not recorded). Recorded by hydrophone in water. Whistles during recording sessions were categorized as stereotyped.

vocal quality. Figure 18 is representative of this group. When the animal was four months old, all of the 126 whistles recorded in air during transport, as well as the 27 recorded by hydrophone after the animal was placed in another tank, were classified as stereotyped. The whistles were still infantile in their lack of complete or multiple loops, but the tremulo had disappeared (Fig. 19).

E. Infant Number 5

Male, conceived in the wild but born in captivity at Marineland of Florida near St. Augustine of parents from the area. Housed in a small tank with mother and, briefly, with an adult-sized male eastern Pacific white-sided dolphin. Recorded intermittently at 25–68 days of age.

This is the only infant in which the development of the multilooped whistle occurred during the period when the animal was being recorded. For this reason,

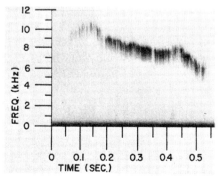

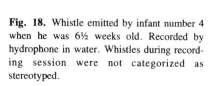

Fig. 18. Whistle emitted by infant number 4 when he was 6½ weeks old. Recorded by hydrophone in water. Whistles during recording session were not categorized as stereotyped.

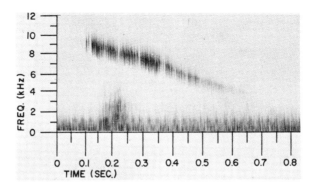

Fig. 19. Whistle emitted by infant number 4 when he was 4 months old. Recorded by microphone in air. Whistles during recording session were categorized as stereotyped.

688 of the whistles were subjected to rather intensive analysis.

All of the whistles (106) recorded under normative conditions when the animal was 25 days old were variable and tremulous; and they exhibited little frequency modulation (Figs. 20 and 21). Whistles (60) recorded when the animal was 26 and 32 days old, both under normative conditions and while the water in the tank was being lowered, showed little if any change. However, many of the analyzed whistles (469) of this animal recorded on day 40, both when undisturbed and while the water was being lowered, were suggestive of having a loop structure (Figs. 22 and 23). Some were suggestive of having two loops despite the high degree of tremulo. At 45 days of age the loops were still more apparent (Figs. 24–26). These whistles (20) were still too tremulous and variable to classify as stereotyped but some were approaching mature multilooped

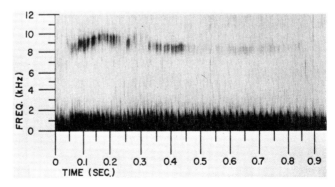

Fig. 20. Whistle emitted by infant number 5 when he was 25 days old. Recorded by hydrophone in water. Whistles during recording session were not categorized as stereotyped, as exemplified by comparison with Fig. 21.

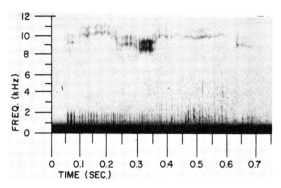

Fig. 21. Whistle emitted by infant number 5 when he was 25 days old. Recorded by hydrophone in water. Whistles during recording session were not categorized as stereotyped, as exemplified by comparison with Fig. 20.

whistles (Fig. 26, for example). At 68 days of age, the whistles (33) were stereotyped as well as multilooped (Fig. 27).

This is the only infant Atlantic bottlenosed dolphin for which we have had the opportunity to observe the development of the whistle in a restricted acoustic environment. The fact that the infant heard the mother's whistle almost exclusively may possibly have influenced the rapid development of the whistle. Again there was a close similarity of the infant's whistle to that of the mother (Fig. 28), a 244-cm standard length animal. The similarity to the mother may be

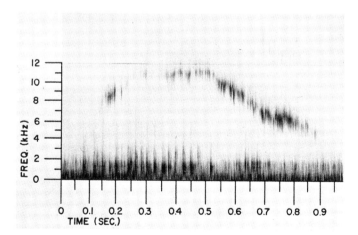

Fig. 22. Whistle emitted by infant number 5 when he was 40 days old. Recorded by hydrophone in water. Whistles during recording session were not categorized as stereotyped, as exemplified by comparison with Fig. 23.

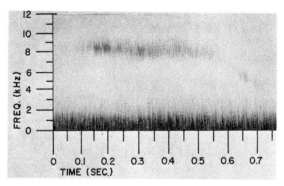

Fig. 23. Whistle emitted by infant number 5 when he was 40 days old. Recorded by hydrophone in water. Whistles during recording session were not categorized as stereotyped, as exemplified by comparison with Fig. 22.

the result of learning or genetics or may only be a coincidence. However, the developing whistles of the infants that we have followed in communities have shown no particular resemblance to those of the mother. It should perhaps be noted that the whistle of the eastern Pacific white-sided dolphin (Fig. 29) to which the infant was briefly exposed was of the same general pattern as that of the mother. If the infant then was learning to imitate the only type of whistle to which it was exposed, there should have arisen no basic conflict in the auditory template from his brief exposure to the white-sided dolphin.

F. Infant Number 6

Male, born in the wild and captured with four other animals from the same herd off Destin, Florida. Recorded at Florida's Gulfarium at Ft. Walton Beach

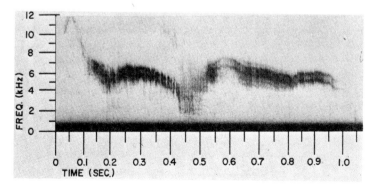

Fig. 24. Whistle emitted by infant number 5 when he was 45 days old. Recorded by hydrophone in water. Whistles during recording session were not categorized as stereotyped, as exemplified by comparison with Fig. 25.

The Whistle of the Atlantic Bottlenosed Dolphin

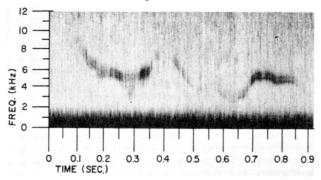

Fig. 25. Whistle emitted by infant number 5 when he was 45 days old. Recorded by hydrophone in water. Whistles during recording session were not categorized as stereotyped, as exemplified by comparison with Fig. 24.

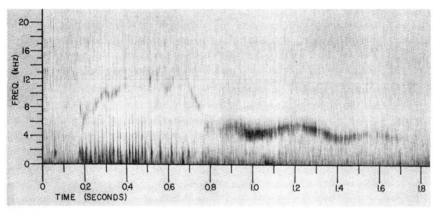

Fig. 26. Whistle emitted by infant number 5 when he was 45 days old. Recorded by hydrophone in water. Whistles during recording session were not categorized as stereotyped. Note similarity between this particular whistle and the final stereotyped multilooped whistle shown in Fig. 27.

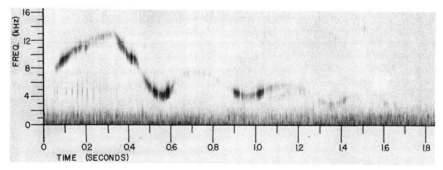

Fig. 27. Whistle of four loops emitted by infant number 5 when he was 68 days old. Recorded by microphone in air. Whistles during recording session were categorized as stereotyped. Note similarity to mother's whistle shown in Fig. 28.

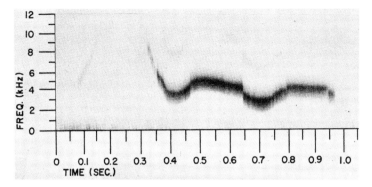

Fig. 28. Whistle of three loops emitted by mother of infant number 5 when she was 244 cm standard length. Whistle recorded by microphone in air. Over a 2-year recording period, the whistles of this adult were categorized as 99% stereotyped.

for a period of four days immediately following capture. Size at capture was 152 cm in standard length, and age estimated from an infant growth curve was 9 months.

The animal was recorded only briefly, but under a variety of circumstances. These included normative conditions, stranding on the bottom of the tank, placement in a strange tank in isolation, and introduction to a new tankmate. The whistle was of interest in two respects. The many breaks in time that occurred while the frequencies were increasing and decreasing may indicate that the animal was only just learning to sweep this wide frequency range. Also, out of the 1051 whistles analyzed, there was only one 1½-looped and one 2-looped whistle. Thus, it is possible that the animal was just learning to cycle the sound loops. Of the remaining whistles, 998 were single-looped of the type shown in Fig. 30, and 51 were partials of less than a complete sound loop.

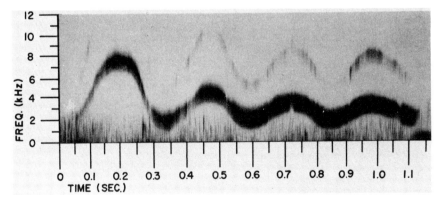

Fig. 29. Whistle of four loops emitted by the eastern Pacific white-sided dolphin to which infant number 5 was briefly exposed. Whistle recorded by microphone in air.

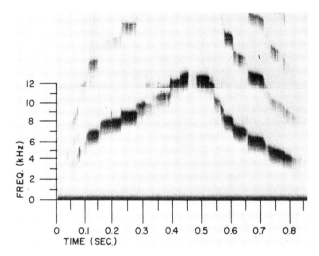

Fig. 30. Whistle emitted by infant number 6 when he was about nine months old (152 cm standard length). Recorded by hydrophone in water. Whistles during period of recording sessions were categorized as stereotyped.

The animal's whistles were classified as 99% stereotyped. We include both partials (the introduction portion of the stereotyped whistle) and multilooped whistles within our definition of stereotypy unless they are distorted (see Caldwell *et al.*, 1973).

G. Infant Number 7

Male, born in the wild and captured off the mouth of the Suwannee River, Florida. Recorded at Marineland of Florida from about two weeks following capture to present. Size at capture was 155 cm in standard length, and age estimated from an infant growth curve was 9½ months.

This animal was precocious in whistle development in that he was emitting stereotyped multilooped whistles of up to nine loops when he was first recorded at approximately 9½ months of age. Placed in isolation after 9 months in captivity, the animal basically retained its same whistle type for about 4½ years (Figs. 31 and 32).

The animal has presently been acoustically isolated from other dolphins for about six years. He has at times demonstrated fairly precise mimicry of pure-tone sounds after being exposed to them in experimental situations (M.C. Caldwell and D.K. Caldwell, 1972). We have not observed such modifications of the stereotyped whistles of animals maintained under more normal conditions. For instance, we have followed community tank animals at Marineland of the Pacific for two years and at Marineland of Florida for up to five years. None of those

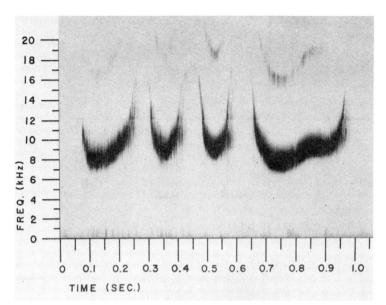

Fig. 31. Whistle of four loops emitted by infant number 7 when he was about 9½ months old (155 cm standard length). Recorded on July 31, 1968, by microphone in air during venipuncture. Whistles during period of recording sessions were categorized as stereotyped. Note basic similarity to whistle of same animal recorded 4½ years later (Fig. 32) under entirely different circumstances.

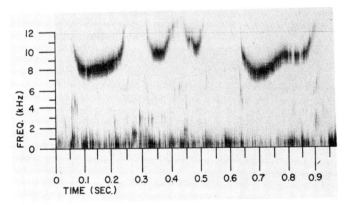

Fig. 32. Whistle of four loops emitted by infant number 7 when he was a juvenile over 5 years old. Recorded by hydrophone in water on January 5, 1973. Whistles during recording sessions were categorized as stereotyped.

animals shifted the stereotyped whistle once the mature whistle pattern had developed.

H. Infant Number 8

Male, born in the wild and captured off Destin, Florida. Recorded at Florida's Gulfarium immediately following capture for a period of 12 days. Size at capture was 158 cm in standard length, and age estimated from an infant growth curve was 10 months.

This animal was recorded following removal of a tankmate, when stranded in the tank, while being fed, and when undisturbed both with and without a tankmate. Several hundred (1013) whistles from these assorted recording conditions were analyzed, and 99% of these were classified as stereotyped. This characteristic whistle has a well-defined loop structure, although none of the sound loops were ever recycled (Fig. 33).

I. Infant Number 9

Male, born in the wild and captured off Steinhatchee, Florida. Recorded at Marineland of Florida shortly after capture for a period of three weeks. Size at capture was 164 cm in standard length, and age estimated from an infant growth curve was 11 months.

The animal was recorded undisturbed, when the water in the tank was being lowered, and in air while being moved. All of the whistles (100) analyzed from

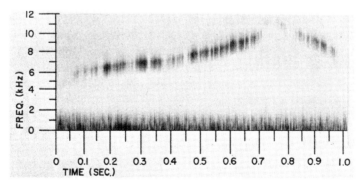

Fig. 33. Whistle emitted by infant number 8 when he was about 10 months old (158 cm standard length). Recorded by hydrophone in water. Whistles during recording sessions were categorized as stereotyped.

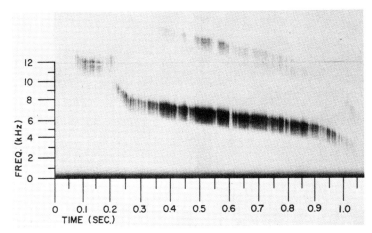

Fig. 34. Whistle emitted by infant number 9 when he was about 11 months old (165 cm standard length). Recorded by hydrophone in water. Whistles during recording sessions were categorized as stereotyped.

these sessions were designated as stereotyped. They consisted of a single sound loop and were frequency-modulated (Fig. 34).

J. Infant Number 10

Female, a third-generation animal born in captivity at Marineland of Florida of bloodlines from the northeastern Florida region. Housed in tank with mother only, but in acoustical contact at various times with about 20 other *Tursiops truncatus* and two (an adult female and a juvenile male) eastern Pacific short-finned pilot whales *(Globicephala macrorhynchus cf. scammoni)*. Recorded intermittently at Marineland of Florida from day of birth through 11½ months of age.

Whistles recorded from this animal on the day of birth and the following two days were variable and tremulous, and exhibited little frequency modulation (Fig. 35). Recorded again when 43 days old while being held out of water for a vaccination, she emitted 65 whistles. They were not stereotyped but were noted as "becoming characteristic." They were all designated as single loops or partials thereof. The illustrated sound (Fig. 36) is fairly typical of this episode. It illustrates the breaking up of the whistle into various frequencies. This is more commonly seen in younger animals under stress, but also occurs in mature animals if they are in great pain. The whistle becomes screechy or strident to the human ear. The patterning may break down completely, particularly in the higher frequency portion of the sound loop.

The Whistle of the Atlantic Bottlenosed Dolphin

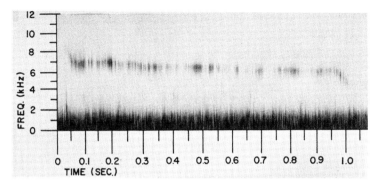

Fig. 35. Whistle emitted by infant number 10 when she was two days old. Recorded by hydrophone in water. Whistles during recording session were not categorized as stereotyped.

At 7½ months of age this animal had developed a stereotyped whistle (Figs. 37 and 38). The breakdown of the stereotyped whistle is shown again in Figure 39 when the animal was recorded in air under the duress of venipuncture. Conversely, of 96 recorded whistles analyzed of this animal during the last hour prior to death, when she was listing on one side, bumping into the side of her tank, and being supported by her mother, 94 were essentially the same stereotyped, single-looped whistle (Fig. 40), one was a stereotyped partial thereof and only one was aberrant (flattened). The animal was by then 11½ months old and her whistle was still essentially the same as at 7½ and 8 months of age (Figs. 37 and 38). No whistles were recorded that consisted of more than a single loop.

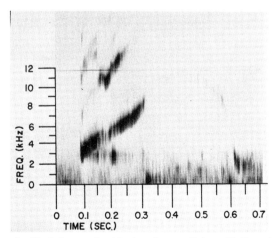

Fig. 36. Whistle emitted by infant number 10 when she was 43 days old. Recorded by microphone in air. Whistles during recording session were not categorized as stereotyped.

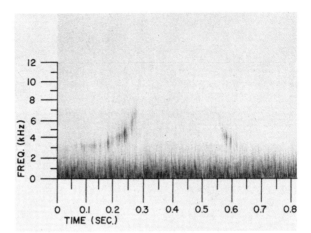

Fig. 37. Whistle emitted by infant number 10 when she was 7½ months old. Recorded by hydrophone in water. Whistles during this and subsequent recording sessions were categorized as stereotyped.

K. Infant Number 11

Female, conceived in the wild and born in captivity at Marineland of Florida of parents from the region. Housed in tank with mother and two other mature female *Tursiops truncatus*. Recorded successfully only once—at six weeks of age when temporarily housed with mother only.

The analysis is based upon 79 whistles recorded under normative conditions, some while the mother was being fed. The whistles were classified as stereotyped in spite of some remaining vocal quaver (Fig. 41). None consisted of more than the single sound loop.

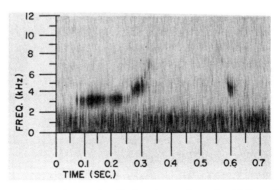

Fig. 38. Whistle emitted by infant number 10 when she was 8 months old. Recorded by hydrophone in water. Whistles during recording session were categorized as stereotyped.

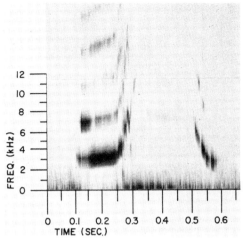

Fig. 39. Whistle emitted by infant number 10 when she was 10½ months old. Recorded by microphone in air during venipuncture. Whistles during recording session were categorized as stereotyped, but this example indicates manner in which the normal stereotyped whistle (Figs. 37, 38, and 40) frequently breaks up under extreme duress.

L. Infant Number 12

Female, born in the wild and captured off Destin, Florida. Recorded at Florida's Gulfarium for a 10-day period after being in captivity for three months. Size when recorded was 156 cm in standard length, and age estimated at that time from an infant growth curve was 9½ months.

The analysis of this animal is based upon 324 whistles and 33 chirps or partials. Only two of these were classified as aberrant. Figure 42 is illustrative of the stereotyped whistles. Recording sessions during which these whistles were emitted took place during periods when the animal was being trained, introduced

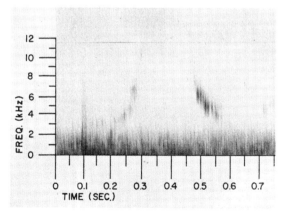

Fig. 40. Whistle emitted by infant number 10 when she was 11½ months old. Recorded by hydrophone in water. Whistles during recording session were categorized as stereotyped although animal was *in extremis*.

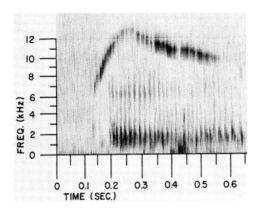

Fig. 41. Whistle emitted by infant number 11 when she was 6 weeks old. Recorded by hydrophone in water. Whistles during recording session were categorized as stereotyped.

to a new tankmate, being stranded in the tank and moved, and when she was undisturbed. None of the whistles consisted of more than one loop.

M. Infant Number 13

Female, born in the wild and captured off Steinhatchee, Florida. Recorded at Marineland of Florida shortly following capture for a 3-week period. Size when recorded was 166 cm in standard length, and the age estimated from an infant growth curve was 11 months.

Whistles from four recording sessions, all when she had been removed from her tank for medication, were analyzed. All 276 of these were stereotyped and were of the single-loop variety, illustrated in Fig. 43, or introductory partials thereof.

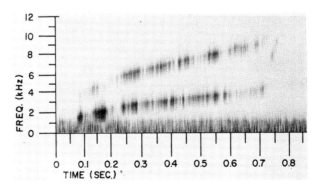

Fig. 42. Whistle emitted by infant number 12 when she was an estimated 9½ months old (156 cm standard length). Recorded by hydrophone in water. Whistles during recording sessions were categorized as stereotyped.

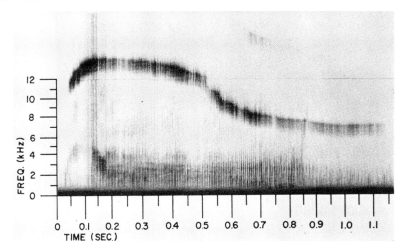

Fig. 43. Whistle emitted by infant number 13 when she was an estimated 11 months old (166 cm standard length). Recorded by microphone in air. Whistles during recording sessions were categorized as stereotyped.

N. Infant Number 14

Female, born in the wild and captured off Steinhatchee, Florida. Recorded at Marineland of Florida for a 4-day period soon after capture. Size when recorded was 168 cm in standard length, and age estimated from an infant growth curve was 11½ months.

The whistles of this animal were analyzed rather extensively as part of another study. All 1453 or 100% of these whistles were designated as signature whistles, although they ranged from less than one through five continuous sound loops. Figure 44 illustrates a whistle of two sound loops. Recordings were made during normative or undisturbed sessions, rather than under conditions potentially disturbing to a naive animal, such as when the tank was being cleaned or when a diver was in the tank.

VII. SUMMARY AND DISCUSSION

This paper is based upon limited data. The conclusions that can be reached are consequently limited in scope. It may well be that the major value of the report is the enunciation of how little is known about the development of the whistle in Atlantic bottlenosed dolphin, as compared to the present state of knowledge in similar areas for many birds.

We start with the premise that each individual Atlantic bottlenosed dolphin

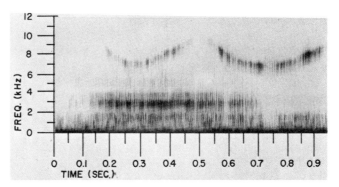

Fig. 44. Whistle emitted by infant number 14 when she was an estimated 11½ months old (168 cm standard length). Recorded by hydrophone in water. Whistles during recording sessions were categorized as stereotyped.

past infancy has a stereotyped or "signature" whistle (M.C. Caldwell and D.K. Caldwell, 1965, manuscript). Figure 45 illustrates an established multilooped whistle of the type that we designate a signature whistle. It will be seen that the sound loops are frequency- and amplitude-modulated and that they are largely repetitive in nature. There is, of course, considerable variability between individuals as to the patterning in the several acoustic parameters; otherwise there would be no signature value to the whistles. Tavolga (1968, p. 201), for instance, figured a whistle which contains a secondary frequency modulation in the sound loop as well as a strong "undertone." Comparisons between these as well as the illustrations of the older animals in this sample, give some idea of the almost infinite variety of ways in which individuality may exist between animals.

The first objective of this study was to determine whether dolphins are born with a whistle and, if so, whether it is stereotyped from birth.

All of the infants did have a whistle when they were first recorded, with three of these recordings having been initiated on the first day of life. As two of the newborn were male and one was female, apparently both sexes are born with the ability to emit a narrow-band sound.

The sounds of the newborn were not stereotyped, however. They were tremulous or quavery and had little frequency modulation. The youngest animal in the sample known to have established a frequency-modulated, stereotyped whistle did so between the ages of 45 and 68 days. The oldest animal that we encountered without a stereotyped whistle was 17½ months of age. In general, both frequency modulation and stereotypy were established by the end of the first year.

Continuous recycling of the sound loops does not normally appear as early as does stereotypy. Only 21% of the yearlings recycled the whistles, whereas almost 80% of the animals were recycling by the end of the juvenile period. In fact, in the sample of 24 adults, one animal was found that even then did not recycle the sound loop. Therefore, there is considerable variability between animals in the ages at which this characteristic may make its appearance.

Some increase in capability for emitting the continuous sound loops and increased durations continues through the subadult stage. The adults then either stabilize or drop off slightly from subadult levels.

We were unable to find any difference between sexes in the postinfantile whistles, nor did we note any differences in the developmental stage. Similarly, no seasonal variations were noted in the infants', nor has there emerged any suggestion of seasonal differences in the whistles of the older animals of either sex (M.C. Caldwell and D.K. Caldwell, manuscript).

In addition to these findings, certain tendencies were noted upon which we can speculate. Under sufficient stress, virtually all postinfantile dolphins may emit whistles which display one or more of the qualities which characterize infantile whistles. These may become tremulous, lose their frequency sweep, break in time, or sometimes show more and stronger overtones. They indicate to us an absence of muscular coordination—only momentary in adults, but undeveloped in infants.

Some of the mature animals in the sample did exhibit a greater loss of stereotypy than that found in the younger age classes. This was particularly true of long-term captives, housed either in isolation or with only one or two other animals. We postulate that these atypical whistles result from the lower levels of arousal more characteristic of older animals, particularly those subject to loss of stimulation, and from the loss of the reinforcing and stabilizing influence of hearing whistles of several other dolphins. In brief, we presently see no reason to

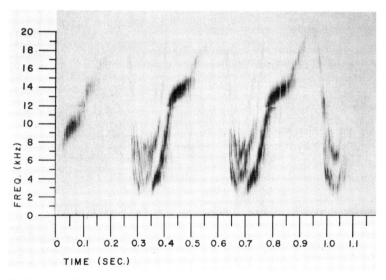

Fig. 45. Signature whistle of three loops emitted by young subadult male (213 cm standard length).

interpret these atypical whistles as positive increments in the vocalization system, but rather as a degradation of the system that would not occur in the natural environment. We have not noted any substantial loss of stereotypy in the mature animals in large communities. They may vocalize less frequently without obvious cause, but this is probably true of most of their activities.

Whistles of younger animals in the sample were also subject to the effects of isolation, but with one exception they were not as marked. The exception is the case of apparent spontaneous mimicry of a pure-tone sound reported for a juvenile long-term isolate (M.C. Caldwell and D.K. Caldwell, 1972).

Finally, as the infant is not born with the stereotyped whistle, we are led to ask what the forces are that shape the final whistle. At one time we assumed that the variations between individuals in their whistles were the direct result of anatomical variations in the sac systems. This may still be a valid interpretation. However, the somewhat singular similarity between the whistle of infant number 5 and that of the mother with whom he was isolated most of the time does give us pause. This similarity could well be only a coincidence, as could the somewhat vague resemblance of the whistle of infant number 1 to the highly vocal eastern Pacific white-sided dolphin in his captive community; yet, we are forced to give more consideration to the potential effects of the acoustic environment in which the dolphin is reared. Based upon the limited information presently available, further speculation serves no useful purpose.

ACKNOWLEDGMENTS

The collection of data for this report necessarily took the cooperation of many people and organizations. For permission to record and observe dolphins both in and out of water, for access to areas not generally open to the public and sometimes not even to all researchers, and for general enthusiastic cooperation and hospitality, we thank the following organizations and the people named: Marineland of Florida near St. Augustine (W. Fred Lyons, B.C. Townsend, Jr., Cecil M. Walker, Jr., and J. Frank Miller); Marineland of the Pacific near Los Angeles (John H. Prescott, David H. Brown, William Monahan); Florida's Gulfarium, Ft. Walton Beach (J.B. Siebenaler, the late Marjorie Siebenaler, and Gregg Siebenaler); Aquatarium, St. Petersburg Beach, Florida (Mike Haslett, Patricia Berger, and Lynn Kephart). William E. Evans, Ruth M. Haugen, and William W. Sutherland, all then of the Lockheed-California Company near Los Angeles, offered considerable patient advice to two neophytes in the early stages of the study over a decade ago, and without their help we never would have gotten the first tape recorder turned on. A number of our student assistants were involved to various degrees in the collection of whistles, sometimes under very trying conditions, and the analyses of the tapes. These included Nicholas R.

Hall, Hazel I. Hall, and Robert W. Hult. Their help was then and is now much appreciated. To any that we may have overlooked, we offer our sincere apologies. Be assured that our thanks to you are no less great than to those named above. Harry Hollien of the University of Florida made a number of helpful comments during part of the study and on an early draft of the manuscript.

Much valued financial support for various phases of this study came primarily from the Office of Naval Research (contracts N00014-67-C-0358 and N00014-70-C-0178), but also in part from the National Science Foundation (grant number GB-1189), the National Institute of Mental Health (grant number MH-07509), the National Institute of Neurological Diseases and Stroke (grant number NS-09694), Marineland, Inc., and Biological Systems, Inc.

REFERENCES

Aronson, L. R., 1966, (No title), in: *Whales, Dolphins, and Porpoises* (K. S. Norris, ed.), pp. 542–543, University of California Press, Berkeley.
Caldwell, D. K., and Caldwell, M. C., 1968, The dolphin observed, *Nat. Hist.* **77(8)**:58–65.
Caldwell, D. K., and Caldwell, M. C., 1972, Senses and communication, in: *Mammals of the Sea; Biology and Medicine* (S. H. Ridgway, ed.), pp. 466–502, Charles C. Thomas, Springfield, Illinois.
Caldwell, M. C., and Caldwell, D. K., 1965, Individualized whistle contours in bottlenosed dolphins *(Tursiops truncatus)*, *Nature* **207**:434–435.
Caldwell, M. C., and Caldwell, D. K., 1971, Statistical evidence for individual signature whistles in Pacific whitesided dolphins, *Lagenorhynchus obliquidens*, *Cetology* **3**:1–9.
Caldwell, M. C., and Caldwell, D. K., 1972, Vocal mimicry in the whistle mode by an Atlantic bottlenosed dolphin, *Cetology* **9**:1–8.
Caldwell, M. C., and Caldwell, D. K., The whistle of the Atlantic bottlenosed dolphin *(Tursiops truncatus)*. General description for the species and statistical analysis of stereotypy in individuals (manuscript).
Caldwell, M. C., Caldwell, D. K., and Miller, J. F., 1973, Statistical evidence for individual signature whistles in the spotted dolphin, *Stenella plagiodon*, *Cetology* **16**:1–21.
Dreher, J. J., and Evans, W. E., 1964, Cetacean communication, in: *Marine Bio-Acoustics* (W. N. Tavolga, ed.), pp. 373–393, Pergamon Press, New York.
Essapian, F. S., 1953, The birth and growth of a porpoise, *Nat. Hist.* **62(9)**:392–399.
McBride, A. F., and Kritzler, H., 1951, Observations on pregnancy, parturition, and post-natal behavior in the bottlenose dolphin, *J. Mammal.* **32**:251–266.
Powell, B. A., 1966, Periodicity of vocal activity of captive Atlantic bottlenose dolphins: *Tursiops truncarus*, *Bull. So. Calif. Acad. Sci.* **65**:237–244.
Sergeant, D. E., Caldwell, D. K., and Caldwell, M. C., 1973, Age, growth, and maturity of bottlenosed dolphin *(Tursiops truncatus)* from northeast Florida, *J. Fish. Res. Bd. Can.* **30**:1009–1011 (2 Figs).
Tavolga, W. N., 1968, Marine Animal Data Atlas, U.S. Naval Training Device Center Technical Report NAVTRADEVCEN 1212-2, pp. i–x, 1–239.
Wood, F. G., 1954, Underwater sound production and concurrent behavior of captive porpoises, *Tursiops truncatus* and *Stenella plagiodon*, *Bull. Mar. Sci. Gulf Caribb.* **3**:120–133 (for 1953).

Chapter 12

MYSTICETE SOUNDS

Thomas J. Thompson, Howard E. Winn, and Paul J. Perkins

Department of Zoology and Graduate School of Oceanography
University of Rhode Island
Kingston, Rhode Island 02881

I. INTRODUCTION

Historic literature contains many allusions to mysterious sounds heard at sea, and these tales may comprise the earliest references to mysticete sounds. The "songs" that sirens sang and the sounds of "ghosts" on ships were perhaps partially based on the sounds of mysticete whales. Many of the stories were probably derived from experiences of sailors traveling by shallow banks and islands in the tropics (L. Winn, personal communication). Much of this lore may be attributable to the humpback whale song, which is emitted during the mating and calving season and could easily be heard through the wooden hull of sailing vessels. Aldrich (1889) stated that whaling captains listened for "singing" whales during their expeditions. Whales included in this subjective account were bowhead, right, humpback, and devil-fish (gray whale), all capable of producing various sounds. On occasion, sounds were heard from live stranded whales. Little else was reported until hydrophones were extensively employed during World War II and various "groans" and "moans" were heard. Although the sound source was usually not known, it was often suspected that at least some of these sounds were produced by whales. However, most hydrophones were incapable of detecting very low frequency sounds below 100 Hz.

In 1951 at various SOFAR (sound fixing and ranging) stations around the world, improved equipment was installed that could detect and record sounds down to and below 20 Hz. Subsequently, many 20-Hz moans were heard and their sources tracked. Several explanations of the source of these sounds were proposed, including distant waves crashing on a beach, meteorological and seismic events, mechanical noise, and even the heartbeats of whales (Walker,

1963). Finally, in 1960, Schevill positively identified the 20-Hz source as the fin whale, *Balaenoptera physalus* (Schevill *et al.*, 1964). The description of frequency and temporal patterning of the 20-Hz signal varied with different authors (Patterson and Hamilton, 1964; Schevill *et al.*, 1964; Northrup *et al.*, 1971; Kibblewhite *et al.*, 1967), and it is now believed that the blue whale, *Balaenoptera musculus,* may also be a possible source of these enigmatic signals (Cummings and Thompson, unpublished manuscript).

As recently as 1962 (Schevill and Watkins, 1962; Schevill *et al.*, 1962), only three of the 11 mysticete species were reliably recorded. These were the fin whale, the northern right whale, *Eubaleana glacialis glacialis,* and the humpback, *Megaptera novaengliae.* Many recordings were made in the western North Atlantic, where the complex sounds of the humpback are presumably not produced, and only simple moans had been reported for this species. However, it was generally known at this time that the humpback produced a variety of sounds (Schevill and Watkins, 1962). This created an added interest in the sounds of baleen whales. Several years later Payne and McVay (1971) and Winn *et al.* (1971) reported on the long, complicated, and repeated "song" of humpbacks produced in the winter (tropical) portion of their range. The humpback call may be the most complex call in the animal kingdom, and yet its acoustic and biological analysis has just begun.

Considerable progress has been made in the last 15 years in documenting mysticete sounds, although not without difficulty. Attendant problems of oceanic ambient noise levels, equipment deficiency, evasive behavior of the whales, and interspecific sound contamination have all plagued bioacousticians. Conclusions on sources of sounds often conservatively rest in the phrase "in the presence of." Additionally, many balaenopterids are especially difficult to visually identify in the field. In one case an investigator necessarily dived around and photographed a Bryde's whale *Balaenoptera edeni* in order to positively identify it (Cummings and Thompson, unpublished manuscript).

Behavioral correlations to emitted sounds are often limited to surface behaviors (e.g., blowing, breaching, tail slapping), and it is not known whether adventitious sounds produced in some of these behaviors have any biological meaning. In many instances an animal observed close to the recording gear may be silent, while an unseen distant animal may be loudly broadcasting. These observational difficulties are obvious.

Some mysticetes can predictably be found within portions of their respective feeding and breeding ranges (gray, humpback, southern right, and bowhead). Recordings from other species are often opportunistic, relying on chance encounters. Sporadic, low-level, or transient signals are difficult to record but may comprise a substantial portion of the vocal repertoire of some species (e.g., minke whale *Balaenoptera acutorostrata,* Winn and Perkins, 1976).

Because of the inherent difficulties in this research, each mysticete encounter requires careful analysis. In time, questionable or unknown sources may be identified as the number of encounters increases and the biology of these animals is better understood. This chapter reviews the advances made in the last 15 years of mysticete bioacoustics and some original data is presented.

II. METHODS

The techniques and equipment used to record sounds vary among investigators and change with progressive electronic sophistication. The required frequency response should be flat over the entire range of emitted signals. Because of the increasing evidence that high-frequency signals are produced by mysticetes (see Section III), investigators should preferably utilize broad-band systems to optimize the quality of the recording. A flexible system should be used in preparation for chance encounters with any other cetacean species for which there are little or no acoustic data.

To maximize the signal-to-noise ratio (S/N) for low-frequency mysticete sounds, Cummings *et al.* (1972) recommended that hydrophone response be "rolled off" to reduce the reception of low frequency noise caused by normal sea conditions. Movement of the hydrophone and cable flutter, caused by wave action, drifting, and pitching of a ship, can be minimized by using various damping techniques. Localization of mysticete sound sources has been accomplished using directional hydrophones and various hydrophone arrays. Different methods of recording, localization, and subsequent analysis of bioacoustic data have been described in several reports (Watkins, 1966, 1967a, 1976; Tyrrell, 1964; Gales, 1966; Perkins, 1966a, Poulter, 1971).

On each recording, a separate channel should be allotted for tape identification and verbal comments on behavior and information concerning conditions, time, and position. Calibration tones are always required on each tape to check and/or correct for frequency response in the recording system. A requirement of utmost importance is the presence of experienced and attentive observers who can correctly identify marine mammals. Recently a series of field guides has been published to aid in proper identification (Leatherwood *et al.*, 1976; Katona *et al.*, 1975). The presence of more than one species of marine mammal can greatly confuse interpretation of received signals. Thus, several observers are needed to scan 360° around a recording vessel to account for any circumstance. Additionally, continued recording during both day and night provides important biological information.

III. SPECIES ACCOUNTS

In reviewing the literature on vocalizations, we will treat each mysticete species separately. In the case of the minke, sei, and fin whales, we have added new bioacoustic information. Each species is presented in order according to the following taxonomic classification.

A. Mysticete Taxonomy

Class Mammalia
  Order Cetacea
    Suborder Odontocoeti, toothed whales, porpoises
    Suborder Mysticocoeti, baleen whales
      Family Balaenidae
        Balaena mysticetus, bowhead
        Eubalaena glacialis glacialis, northern right whale
        Eubalaena glacialis australis, southern right whale
        Caperea marginata, pygmy right whale
      Family Eschrichtiidae
        Eschrichtius robustus, gray whale
      Family Balaenopteridae
        Balaenoptera musculus, blue whale
        Balaenoptera physalus, fin whale
        Balaenoptera edeni, Bryde's whale
        Balaenoptera borealis, sei whale
        Balaenoptera acutorostrata, minke whale
        Megaptera novaengliae, humpback whale

B. Family Balaenidae

1. Bowhead: *Balaena mysticetus* (Linnaeus 1758)

The bowhead has been recorded by Poulter (1966, 1968, 1971) in the Arctic. Its signals consist primarily of ascending or descending low-frequency moans. The moans are narrow-frequency signals typically 1–2 sec in duration. Poulter's published spectrograms are difficult to analyze since most of the obvious signals present belong to the bearded seal, *Erignathus barbatus*. Poulter (1971) stated that certain signals of the bowhead and bearded seal are similar. Certainly, acoustic identification of the bowhead in the Arctic is difficult at best, due to loud and numerous bearded seal vocalizations which may interfere with correct signal assignment. Further information on the biology of this species is given by McVay (1973).

2. Northern Right Whale: *Eubalaena glacialis glacialis* (Borowski 1781)

We will treat the northern and southern subspecies of right whales separately; although the vocalizations of these animals are basically the same, differences may become evident as more recordings are analyzed. The northern right whale was one of the first mysticetes to be recorded (Schevill *et al.*, 1962), although there now exist more recordings and behavioral analyses of the southern right whale (Cummings *et al.*, 1972; Payne and Payne, 1971).

Low-frequency moans below 400 Hz with slight frequency modulation have been reported by Schevill and Watkins (1962) and by Schevill *et al.* (1962). Figure 1 depicts a spectrogram of this sound taken from the disk by Schevill and Watkins (1972) and recorded off Cape Cod. In many subsequent encounters, despite ideal recording conditions and using equipment sensitive to 150 kHz, no sounds suggestive of echolocation have been recorded for the right whale (Watkins and Schevill, 1976). Adventitious sounds produced from "baleen rattle" as the whale feeds on the surface have been described (Watkins and Schevill, 1976), although it appears unlikely that these sounds are communicative or biologically important. However, moans which are emitted sporadically have been regarded as communicative (Watkins and Schevill, 1976) (see Section V).

Cummings and Phillipi (1970) reported that northern right whales off the coast of Newfoundland produced repetitive patterns or "stanzas" lasting 11–14 min. Sounds within the "stanzas" consisted of brief pulses, blips, and moans in the frequency range of 20–175 Hz. This report has met with some dispute (Payne and Payne, 1971) for the following reasons: (1) Cummings and Phillipi used equipment sensitive only up to 175 Hz and thus were unable to record possible sounds above that frequency; (2) the basis for their "tentative identification" was not specified; and (3) comparisons with southern right whales recorded by Payne and Payne (1971) agreed well with known northern right whale sounds published by Schevill and Watkins (1962), but differed considerably from those of Cummings and Phillipi (1970). Thus, it has been suggested that the repetitive sounds recorded and identified as right whale sounds were the lower-frequency sounds produced in the humpback whale song which is highly redundant and repeated (Payne and Payne, 1971). However, the complex humpback whale song is only known to be emitted in the tropical portion of their range during the winter mating and calving season, although there is strong evidence that they are making sounds during at least part of their southern migration (Payne and McVay, 1971; Winn *et al.*, 1971; Winn and Winn, 1978). An encounter with a calling humpback whale in late December off the coast of Newfoundland may be significant, although it is possible that humpbacks could start their song in December as they are about to migrate. However, the evidence presented to dispute the claim that right whales do not produce a repetitive "song" is inconclusive. Further investigations of long recording sessions (an hour or more) with wide-band equipment may be needed to determine the existence of a repetitive signal. Cummings (personal communication) is gathering data that will show the repetitive stanzas were produced by humpback whales.

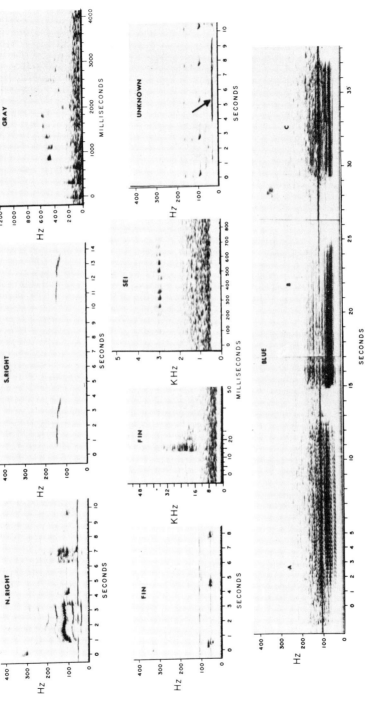

Fig. 1. A composite of various narrow-band spectrographs from mysticete recordings. The continuous narrow frequency band lines on the Northern right, fin, and blue whale illustrations are a result of engine noise. Note the 350-Hz pulse preceding part "C" on the blue whale spectrograph. An arrow indicates the unknown 35-Hz sound of possible mysticete origin. The 100-Hz short moans on the same "unknown" spectrograph may be a resonant harmonic of 20-Hz signals. The first of the two fin whale illustrations is a typical "20-Hz signal" and the other shows one of a train of clicks. The Southern right whale sound, courtesy of Capitol Records Inc., was recorded by R. Payne. See text for other credits and circumstances. The effective analyzing bandwidths were: 2 Hz for the several 0- to 400-Hz scales, 6 Hz for the gray whale spectrograph, 240 Hz for the fin whale click, and 20 Hz for the sei whale pulses. A 4800-Hz high-pass filter was used for the fin whale click, and a 300 high-pass used for the sei whale.

3. Southern Right Whale: *Eubalaena glacialis australis*

The bioacoustic information on the southern right whale has been greatly increased by the fact that a concentration of animals is found in ideal locations along the coast of the Valdes Peninsula in Patagonia, Argentina. These whales migrate from the south during the southern hemisphere winter and enter protected bays, Golfo Nuevo and Golfo San Jose, to mate (Gilmore, 1969; Cummings *et al.*, 1972; Payne and Payne, 1971). Animals can be observed from cliffs overlooking the bays, and thus it may be possible to correlate sound production with various behaviors (Payne, 1972).

Cummings *et al.* (1972) reported that the most common sound produced by these whales were "belch-like" sounds averaging 1.4 sec (range 0.9–2.2) in duration with principal energy at 235 Hz. The frequency of these sounds ranged from 30 to 2200 Hz with the majority at 500 Hz. Moans constituted a second classification which was further divided into simple or complex moans. Simple moans ranged from 0.6–1.6 sec in duration within a narrow band frequency centered at 160 Hz (range 70–320 Hz). Complexity in moans referred to frequency modulations, overtones, and longer durations. Principal energy for these complex moans centered at 235 Hz (range 30–1250 Hz) and lasted from 0.2 to 4.1 sec. Other miscellaneous sounds recorded include pulses ranging from 20 to 2100 Hz, lasting only 0.06 sec and often associated with moans. No conclusive evidence of diurnal periodicity in vocalizations was found, and there was no evidence of repetitive "stanzas" as previously reported for northern right whales. This latter case may be due to seasonal, environmental, or geographic differences, and/or a case of mistaken source identity (see previous section).

In the following year the same authors (Cummings *et al.*, 1974) returned to Patagonia and made several new observations. Twenty-five to 30 southern right whales were seen 250 km north of Valdes during the breeding season, resulting in a substantial increase to the population estimate of the area. The authors stated that the sounds "were most numerous, diverse, and spectacular when the animals were courting or apparently copulating." The "belch-like" sounds were less common than in the previous year and were replaced by "bellowings and moans of rising pitch." Over 1750 sounds were analyzed, including one strong pulse ranging from 50 to 2200 Hz resembling a reverberating gunshot. This report also included a description of the airborne and underwater sounds produced by a right whale's blow, as did Watkins' (1967*b*) report for the humpback whale.

Southern right whale sounds were recorded and reported by Payne (1972) and Payne and Payne (1971) in the same location. Some of the sounds were very similar (and a few nearly exact) to the sounds of moans published by Schevill and Watkins (1962) for the northern right whale. While daytime signals were produced infrequently, at one sound per half hour, there was an apparent increase in vocal activity at night, with 1–15 sounds per minute. Most sounds ranged between 50 and 500 Hz, with occasional sounds having energy up to 1500 Hz. A typical moan from a southern right whale, taken from a long-playing album (Payne, 1977; Capitol Records, Inc., Survival Anglia Ltd.) is illustrated in Fig. 1.

4. Pygmy Right Whale: *Caperea marginata* (Gray 1864)

It is not known whether this species produces any vocalizations. The occurrence of this particular species, known only from the southern hemisphere, is reviewed by Ross *et al.* (1975).

C. Family Eschrichtiidae

1. Gray Whale: *Eschrichtius robustus* (Lilljeborg)

The gray whale *Eschrichtius robustus* has been subjected to close bioacoustic scrutiny because of its coastal migratory behavior along California and Baja. Attempts to record these animals were unsuccessful until Asa-Dorian recorded "clicks" from below 200 Hz to over 3000 Hz in 1955 (Wenz, 1964; Gales, 1966). The history of obtaining these vocalizations has been carefully examined since skepticism and controversy followed the investigations even 10 years after the original recording. This skepticism was not unfounded. Rasmussen and Head (1965) spent over two months conducting acoustic field studies at several locations near San Diego and Baja, California, and concluded that no subsurface sounds could be undeniably attributed to the gray whale. In their study, approximately 200 whales were acoustically monitored under a variety of conditions. But there was contradictory evidence that gray whales do produce croaker-like grunts, low-frequency rumbles from 40 to 700 Hz (Eberhardt and Evans, 1962), and pulses and low-pitched grunting (Painter, 1963), as well as the clicks recorded by Asa-Dorian.

J. F. Fish *et al.* (1974) tabulated the sounds of the gray whale and included three more reports of vocalizations (Asa-Dorian and Perkins, 1967; Cummings *et al.*, 1968; Poulter, 1968). These reports extended the repertoire to include respectively: (1) "echolocation-like pulses, variable whistles," (2) "moans, bubble-type sounds, knocks," and (3) "cries, rasping, chirps, and bongs." Cummings *et al.* (1968) stated that 87% of all utterances were moans (20–200 Hz), and the mean duration of 155 moans was 1.54 sec. Knocks extended to 350 Hz and bubble-type sounds to 305 Hz averaging 0.7 sec. The range of received sound-pressure level (SPL) for all the sounds averaged between 138 and 152 dB re 1 μPa at 1 m (Cummings *et al.*, 1968).

The echolocation-like pulses reported by Asa-Dorian and Perkins (1967) range from 70 to 3000 Hz (with most energy at 400–800). These pulses are 10–15 msec in duration, with 150–350 msec intervals, and have been characterized as similar to the sound made by running a thumbnail across a comb (Fig. 1).

The capture and subsequent captivity of a yearling gray whale, "Gigi," enabled J.F. Fish *et al.* (1974) to definitely assign acoustic emissions to the gray whale. A "metallic-sounding pulsed signal" was the most common sound recorded and occurred nearly every time the trainer touched its back. This signal consisted of 8–14 pulses in bursts up to 2 sec in duration in a frequency range from below 100 Hz to over 10 kHz, with most energy at 1.4 kHz. This signal was emitted up to 5 times a minute. "Moans" similar to those reported by Cummings *et al.* (1968) were also recorded by J. F. Fish *et al.* (1974) and were described between 100 and 200 Hz with a secondary peak of energy at 115 kHz lasting over 1 sec. Three other sounds recorded from Gigi were: (1) a "grunt-like" sound at 200–400 Hz (second peak at 1.6 kHz) for 0.2 sec, (2) a "blowhole rumble," and (3) a long "metallic-sounding pulse train" which merged into a long, low-frequency groan.

On three separate occasions J. F. Fish *et al.* (1974) recorded click trains from gray whales. The first encounter in which the click trains were heard was at the time of Gigi's release. Although this signal was unlike any previously recorded from Gigi, identical signals occurred in subsequent encounters of feeding gray whales in Wickaninnish Bay, Vancouver Island. These click trains occupy a frequency range of 2–6 kHz centered at 3.4–4.0 kHz. The number of clicks per train varied from 1 to 833, with repetition rates from 9.5 to 36.0 sec in the first encounter (i.e., Gigi), and 8–40/sec at Wickaninnish Bay with an average click duration of under 2 msec. Further support for the existence of click trains was provided by Norris *et al.* (1977), based on recordings of gray whales in southern Baja. They also described intense, sporadic clicks or a "loud bang" which occurred singly, or occasionally up to 2/sec, from partially stranded male gray whale calves. These clicks differ from those reported by J. F. Fish *et al.* (1974) because of the broader frequency band (energy over 20 kHz), longer duration (0.25 sec compared to 1–2 msec), and repetition rate observed. Minor differences found between clicks can often be attributed to differences in recording and analyzing techniques. "Low resonant pulses" were qualitatively described from a stranded calf as a second or less in duration, emitted each 2–3 sec.

The literature on gray whale vocalizations indicates that a wide range of sounds are produced under different behavioral contexts. Speculation on the functions of these sounds has been made (see Section V). Nevertheless, patience, proper recording equipment, and, in many cases, special circumstances such as obstructing the path of a whale (Asa-Dorian and Perkins, 1967) are required to obtain signals.

D. Family Balaenopteridae

1. Blue Whale: *Balaenoptera musculus* (Linnaeus 1758)

Poulter (1968) first described signals produced by the blue whale consisting of "clicks or groups of clicks which merge together to form buzzes, rasps, etc., with some rather low frequency intermittent tones distributed through them." Poulter's sonograms did not include frequency or time references, and no information concerning the encounter(s) was given. Thus, the source is in doubt. In 1971, two publications expressed confidence in attributing underwater sounds to the blue whale. Cummings and Thompson (1971a,b) recorded a series of repeated signals from two of the four blue whales they encountered at Guafo Island, Chile (74°40'W, 43°36'S) on May 30 and 31, 1970. Their recording apparatus had a response of ± 5 dB from 25 Hz to 18 kHz, with the lowest frequency response attenuated or "rolled downward" beginning with 3 dB at 12 Hz to reduce noise. Changes in amplitude correlated with the movements of the animals. Four estimates of levels at the source averaged 188 dB re 1 μPa (= 88 dB re 1 μbar) at 1 m, the most powerful sounds produced by any living source.

These sounds are low-frequency moans between 12.5 and 200 Hz with an average duration of 36.5 sec (range 34.7–38.1). These long signals are heavily amplitude-modulated. They consist of three parts, labeled A, B, and C in Fig. 1, separated by: (1) degree of modulation (part A with a rate of 3.85/sec and parts B and C at 7.7/sec); and (2) presence of a brief interval between parts B and C in both animals and between A and B in only one. A 390-Hz pulse of 0.5–1 sec duration always preceded part C in all 27 signals analyzed. The energy in these three-part moans was strongest in the ⅓ octave bands at 20, 25, and 31.5 Hz. Intervals between the long three-part moans were very consistent, with medians of 100 and 106 sec for the two animals. Longer intervals evidently correlated with blowing at the surface.

Beamish and Mitchell (1971) recorded ultrasonic signals in the presence of a blue whale encountered while apparently feeding off Sable Island Bank (43°24'N, 60°24'W) on May 22, 1969. The response of the system was flat from 0.1 Hz to 150 kHz but useful beyond 250 kHz. Clicks produced within a frequency range of 21–31 kHz were recorded with a narrow spectral peak of energy at 25 kHz. Approximately 5000 of these clicks were recorded, and the amplitude levels increased as the animal approached the hydrophone. The level extrapolated to the source was calculated as 159.2 dB re 1 μPa at 1 m. The authors suggested that these clicks may have served to echolocate zooplankton during feeding. The possibility of interspecific sound contamination by sei whales and white-sided dolphins, *Lagenorhynchus acutus*, existed, but the authors believed that the clicks were transmitted only by the blue whale. No low-frequency signals were reported in this account. Beamish (1974) later reported that a blue whale entrapped in ice produced clicks which he could hear using a stethoscope placed on the animal's back.

The blue whale sounds recorded from Chile have never been recorded elsewhere, despite their levels and hours of monitoring oceanic sounds. Cummings and Thompson (unpublished manuscript) suggested that a seasonal occurrence of blue whales may correlate with a long, multicomponent 20-Hz signal common to the northeast Pacific and that variation from the Chilean signals may relate to geographic separation or dialects. Some of these 20-Hz signals have been attributed to the fin whale *Balaenoptera physalus*.

2. Fin Whale: *Balaenoptera physalus* (Linnaeus 1758)

The fin whale, second only to the blue whale in size, is found in all oceans. Sounds reported from this species have been broadly termed "20-Hz signals" (see Section I) despite variation in signal frequencies. Because of the intensity of these sounds, transmission distances of up to hundreds of miles have been theorized (Payne and Webb, 1971). However, despite calculated source levels of 170–185 dB re 1 μPa at 1 m (Walker, 1963), hydrophone cable noise may greatly interfere with reception. Thus, fixed hydrophones have primarily accounted for detection; Fig. 1 depicts a typical sonogram of these subsonic "20-Hz signals." Various signal descriptions from several investigators are listed in Table I. Note that doublets are often recorded with stereotyped temporal characteristics. A doublet labeled "22–15 sec" (Patterson and Hamilton, 1964) is characterized by a 22-sec interval between doublets and 15 sec between the two pulses. Each pulse or moan typically lasts approximately 1 sec.

No assumption is made that *all* of the signals listed in Table I are fin whale vocalizations, although it is well documented that fin whales do produce these kinds of signals (Schevill *et al.*, 1964). Perhaps other balaenopterid species are also involved, as most mysticetes produce low-frequency sounds.

In cases where several fin whales are calling, the "20-Hz" signals appear to differ slightly in frequency. Perhaps individuals with 17-, 18.5-, 19-, 20-, and 40-Hz signals, for example, utilize these specific frequencies for individual identification. However, it seems doubtful that individuals could discriminate differences of only 1 Hz.

Some information is available on possible high-frequency sounds produced by fin whales. Wright (1962) described an account of a fin whale approaching the ship within 20 yd and producing an "alarm" sound resembling an "echo sounder running fast." Presumably a superheterodyne receiver tuned to a single frequency between 20 kHz and 39 kHz (ultrasonic signals converted to audible range) was used during this wartime ASDIC/sonar operation. Schevill (1964) discounted Wright's account as a misidentification of a sperm whale. In an encounter of five fin whales 125 miles east of Bermuda, Perkins (1966*b*) reported that 4 min of intermittent and varied sounds were recorded while two fin whales

Table I. Review of 20-Hz Literature

		Signal descriptions			
Author and date	Location	Frequency (Hz)	Duration (sec)	Interval	Notes
Schevill and Watkins, 1962	42°21'N; 70°04'W	40–75			"Moans"
Walker, 1963	Cont. Shelf New England	Narrow band centered at 20	Approx. 1 sec	10 sec	Uninterrupted trains 6–25 min; silent intervals 2–3 min; sources tracked at 2–3 and up to 7–8 knots
Patterson and Hamilton, 1964	Bermuda (Nov.–Mar.)				"Blips" recorded on drum; sources tracked at 1–4 knots
		(A) Close to 20	Approx. 1 sec	12 sec	(A) Accounts for 50–60% of season's activity
		(B) Doublets of varied amplitude; 1st pulse 25–28 Hz sweeping to 20 Hz; 2nd pulse of smaller amplitude constant at 18–19 Hz	Approx. 1 sec	22–15 sec doublet	(B) 20–30%; 140-Hz tone often precedes larger amplitude pulse by 1 sec, detectable only when close to hydrophone; source level 173–181 db re 1 μPa
		(C) Variety of doublets		15–12 sec, 19–11 sec, 20–13 sec, 9–12 sec, 12–8 sec	(C) 15–30%; signals present for several hours, typically 15 min pulsing with pause from 1 to 3 min
Schevill et al., 1964					Attributed sounds to fin whale *B. physalus*; have not found the 140-Hz signal reported above.

Reference	Location	Frequency	Duration	Interval	Comments
Weston and Black, 1965	S. Norwegian Sea	(A) 6–12 Hz (B) 20 Hz "peaked sharply" (C) 16 Hz (D) 23–18 Hz	2.5 sec 1 sec some 1.8 sec, some 1.0 sec 23 sec	Regular, 16 sec Not as regular, 7–2 sec Regular, 1 min	(A) Precedes a "grunt" at 50–100 Hz of 2 sec (B) Pulse trains lasting several min termed 20-Hz pulses (C) Total duration 2 min; 13 pulse only; grunt heard in middle of series; "quasi 20 Hz." (D) "Moans"; typical train contained 6 pulses (varied from 1–8).
Kibblewhite et al., 1967	New Zealand	(A) 25-Hz peak (B) 23.6-Hz peak (C) 31.5-Hz source, x; 25-Hz source, y	5–10 sec 20 sec 20 sec	Notably regular, bet. 2–10 min (4–5 min most common) 2–10 min (3–5 min most common) 157 sec bet. doublets (2 sources)	(A) "5-sec pulse" (B) "20-sec pulse." } Resemble Weston and Black's "20-Hz moan" (C) "Three King's Pulse;" similar to 20-sec pulse
Cummings and Thompson, unpublished	Northeast Pacific	19.7 Hz, 1st pulse; 22.2–19.8, 2nd pulse	19 sec	23 sec bet. pulses 1–3 min bet. doublets	
Northrup et al., 1968	Central Pacific	27–18 Hz, 1st pulse; 44–18 Hz, 2nd pulse	1 sec	16–20 sec doublets	
Northrup et al., 1971	Midway Is., Pacific	25 Hz, 1st pulse; 25–23 Hz, 2nd pulse	12.8 sec (1st); 11.5 sec (2nd)	1 sec. bet. pulses 59–70 sec. bet. doublets	50–100 pulses separated by 1–5 min; several hours/sessions not seen in summer months. Trains lasted 11–22 min with 3.7–4.6 silence; similar to "Three King's Pulses" of Kibblewhite et al., 1967
Cummings and Thompson, unpublished	Gulf of California	68 Hz, 1st pulse; 34 Hz, 2nd pulse (15–95 Hz.) $\bar{x} = 58.5$	1.5 sec, 1.3 sec $\bar{x} = 0.8$ sec	1.6–2.2 times/min	Fin whale identified 68 "moans" from unidentified balaenopterids

were 25 m from the ship. These sounds were described as "chirps and whistles .... from 1500 to 2500 Hz, occasionally reaching 5000 Hz." The durations of these sounds range from less than 50 msec to over 600 msec. These signals have yet to be verified, and it is not certain that the animals were not sei whales. However, few recordings have been made in the temperate winter habitat of this species.

To add to these high-frequency signals, we now report on a long train of high-frequency pulses recorded as two fin whales approached within 50 yd of a ship during an encounter off Nova Scotia in September 1971. Figure 1 depicts one pulse from a train of 28 wide-band pulses in the 16- to 28-kHz range. Each pulse, composed of two to three parts, ranges from 3 to 3.4 msec in duration, with pulse intervals ranging from 250 to 336 msec. Train duration was 8.8 sec. No other animals were seen during this encounter. The ultrasonic information was received on a separate channel using a TR-128 WQM hydrophone (flat response from 20 to 167 kHz) and Precision Instrument recorder (50 Hz to 160 kHz). We believe that these high-frequency signals merit further investigation. An apparent prerequisite for obtaining these high-frequency sounds is close proximity to the whales due to low signal levels.

3. Bryde's Whale: *Balaenoptera edeni* (Anderson)

The Bryde's whale is generally restricted to warm waters and is found between 40°N and 40°S (Nishiwaki, 1972). This species can be identified from other balaenopterids by the presence of three ridges on the dorsal surface of the head, but is difficult to distinguish from the sei or fin whale.

Sounds produced by this species were described by Thompson and Cummings (1969) and Cummings and Thompson (unpublished manuscript). While off Loreto, Mexico, in the Gulf of California, 288 moans in 50 min were recorded as one of two whales remained near the boat and was positively identified with underwater photography. The signals recorded varied in frequency and duration. Ninety-three analyzed moans ranged from 70 to 245 Hz and durations from 0.2 to 1.5 sec with an average of 0.42 sec. Signals were produced sporadically at intervals ranging from 0.2 to 9 min. Frequency modulations consisted of only small shifts (median 15.2 Hz) upward, downward, or in combination in about three quarters of the moans.

A whale of uncertain identification, encountered the same day by these authors, was assumed to be a Bryde's whale based on comparisons of the known Bryde's vocalizations obtained earlier. In this contact 35 moans were recorded, analyzed, and described with a mean frequency of 132 Hz and a mean duration of 0.4 sec.

In the first encounter, the presence and sounds of saddleback dolphins *Delphinus delphis* were readily identifiable without the occurrence of moans. The amplitude of the low-frequency signals was correlated with the proximity of the Bryde's whale to the ship. The equipment used was sufficiently broad band (over-all response ± 5 dB from 25 Hz to 18 kHz) to account for any likely signals from the whale. This fortunate encounter with a relatively rare species has also extended the reported distribution of this whale to include the Gulf of California.

The only other published report of Bryde's whale sounds is by Beamish and Mitchell (1973). While in the South Atlantic (30°07'W, 14°55'S) Beamish heard a single series of "short pulse length audio frequency clicks" and observed a Bryde's whale approaching the hydrophone. Identification was verified by photographs showing the three dorsal head ridges, and no other cetacean species was seen at this time. Unfortunately, no recording was made during the brief encounter. These sounds were described as similar to the short pulse length audio frequency sounds recorded from minke whales (Beamish and Mitchell, 1973), supporting the contention of mysticete origin of these signals.

4. Sei Whale: *Balaenoptera borealis* (Lesson 1828)

Previous to the present account, no published report of sounds from the sei whale existed. The sei whale can easily be misidentified and confused at a distance with either the fin whale or Bryde's whale. However, in one encounter (Winn and Perkins) on August 8, 1968, between Nova Scotia and Newfoundland (44°49'N, 56°28'W), tens of animals were sighted; positive identification was based on the characteristic surfacing behavior, shape and position of the dorsal fin, and the lack of white on the right jaw and baleen (see Leatherwood *et al.*, 1976). This encounter was also well north of the reported range of the tropical and subtropical Bryde's whale (Nishiwaki, 1972).

The recording system used was an AN/UNQ 7a tape recorder and Hydro Products R-130 hydrophone which is flat between 50 Hz and 7500 Hz at 3¾ in./sec (Perkins, 1966a). At times a few animals approached close to the ship. During one of these close encounters, a sei whale emitted a sonic burst of 7–10 metallic pulses with peak energy at 3 kHz. The train of pulses lasted 0.7 sec with each pulse 4 msec in duration (Fig. 1). An effort to record 20-Hz signals (commonly attributed to the fin whale) was made, but low-frequency cable noise masked possible signals. Other sounds recorded in the presence of sei whales, including low-amplitude sonic clicks, may have been produced by common dolphins *Delphinus delphis* seen 5 min after the encounter. However, metallic pulses recorded as the sei whale approached the ship do not resemble any known *Delphinus* signals (Busnel and Dziedzic, 1966).

5. Minke Whale: *Balaenoptera acutorostrata* (Lacepede 1804)

The minke or little piked whale is the smallest of the Balaenopteridae; the largest reported individual was 10.2 m (Nishiwaki, 1966). Comments on this species' wide distribution from the poles to tropical waters were given by Winn and Perkins (1976).

Schevill and Watkins (1972) first reported on low-frequency sounds from two minke whales encountered off Ross Island, Antarctica, in the proximity of breathing holes in the ice pack. The equipment they used was flat from 30 Hz to 30 kHz. Several sounds consisting of downward sweeps in frequency ranged from 130 to 115 Hz at the beginning and dropped to about 60 Hz. The duration of these loud sounds (165 dB re 1 μPa at 1 m above ambient) lasted from 0.2 to 0.3 sec and had no obvious repetitive pattern as the intervals between sounds varied from 8 to 97 sec.

Low-frequency "grunts" with variable frequencies between 80 and 140 Hz and durations of 165–320 msec were reported by Winn and Perkins (1976). The frequency and duration of these grunts are comparable to those reported above by Schevill and Watkins (1972), but they lack a downward frequency shift and are often found in trains with irregular or regular intervals at 2.1–2.3 pulses/sec.

The most distinctive vocalization of the minke whale is the so-called "thump train," the source of which eluded investigators for several years. It has been referred to as the "A-train" frequently recorded on fixed hydrophone systems. Minke whales have an invisible blow and are difficult at times to see, although these animals are notoriously curious and often approach a stationary ship. It was in this condition that many thump trains were recorded by Winn and Perkins (1976) during several encounters below 100 Hz to at least 800 Hz with maximum energy between 100 and 200 Hz. Subsequent to this report, other recordings have shown peak frequencies as high as 2 kHz (Fig. 2B). Individual thumps range from 50 to 70 msec with relatively constant repetition rates. Each train may last over 1 min. We now report that there is potential signature information within this signal, since frequency composition and repetition rates vary substantially between individuals; Fig. 2 shows sections of thump trains from different animals demonstrating this hypothesis. Further evidence of possible signature comes from recordings in which thump trains of various repetition and spectral characteristics from two or three animals obviously overlapped. Thus it is unlikely that only one animal was responsible for the different signals. Detailed investigations of signature as well as geographic differences in minke vocalizations are presently being conducted and will be reported separately.

Other sounds reported by Winn and Perkins (1976) include short trains and occasional pings or clicks in three general classes from 3.3 to 3.8 kHz, 5.5 to 7.2 kHz, and 10.2 to 12 kHz. All ranged from 0.5 to 1.0 msec and had peak energy

---

Fig. 2. Sections of various minke whale thump trains demonstrating spectral and repetition rate differences from separate individuals. (A) Two sections recorded from a single encounter with 2 animals; (B) an unusually high frequency (2 kHz) thump train; (C) further examples from 2 individuals from different geographic locations; 1 — the amplitude trace shows the total number of thumps in the train of the similarly numbered spectrographic section (total duration approx. 14 sec). The effective analyzing bandwidth for A and C was 6 Hz; B was 20 Hz. Section C-2 had a 450-Hz low-pass filter.

Mysticete Sounds

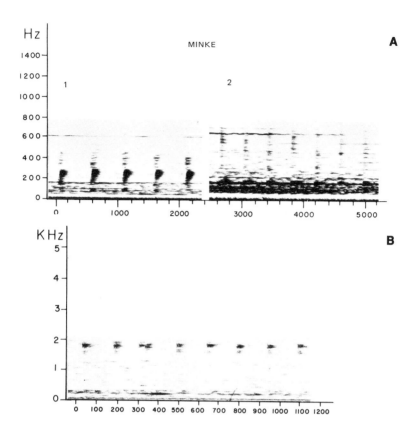

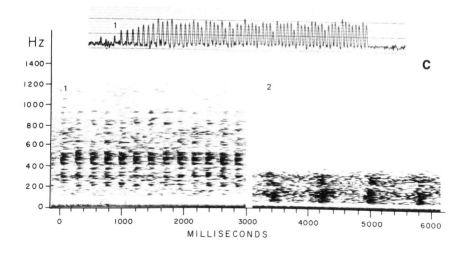

at 5–6 kHz but with energy extending over 20 kHz. Ratchet-like pulses centered at 850 Hz were composed of single- and multipulsed units, lasting 1–6 msec for the single pulses and 25–30 msec for the multipulses.

Beamish and Mitchell (1973) also reported on high-frequency signals from a minke whale which apparently emitted 200 clicks in 50 distinct series. The repetition rate was consistent, with a mean of 6.75 ± 1.02 (SD) clicks per second. The principal energy of these clicks occupied a frequency band between 4 and 7.5 kHz.

It is obvious that there is considerable variation in the vocal repertoire of minke whales and that the information on high-frequency "clicks" produced by this species encourages further investigation of these signals in other mysticetes.

6. Humpback Whale: *Megaptera novaeangliae* (Borowski 1781)

Unquestionably the most vocal mysticete species is the humpback whale, whose repertoire in the tropical or winter portion of its range is organized into a complex, repeating song. This song contrasts with the limited repertoire of these animals when feeding during summer months in higher latitudes.

Several investigators (M. P. Fish, 1949; Schreiber, 1952; Griffin, 1955; Kibblewhite *et al.*, 1967; Poulter, 1968) reported on sounds believed to be humpback vocalizations, although Schevill and Watkins (1962) were the first to publish a spectrogram of a few low-frequency signals. Tavolga (1968) then reported on several examples of these humpback signals which include vocalizations from 150 to 800 Hz at 1–1.5 sec and "cries or squeals" which sometimes reached 2 kHz at 0.5-sec duration. Most of these sounds were recorded off Bermuda by B. Patterson. Two accounts by Levenson (1969, 1972), also recording in Bermuda, extended the reported range of humpback signals to occasional 4- and 8-kHz levels. The two most common sounds, "low grunts" from 120 to 250 Hz and "squeals" around 1600 Hz, were associated with "chirps, whistles, wails, and turkey-like" sounds ranging from 500 to 1650 Hz. Source levels were calculated at an average value of 155.4 dB re 1 μPa at 1 m within a range of 144.3–174.4 dB.

Humpbacks are also known to produce a "wheezing" blow at the surface which is acoustically coupled to the water and differs substantially from a normal blow (Watkins, 1967*b*). This signal may have a communicative function.

The song of the humpback (Anon., 1969; Payne and McVay, 1971; Winn *et al.*, 1971) may be described as a repeating, complex series of sounds in frequencies generally less than 4 kHz. Songs are organized into several themes composed of repeating phrases and syllables (or units, as in Payne and McVay, 1971) within a phrase. Themes and phrases follow a rigid or fixed order and are monotonously repeated. Figure 3 is a continuous spectrogram of an entire song with phonetic labeling of sounds taken from Winn *et al.* (1971). Temporal analysis indicates that songs may last from 7 to 36 min, but in each case themes are presented in the same fixed order. The difference in song duration simply reflects the number of times phrases are repeated within each theme. A detailed temporal analysis is under separate study (Thompson and Winn, 1977). Although songs are continuously emitted, Winn *et al.* (1971) correlated surfacing with a sound known as the surface ratchet which conveniently marks the end of the song. Immediately the animal begins the same sequence over again. This pattern differs from bird song, which has long intersong intervals and a variable sequence of notes within the song.

Generally, the type of song produced (designated by differences in sounds and phrase composition) is the same on specific mating and calving banks, with some variation between individuals. Dialect analysis (Winn and Winn, 1978) between banks is difficult due to overlapping song types, inadequate sampling, and annual changes in song format (Payne and Payne, 1978; Winn and Winn, 1978). However, differences do exist in song formats and the possibility of dialects merits further investigation. Analysis and comparison of songs from the Pacific with those of Winn *et al.* (1971) in the West Indies, and those recorded by the Paynes in Bermuda, may provide more information about species variability and generalized properties of the humpback song. It is clear that phrases found in songs from New Zealand, Hawaii, and California have not been found in those from the western North Atlantic (Winn, unpublished).

Winn *et al.* (1971) also reported on sonic pulses or clicks in the 2- to 7-kHz region and white noise blasts recorded from nonsinging animals. Analysis of individual sounds produced by entrapped humpbacks in Newfoundland has been completed by Winn, Beamish, and Perkins (1978). A wide variety of sounds were recorded from a female and a male under these conditions, but the sounds were irregular and not organized into a song.

The functions of humpback vocalizations have received speculative comment (see Section V) and remain as some of the most intriguing questions in cetology.

IV. MISCELLANEOUS SOUNDS

Due to observational difficulties, many sounds are recorded with no information available concerning source. Occasionally these unknown sounds can be tracked, as with the 20-Hz fin whale signals. One such sound is a long 35-Hz moan of considerable amplitude (Fig. 1). This narrow frequency band sound is longer than the typical 1-sec "20-Hz" signals of fin whales and may originate from the blue whale. The same signal is on the record produced by Payne (1977), Capitol Records, Inc., Survival Anglia Ltd., recorded by B. Patterson in Bermuda. If these sounds originate from blue whales, perhaps we will eventually learn more of this species' little known winter distributions in the

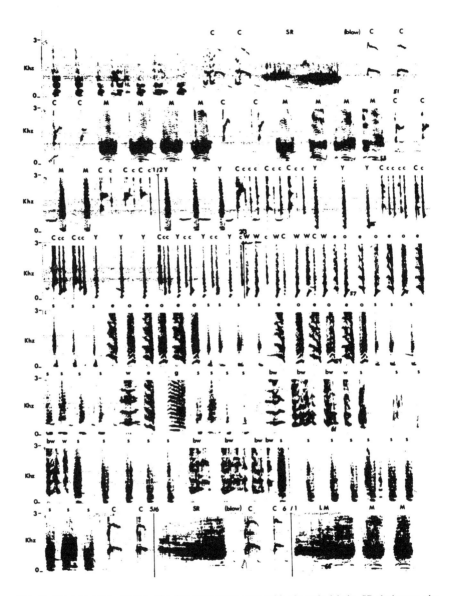

Fig. 3. Spectrographic display of a full humpback song with phonetic labels; SR designates the "surface ratchet" which marks the end of the song. (From Winn *et al.*, 1971.)

Mysticete Sounds

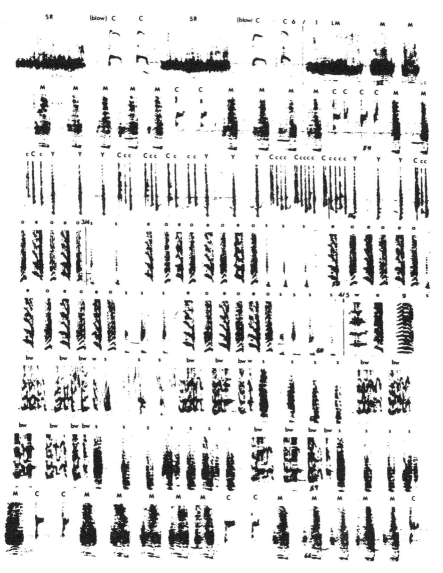

Fig. 3. *(continued).*

Atlantic. However, the documented signals from blue whales off Chile (Cummings and Thompson, 1971a) are not the same (see Section III.D.1). If geographic differences in blue whale signals do occur, this information may support dialect hypotheses in other mysticete species.

An occasional sound detected only in the Pacific has caused considerable interest for many years. This apparently biological sound is known as the "boing" (Wenz, 1964), with a fundamental frequency of about 100 Hz with variable durations up to 4.5 sec and intervals between 8 and 40 sec. The "boing" appears mechanically produced with strong pulsed modulations. The possibility that this sound originates from a mysticete species has not been dispelled.

Discerning biological signals from ambient noise is often difficult, particularly since many naturally occurring phenomena, such as the movement of ice packs, produce strange combinations of sounds. However, most ambient sounds are well documented. Searching for and identifying sources of unknown sounds may likely extend the knowledge of distribution and occurrence of several species.

V. DISCUSSION

It is important to bear in mind the limitations imposed by difficulties inherent in mysticete bioacoustic research. It is possible that some of the sounds reported in this review were not produced by the specified mysticetes. It is also highly probable that many mysticete sounds are yet to be recorded, and the wide variability in the vocal repertoires of the humpback and minke, for example, lend support to this claim. Very little is known about the sounds of some mysticete species, such as the sei and Bryde's and nothing is known of sounds, if any, of the pygmy right whale *Caperea marginata*. Analysis of the complex humpback whale song is an example of a problem which has yet to be fully solved. The humpback has shown an annual change in its song format (Payne and Payne, 1979; verified by Winn and Winn, 1978) by varying sound sequences, although the sounds are generally similar from year to year. Thus, yearly continuity in recording humpbacks is necessary to define the general properties of the song and range of acoustic and thematic variability.

A broad classification of mysticete sounds will be presented here. The first group includes the typical low-frequency moans from 0.4 to 36 sec long, with fundamental frequencies from 12 to 500 Hz but usually between 20 and 200 Hz. Moans may either contain strong harmonic structure or be pure tone, such as the 20-Hz signals. All but the sei and minke whales are known to produce these sounds. The second classification includes the gruntlike thumps and knocks of shorter duration. The humpback, southern and northern right whale, bowhead, gray, fin, and minke are known to produce these sounds, which range in duration from 50 to 500 msec with major energy between 40 and 200 Hz. The third group contains chirps, cries, and whistles at frequencies above 1000 Hz. Chirps are generally pulses producing short (50–100 msec) discrete tones which change frequency rapidly and are not harmonically related, whereas cries and whistles are pure tonal with or without harmonics. The fourth and last group of sounds is the clicks or pulses which generally last from 0.5 to 5 msec and reach as high as 20–30 kHz in the blue whale (Beamish and Mitchell, 1971). Clicklike sounds of either pure frequency or broad band are reported from minke, gray, humpback, sei, Bryde's, and fin whales. Perhaps as the number of acoustic encounters increases, we can more fully document cases of clicklike vocalizations.

Species-specific signals have some utility in population assessment and distribution. For instance, Winn et al. (1975) estimated humpback populations in the West Indies by supplementing visual sightings with acoustic detection of humpback sounds from animals beyond visual range. Levenson and Leapley (1976) were able to verify the wide range and distribution of humpbacks in the Caribbean using sonobuoys with known distance of detection, deployed from an aircraft. The characteristic thump train of the minke whale recorded on a fixed hydrophone in Antigua has extended the previously known distribution of this species to lower latitudes (Winn, unpublished). Thus, important biological information on specific species can be obtained by listening. Fixed hydrophone stations throughout the oceans may provide an increasing amount of data concerning the biology of several species of whales.

Besides species recognition there is increasing evidence of individual or signature information within species, as has been suggested with the click vocalizations of sperm whales, *Physeter catodon* (Backus and Schevill, 1966) and other odontocetes (Caldwell *et al.*, 1972). We have presented evidence herein of possible individual frequency and repetition rate differences in the minke thump trains and have noted consistent frequency distinction in the possibly antiphonal 20-Hz signals of fin whales. Recently, Hafner *et al.* (1979) showed significant discrimination between individual humpbacks by using multivariate techniques on measurements of the high-frequency "cries" in the humpback songs. Certainly, signature information within vocalizations would have prominent consequences in developing models of social organization.

Interspecific communication has been shown by Cummings and Thompson (1971c) by playing back killer whale *(Orcinus orca)* sounds to gray whales during their migration. Killer whales are known predators of gray whales, and the experimental results showed statistically significant avoidance by gray whales when *Orcinus* sounds are broadcast. The same experiment with southern right whales failed to elicit the same reaction, although playback level was lower (Cummings *et al.*, 1972).

Evidence has been presented that blue, fin, and minke whales produce transient "click" energy in the ultrasonic (20-kHz) range. Species for which sonic pulses have been reported include the humpback, minke, and gray whales.

However, in no case has it been shown that echolocation is utilized by any of these species (see Norris, 1969; Beamish, 1974). It is hard to imagine a signal designed to detect food organisms as small as plankton, although Beamish and Mitchell (1971) make an interesting case regarding the blue whale 25-kHz signals. The high-frequency sounds of the minke whale may serve to detect fish, which comprise part of the diet of several mysticete species (Beamish and Mitchell, 1973). Certainly experimental evidence is lacking which could demonstrate echolocation abilities, although recently Beamish (1978) showed that a humpback whale tethered to shore could only visually navigate in a maze constructed of poles anchored in an embayment. When blindfolded, the humpback ran into the obstacles. However, the powerful low-frequency sounds of mysticetes are thought to communicate information on bathymetry since resounding whale echoes can often be detected from pinnacles and banks. Echoes such as these from large reflective bodies may well be the only type of "echolocation" in mysticetes.

Some of the most fundamental biological descriptions, such as mode of sound production, are lacking for mysticetes. Vocal cords typically seen in other mammals are absent, but there is anatomical complexity in the larynx and respiratory airways of mysticetes (see Hosakowa, 1950) which might be used for vocalizations. Cetaceans produce the loudest signals of any living source: up to 188 dB re 1 μPa at 1 m registered from the blue whale (Cummings and Thompson, 1971a). Even the relatively small minke has been recorded at 152.6 dB re 1 μPa at 1 m (Beamish and Mitchell, 1973). A compilation of source levels measured from cetaceans is presented by Fish and Turl (1976). High-amplitude, low-frequency sounds typical of mysticetes will transmit tens, if not hundreds, of miles under certain conditions. Thus, Payne and Webb (1971) have redefined the term "herd" of animals from those in close proximity to those in a wide acoustic range of "tenuous contact." If this is the case, investigations of herd or social structure become extremely difficult.

The exact hearing capabilities of mysticetes is difficult to establish, whereas those of a captive odontocete, *Tursiops truncatus,* have been measured (see Johnson, 1967). Anecdotal evidence is available suggesting that mysticetes respond to ship noise, sonar pinging, etc. Meaningful playback experiments with mysticetes are lacking due to technically strict requirements for large and powerful transducers to produce accurate signals. Anatomically, the mysticete ear has been studied and reviewed (Fraser and Purves, 1954, 1960; Reysenbach de Haan, 1957, 1966; Dudok van Heel, 1962; Purves, 1966). Recently, Fleischer (1976) compared cochlear morphometrics in extinct and extant cetaceans and concluded that mysticete cochlea have structurally evolved for sensitivity to low-frequency sounds as compared to odontocetes' high-frequency sensitivity; although mysticete hearing for high frequency is probably very good.

We have already mentioned that some mysticete sounds may serve to maintain contact with a herd miles away. It has also been suggested that 20-Hz signals of fin whales may be produced to indicate the presence of food (Payne and Webb, 1971). In a visually limited, three-dimensional environment, acoustic signals may provide information about position, behavioral state (reproductive, feeding, alarm, etc.), signature, and population size (e.g., a humpback confusion chorus on a densely populated or crowded bank). Singing humpbacks are presumed to be isolated, sexually active males (Winn et al., 1971) attempting to attract a mate. Testing this hypothesis is possible by using a biopsy sampler to cytologically sex individuals at sea (Winn et al., 1973). No obvious sexual dimorphism is evident from the surface, and knowledge of sex would greatly facilitate social analysis. In the case of the humpback, if only males sing, the acoustic census technique could be refined by knowing the ratio of singing to nonsinging animals (Winn et al., 1975).

The sporadic low-frequency moans of the northern right whale appear to be communicative. These sounds are heard as the whale moves to reposition for feeding on a slick of plankton (Watkins and Schevill, 1976). A specific behavior correlated with a specific sound has been observed with the humpback "surface ratchet" (Winn et al., 1971). Norris et al. (1977) also describe sharp clicks or pulses consistently emitted when gray whale calves were reunited with their mothers. Observational problems have restricted behavioral interpretation of sounds to a minimum.

Marked periodicity in sound production has been noted for the southern right whale (Payne and Payne, 1971), the gray whale (Painter, 1963), and possibly for humpbacks in Bermuda (Payne and McVay, 1971; Al Seidl, personal communication). The increase in vocalizations during the night may simply be a function of reduced visibility. We have not been able to verify this periodicity in the humpback whale.

The presence of a boat affects whales' behavior to a degree. It is well known, for instance, that a minke whale is attracted to a ship and is most likely to phonate on its first approach. Sonobuoys deployed from an aircraft offer a means to record without the interference of a ship.

In summary, various functions of mysticete vocalizations that have been advanced can be enumerated as follows: (1) maintenance of interindividual distance; (2) species and individual recognition; (3) contextual information (feeding, alarm, courtship, etc.); (4) maintenance of social organization or herd structure; and (5) the location of banks and pinnacles or other oceanographic features. Investigators can use these various vocalizations to: (1) identify species, and possibly individuals; (2) locate and track animals; (3) monitor population levels; (4) identify substocks if dialects exist, and (5) to further understand more of the animals' behavior and life history.

In this chapter we have assigned and characterized mysticete sounds according to species. Species-specific sounds are documented in several of the baleen whales. While we have physically and qualitatively described these sounds produced, it is difficult to ascribe a function to any given sound. Each sound may provide multiple functions which differ in various behavioral and/or environmental contexts. The inherent difficulties of mysticete behavioral and bioacoustic research are further compounded by the relatively sparse populations of some of these mysticetes (e.g., the northern right whale, blue whale). Periodic compilations of acquired data from all cetacean investigations will undoubtedly facilitate an understanding of cetacean life histories and ecological requirements. This knowledge may aid in man's attempt to further the conservation and management of these animals.

ACKNOWLEDGMENTS

This review of the literature and the presentation of original data was supported by ONR contract N00014-76-C-0226 to the University of Rhode Island. We wish to thank Ms Marilyn Nigrelli for typing the manuscript.

REFERENCES

Aldrich, H. L., 1889, *Arctic, Alaska, and Siberia,* Rand, McNally, Chicago, pp. 32–35.
Anon., 1969, Singing whales, *Nature* **224(5216)**:217.
Asa-Dorian, P. V., and Perkins, P. J., 1967, The controversial production of sound by the California gray whale, *Eschrichtius gibbosus, Nor. Hvalfangst-Tid.* **4**:74–77.
Backus, R. H., and Schevill, W. E., 1966, *Physeter* clicks, in: *Whales, Dolphins, and Porpoises* (K. S. Norris, ed.), pp. 510–528, University of California Press, Berkeley.
Beamish, P., 1974, Whale acoustics, *J. Can. Acoust. Assoc.* **2(4)**:8–12.
Beamish, P., 1978, Evidence that a captive humpback whale (*Megaptera novaeangliae*) does not use sonar, *Deep-Sea Res.* **25(5)**:469–472.
Beamish, P., and Mitchell, E., 1971, Ultrasonic sounds recorded in the presence of a blue whale (*Balaenoptera musculus*), *Deep-Sea Res.* **18**:803–809.
Beamish, P., and Mitchell, E., 1973, Short pulse length audio frequency sounds recorded in the presence of a minke whale (*Balaenoptera acutorostrata*), *Deep-Sea Res.* **20**:375–386.
Busnel, R-G., and Dziedzic, A., 1966, Acoustic signals of the pilot whale *Globicephala melaena* and of the porpoises *Delphinus delphis* and *Phocoena phocoena*, in: *Whales, Dolphins, and Porpoises* (K. S. Norris, ed.), pp. 607–648, University of California Press, Berkeley.
Caldwell, M. C., Caldwell, D. K., and Hall, N. R., 1972, Ability of an Atlantic Bottlenosed Dolphin (*Tursiops truncatus*) to Discriminate Between and Potentially Identify to Individual, the Whistles of Another Species, the Common Dolphin *(Delphinus delphis)*, Marineland Res. Lab., Tech. Rep. 9, 7 pp.
Cummings, W. C., and Phillipi, L. A., 1970, Whale Phonations in Repetitive Stanzas, Naval Undersea Res. and Dev. Center, NUC TP 196, San Diego, 4 pp.
Cummings, W. C., and Thompson, P. O., 1971a, Underwater sounds from the blue whale, *Balaenoptera musculus, J. Acoust. Soc. Am.* **50**(4 pt. 2):1193–1198.
Cummings, W. C., and Thompson, P. O., 1971b, Bioacoustics of marine animals; R/V *Hero* cruise 70-3, *Antarc. J. U.S.,* **6(5)**:158–160.

Cummings, W. C., and Thompson, P. O., 1971c, Gray whales, *Eschrichtius robustus,* avoid the underwater sounds of killer whales, *Orcinus orca, Fish. Bull.* **69(3)**:525–530.

Cummings, W. C., and Thompson, P. O., Sounds from Bryde's and finback whales in the Gulf of California (unpublished manuscript).

Cummings, W. C., Thompson, P. O., and Cook, R., 1968, Underwater sounds of migrating gray whales, *Eschrichtius glaucus* (Cope), *J. Acoust. Soc. Am.* **44(5)**:1278–1281.

Cummings, W. C., Fish, J. F., and Thompson, P. O., 1972, Sound production and other behavior of southern right whales, *Eubalena glacialis, Trans. San Diego Soc. Nat. Hist.* **17(1)**:1–14.

Cummings, W. C., Thompson, P. O., and Fish, J. F., 1974, Behavior of southern right whales: *RV Hero* cruise 72-3, *Antarct. J. U.S.,* March–April 1974, 33–38.

Dudok van Heel, W. H., 1962, Sound and cetacea, *Neth. J. Sea Res.* **1(4)**:407–507.

Eberhardt, R. L., and Evans, W. E., 1962, Sound activity of the California gray whale, *Eschrichtius glaucus, J. Audio Eng. Soc.* **10(4)**:324–328.

Fish, J., and Turl, C. W., 1976, Acoustic Source Levels of Four Species of Small Whales, Naval Undersea Center, NUC TP 547, 14 pp.

Fish, J. F., Sumich, J. L., and Lingle, G. L., 1974, Sounds produced by the gray whale, *Eschrichtius robustus, Mar. Fish Rev.* **36(4)**:38–45.

Fish, M. P., 1949, Marine Mammals of the Pacific with Particular Reference to the Production of Underwater Sound, Office of Naval Res., Tech. Rep. No. 8, Ref. no. 49-30, 69 pp.

Fleischer, G., 1976, Hearing in extinct cetaceans as determined by cochlear structure, *J. Paleontol.* **50(1)**:133–152.

Fraser, F. C., and Purves, P. E., 1954, Hearing in cetaceans, *Bull. Br. Mus. (Nat. Hist.)* **2**:103–116.

Fraser, F. C., and Purves, P. E., 1960, Hearing in cetaceans, *Bull. Br. Mus. (Nat. Hist.)* **7**:1–140.

Gales, R. S., 1966, Pickup, analysis, and interpretation of underwater acoustic data, in: *Whales, Dolphins, and Porpoises* (K. S. Norris, ed.), pp. 435–444, University of California Press, Berkeley.

Gilmore, R., 1969, Populations, distribution, and behavior of whales in the western South Atlantic; Cruise 69-3 of *R/V Hero, Antarct. J. U.S.* **IV(6)**:307–308.

Griffin, D. R., 1955, Hearing and acoustic orientation in marine animals, *Deep-Sea Res.* **3**(Suppl.):406–417.

Hafner, G. W., Hamilton, C. L., Steiner, W. W., Thompson, T. J., and Winn, H. E., 1979, Signature information in the song of the humpback whale, (in preparation).

Hosakowa, H., 1950, On the cetacean larynx with special remarks on the laryngeal sac of the sei whale and the aryteno-epiglottideal tube of the sperm whale, *Sci. Rep. Whales Res. Inst.* **3**:23–62.

Johnson, C. S., 1967, Sound detection thresholds in marine mammals, in: *Marine Bio-Acoustics* (W. N. Tavolga, ed.), Vol. 2, pp. 247–260, Pergamon Press, New York.

Katona, S., Richardson, D., and Hazard, R., 1975, *A Field Guide to the Whales and Seals of the Gulf of Maine* pp. 1–63, Maine Coast Printers, Rockland.

Kibblewhite, A. C., Denham, R. N., and Barnes, D. J., 1967, Unusual low-frequency signals observed in New Zealand waters, *J. Acoust. Soc. Am.* **41(3)**:644–655.

Leatherwood, S., Caldwell, D. K., and Winn, H. E., 1976, Whales, Dolphins, and Porpoises of the Western North Atlantic, A Guide to Their Identification, NOAA Tech. Rep. NMFS CIRC-396, 175 pp.

Levenson, C., 1969, Behavioral, Physical, and Acoustic Characteristics of Humpback Whales *(Megaptera novaeangliae)* at Argus Island, Naval Oceanogr. Office, Informal Rep. 69-54, 13 pp.

Levenson, C., 1972, Characteristics of Sounds Produced by Humpback Whales *(Megaptera novaeangliae),* NAVOCEANO Tech. Note 7700-6-72, 17 pp.

Levenson, C., and Leapley, W. T., 1976, Humpback Whale Distribution in the Eastern Caribbean Determined Acoustically from an Oceanographic Aircraft, NAVOCEANO Tech. Note 3700-46-76, 7 pp.

McVay, S., 1973, Stalking the Arctic whale, *Am. Sci.* **61**:24–37.
Nishiwaki, M., 1972, General biology, in: *Mammals of the Sea* (S. H. Ridgway, ed.), pp. 3–33, Charles C. Thomas, Springfield, Illinois.
Norris, K. S., 1969, The echolocation of marine mammals, in: *The Biology of Marine Mammals* (H. T. Andersen, ed.), pp. 391–423, Academic Press, New York.
Norris, K. S., Goodman, R. M., Villa-Ramirez, B., and Hobbs, L., 1977, Behavior of California gray whale, *Eschrictius robustus,* in Southern Baja California, Mexico, *Fish. Bull.* **75**(1):159–172.
Northrup, J., Cummings, W. C., and Thompson, P. O., 1968, 20-Hz signals observed in the Central Pacific, *J. Acoust. Soc. Am.* **43**(2):383–384.
Northrup, J., Cummings, W. C., and Morrison, M. F., 1971, Underwater 20-Hz signals recorded near Midway Island, *J. Acoust. Soc. Am.* **49**(6 pt. 2):1909–1910.
Painter, D. W., 1963, Ambient noise in a coastal lagoon, *J. Acoust. Soc. Am.* **35**(9):1458–1459.
Patterson, B., and Hamilton, G. R., 1964, Repetitive 20 cycle per second biological hydroacoustic signals at Bermuda, in: *Marine Bio-Acoustics* (W. N. Tavolga, ed.), Vol. 1, pp. 125–146, Pergamon Press, New York.
Payne, R., 1972, The song of the whale, in: *The Marvels of Animal Behavior,* pp. 144–166. National Geographic Society, Washington D.C.
Payne, R., 1977, "Deep Voices," a long playing album produced by R. Payne, Capitol Records, Inc., Survival Anglia Limited.
Payne, R., and McVay, S., 1971, Songs of humpback whales, *Science* **173**:583–597.
Payne, R., and Payne, K., 1971, Underwater sounds of southern right whales, *Zoologica* **4**:159–165.
Payne, K., and Payne, R., 1979, Annual changes in songs of humpback whales, *Z. Tierpsychol.* (in press).
Payne, R., and Webb, D., 1971, Orientation by means of long range acoustic signalling in baleen whales, *Ann. N.Y. Acad. Sci.* **188**:110–142.
Perkins, P. J., 1966a, Passive sonar aids deep-sea research on *Trident, Undersea Tech.* March:3.
Perkins, P. J., 1966b, Communication sounds of finback whales, *Nor. Hvalfangst-Tid.* **10**:199–200.
Poulter, T. C., 1968, Marine mammals, in: *Animal Communication; Techniques of Study and Results of Research* (T. Sebeok, ed.), pp. 405–465, Indiana University Press, Bloomington.
Poulter, T. C., 1970, Vocalization of the gray whales in Laguno Ojo de Liebre (Scammons Lagoon) Baja California, Mexico (unpublished manuscript).
Poulter, T. C., 1971, Recording of marine mammals in the arctic, *U.S. Navy J. Underwater Acoust.* **1**:97–104.
Purves, P. E., 1966, Anatomy and physiology of the outer and middle ear in cetaceans, in: *Whales, Dolphins, and Porpoises* (K. S. Norris, ed.), pp. 320–380, University of California Press, Berkeley.
Rasmussen, R. A., and Head, N. E., 1965, The quiet gray whale (*Eschrichtius glaucus*), *Deep-Sea Res.* **12**:869–877.
Reysenbach de Haan, F. W., 1957, Hearing in whales, *Acta Otolaryngol.* **134**(Suppl.):1–114.
Reysenbach de Haan, F. W., 1966, Listening underwater: Thoughts on sound and cetacean hearing, in: *Whales, Dolphins, and Porpoises* (K. S. Norris, ed.), pp. 583–596, University of California Press, Berkeley.
Ross, G. J. B., Best, P. B., and Donnelly, B. G., 1975, New record of the pygmy right whale (*Caperea marginata*) from South Africa, with comments on distribution, migration, appearance, and behavior, *J. Fish. Res. Board. Can.* **32**(7):1005–1017.
Schevill, W. E., 1964, Underwater sounds of cetaceans, in: *Marine Bio-Acoustics* (W. N. Tavolga, ed.), Vol. 1, pp. 307–316.
Schevill, W. E., and Watkins, W. A., 1962, "Whale and Porpoise Voices," a phonograph record, Woods Hole Oceanographic Inst., Woods Hole, Massachusetts, 24 pp.

Schevill, W. E., and Watkins, W. A., 1972, Intense low-frequency sounds from an Antarctic minke whale, *Balaenoptera acutorostrata, Breviora* **388**:1–8,

Schevill, W. E., Backus, R. H., and Hersey, J. B., 1962, Sound production by marine animals, in: *The Sea* (M. N. Hill, ed.), Vol. 1, pp. 540–562, Interscience Publishers, J. Wiley and Sons, New York.

Schevill, W. E., Watkins, W. A, and Backus, R. H., 1964, The 20 cycle signals and *Balaenoptera* (fin whales), in: *Marine Bio-Acoustics* (W. N. Tavolga, ed.), Vol. 1, pp. 147–152, Pergamon Press, New York.

Schreiber, D. W., 1952, Some sounds of marine life in the Hawaiian area, (abstract), *J. Acoust. Soc. Am.* **24**:116.

Tavolga, W. N., 1968, Marine Animal Data Atlas, Naval Training Device Center, Tech. Rep., NAVTRADEVCEN, 1212-2, 239 pp.

Thompson, P. O., and Cummings, W. C., 1969, Sound production of the finback whale, *Balaenoptera physalus,* and Eden's whale, *B. edeni,* in the Gulf of California, *Proc. Sixth Ann. Calif. Biol. Sonar and Diving Mammals* (abstract), Stanford Res. Inst., p. 109.

Thompson, T. J., and Winn, H. E., 1977, Temporal aspects of the humpback whale song (abstract 147), Anim. Behav. Soc. Meeting, Pennsylvania State University.

Tyrrell, W. A., 1964, Design of acoustic systems for obtaining bio-acoustic data, in: *Marine Bio-Acoustics* (W. N. Tavolga, ed.), Vol. 1, pp. 65–86, Pergamon Press, New York.

Walker, R. A., 1963, Some intense, low-frequency, underwater sounds of wide geographic distribution, apparently of biological origin, *J. Acoust. Soc. Am.* **35(11)**:1816–1824.

Watkins, W. A., 1966, Listening to cetaceans, in: *Whales, Dolphins, and Porpoises* (K. S. Norris, ed.), pp. 471–476, University of California Press, Berkeley.

Watkins, W. A., 1967a, The harmonic interval; fact or artifact in spectral analysis of pulse trains, in: *Marine Bio-Acoustics* (W. N. Tavolga, ed.), Vol. 2, pp. 15–43, Pergamon Press, New York.

Watkins, W. A., 1967b, Airborne sounds of the humpback whale, *Megaptera novaeangliae, J. Mammal.* **48(4)**:573–578.

Watkins, W. A., 1976, Biological sound-source locations by computer analysis of underwater array data, *Deep-Sea Res.* **23**:175–180.

Watkins, W. A., and Schevill, W. E., 1976, Right whale feeding and baleen rattle, *J. Mammal.* **57(1)**:58–66.

Wenz, G. M., 1964, Curious noises and the sonic environment in the ocean, in: *Marine Bio-Acoustics* (W. N. Tavolga, ed.), Vol. 1, pp. 101–119, Pergamon Press, NewYork.

Weston, D. E., and Black, R. I., 1965, Some unusual low-frequency biological noises underwater, *Deep-Sea Res.* **12**:295–298.

Winn, H. E., and Perkins, P. J., 1976, Distributions and sounds of the minke whale, with a review of mysticete sounds, *Cetology* **19**:1–12.

Winn, H. E., and Winn, L. K., 1978, The song of the humpback whale *(Megaptera novaeangliae)* in the West Indies, *Mar. Biol.* **47**:97–114.

Winn, H. E., Perkins, P. J., and Poulter, T. C., 1971, Sounds of the humpback whale, *Proc. Seventh Ann. Conf. Biol. Sonar and Diving Mammals,* Stanford Res. Inst., pp. 39–52.

Winn, H. E., Bischoff, W. L., and Taruski, A. G., 1973, Cytological sexing of cetacea, *Mar. Biol.* **23**:343–346.

Winn, H. E., Edel, R. K., and Taruski, A. G., 1975, Population estimate of the humpback whale *(Megaptera novaeangliae)* in the West Indies by visual and acoustic techniques, *J. Fish. Res. Board Can.* **32(4)**:499–506.

Winn, H. E., Beamish, P., and Perkins, P. J., 1978, Sounds of two entrapped humpback whales *(Megaptera novaeangliae)* in Newfoundland (manuscript).

Wright, B. S., 1962, Notes of North Atlantic whales, *Can. Field Nat.* **76**:62–65.

INDEX

For species information see both common and scientific names.

Acoustic tracking, 3
Activity patterns, 203, 359-362
Age and growth, 158-162, 172, 228-229, 237, 239-244, 255, 257-261, 267, 281-282, 369, 371, 372
Aggression, 80, 166, 222, 251, 278, 280
Agonistic behavior, 76
Alarm call (*see also* Vocalizations), 339
Arctocephalus pusillus (*see also* Cape fur seal), 209

Balaena mysticetus (*see also* Bowhead), 406
Balaenoptera acutorostrata (*see also* Minke whale), 143, 148, 404, 417
  B. borealis (*see also* Sei whale), 417
  B. edeni (*see also* Bryde's whale), 416
  B. musculus (*see also* Blue whale), 292, 296, 404, 412, 421
  B. physalus (*see also* Fin whale), 7, 28, 75, 404, 413

Baleen whales, 8, 33, 55-56, 228, 291-309
Beaked whale, 143-164
Beluga, 77, 273, 279, 311-341
Berardius bairdi (*see also* Beaked whale), 144
Blue whale, 292, 294-296, 298-299, 307-308, 404, 412, 421
Bottlenose dolphins, 3, 6, 15, 165-166, 175, 196, 208-209, 212-214, 220-222, 277, 369
Bottlenose whale, 143-164
Bowhead, 406
Breeding season (*see* Reproduction)
Bryde's whale, 416

California sea lion, 266
Callorhinus ursinus (*see also* Northern fur seal), 266
Calves, 27, 66, 68, 150, 153, 155, 158, 160, 172, 182-183, 186-190, 196, 227, 246, 250, 258-259, 273, 277, 279-280, 338, 370-401
Cape fur seal, 209

Caperea marginata (*see also* Pygmy right whale), 410
Care-giving behavior, 153, 218, 277
Coloration, 46-47, 51-53, 74, 78, 120-124, 172-173, 275
Commensals, 53, 55-56, 79
Common dolphin, 211
Communication (*see also* Sound production; Vocalizations), 167, 218, 345-368, 370-401, 425-427
Competition, 208, 251, 256, 276, 278, 280

Dall's porpoise, 6, 45-83
Deep scattering layer, 16, 17
Delphinapterus leucas (*see also* Beluga), 28, 65, 77, 279, 311-341
Delphinids, 18, 51, 53, 56, 59, 77-78, 165, 167, 172, 175, 218-222, 276
Delphinus, 5, 6, 10, 16, 17, 18, 65, 70, 220
  D. delphis (*see also* Common dolphin), 5, 6, 10, 11, 13, 16, 21, 28, 65-66, 105, 107, 109, 211, 213, 220
Distribution (*see also* Range), 88, 366, 425
Diving, 2, 11, 16, 20-24, 57, 75, 153-154, 161, 269-270, 276-279, 284, 327, 329
Dominance, 80, 245, 273, 276
Dusky dolphin, 5

Echolocation, 116, 278, 291, 299-300, 306, 308-309
Elephant seal, 266
Entrapment, frontispiece, 291-309

Eschrichtius robustus (*see also* Gray whale), 3, 5, 14, 15, 28, 75, 312, 410
Eubalaena australis (*see also* Right whale), 208
  E. glacialis australis (*see also* Right whale), 409
  E. glacialis glacialis (*see also* Right whale), 406
Eumetopias jubata (*see also* Stellar sea lion), 266

False killer whale, 279
Feeding (by species) (*see also* Food),
  Hyperoodon ampullatus, 148, 154-156, 161
  Megaptera novaeangliae, 34
  Orcinus orca, 51
  Phocoenoides, 56-57, 79
  Phocoenoides dalli dalli, 58
  Physeter catodon, 268-270, 272, 276-279, 284
  Sousa, 192-193, 196-197, 200-207, 212-215, 223
  Stenella attenuata, 21
Fin whale, 75, 299, 307, 404, 413
Finless porpoise, 77
Food (*see also* Feeding), 5, 17, 107, 126-127

Geographical distribution
  Hyperoodon ampullatus, 144-145, 149
  Megaptera novaeangliae, 34
  Orcinus orca, 51
  Phocoenoides, 45, 48-50, 78
  Physeter catodon, 230, 242, 262-270
  Sousa, 177, 213
Globicephala (*see also* Pilot whale), 5, 6, 23, 65, 70, 345-367

Globicephala (cont.)
 G. macrorhynchus, 345
 G. macrorhynchus cf. *scammoni*, 28, 392
 G. melaena, 219, 279, 345
 G. scammoni, 220
Grampus griseus, 65
Gray whale, 3, 5, 7, 16, 23, 25, 27, 41, 75, 299, 410
Group composition, 150-153, 161, 182-189, 216, 227-232, 234-271, 273-274, 277, 279, 281-284

Harbor porpoise, 53
Harems, 228, 231, 242, 245, 250, 254, 259, 269, 275, 277, 280-283
Hearing, 3, 77, 426
Herds (*see also* Pods; Schools), 14, 18, 107-111, 152, 339-340, 366
Humpback dolphin, 165-226
Humpback whale, frontispiece, 3, 33-44, 291, 299-308, 403, 407, 420
Hyperoodon ampullatus, 143-164
 H. planifrons, 143

Identification of individuals, 33-44, 76, 174, 182, 184, 365, 418, 425
Indian Ocean bottlenose dolphin, 194-195
Indus dolphin, 222
Inia geoffrensis, 65
Interspecific associations, 111-112

Killer whale, 6, 41, 50-53, 78, 210-211, 223
Kogia, 59

Lagenorhynchus, 65, 70, 221
 L. obliquidens (*see also* White-sided dolphin), 13, 28, 51, 55-56, 65, 219, 376
 L. obscurus (*see also* Dusky dolphin), 5, 6, 28
Learning, 369
Leptonychotes weddelli, 312
Lissodelphis, 70
 L. borealis (*see also* Right whale dolphins), 77, 85-141
 L. peronii, 123

Management, 34, 41, 280, 285
Megaptera novaeangliae (*see also* Humpback whale), 3, 7-8, 28, 34, 300, 403, 407, 420
Migration (*see also* Movements), 34, 43, 49-50, 144-148, 152, 160-161, 177, 180, 228, 250, 261, 263, 266-267, 271-273, 279, 284
Minke whale, 143, 148, 299, 307, 404, 417
Mirounga angustirostris (*see also* Elephant seal), 266
Morphology, 41, 57-68, 79, 127-138, 172, 176, 180, 228-229, 233-234, 237, 244, 249, 257, 277
Mortality, 233-234, 237, 256, 260, 275, 280, 282, 285
Movements (*see also* Migration), 5, 13, 14, 18-20, 34, 88-99, 106, 256, 300
 group, 175, 180, 182, 185, 191, 196-200, 204-207
 seasonal, 48, 50, 78, 279
Mysticetes, 69, 79, 299-300

Neophocaena phocaenoides (*see also* Finless porpoise), 59, 77

North Atlantic pilot whale (*see* Globicephala)
Northern fur seal, 266, 280-281, 283
Northern right whale dolphin, 85

Odontocetes, 51, 55, 57, 65-66, 69-70, 79, 149, 158, 299
Orcinus orca (*see also* Killer whale), 5, 23, 27-28, 50, 210, 312

Parasites, 53, 55, 79, 127, 250, 260-262
Parturition (*see* Reproduction)
Phocoena, 53, 59, 65, 70, 78, 311
  P. dalli, 46
  P. phocoena, 51, 53, 55-56, 77, 311
  P. sinus, 51
Phocoenoides, 46, 53, 56, 59, 65, 69-70, 74-79
  P. dalli (*see also* Dall's porpoise), 6, 28, 45-48, 64, 66, 77
  P. dalli dalli (*see also* Dall's porpoise), 46-47, 50, 55-56, 58, 60-63, 69, 77
  P. dalli truei (*see also* True's porpoise), 46-48, 77
  P. truei, 46, 77
Physeter, 59, 65
  P. catadon (*see also* Sperm whale), 55, 77, 227-289
Pilot whales, 5, 6, 26, 167, 219-220, 279
Pinnipeds, 275-277, 283
Playback experiments, 311-341
Platanista indi (*see also* Indus dolphin), 222
Play behavior, 75, 209
Pods (*see also* Herds; Schools), 42, 227, 258
Polygyny, 273, 275-281, 283

Population assessment, 34, 43, 102, 107, 229, 254, 281, 284-285, 425, 427
Preanal gland, 77
Predation, 50-51, 53, 78, 223
Predators (*see* Predation)
Pregnancy (*see* Reproduction)
Pseudorca crassidens (*see also* False killer whale), 279
Pygmy right whale, 410

Radiotelemetry, 1-31, 176, 307-308
Range, 34, 37, 39, 175-177, 180, 212, 272, 284
Reproduction
  bottlenose dolphins, 220
  Delphinapterus leucas, 279
  Delphinus, 70
  Globicephala melaena, 279
  Hyperoodon ampullatus, 152-153, 155-158, 160-162
  Lagenorhynchus, 70
  Lissodelphis borealis, 124-126
  Megaptera novaeangliae, 34
  Phocoena, 70
  Phocoenoides, 65-70, 79-80
  Physeter catodon, 228-229, 231, 233-234, 243-244, 246-251, 254-260, 267-268, 272-282

  Sousa, 189-191, 219-223
  Stenella coeruleoalba, 221
  Tursiops aduncus, 221
  T. truncatus, 221
Reproductive rates (*see* Reproduction)
Respiration, 4-5, 7, 20-23, 33, 35, 73-74, 153, 172, 192, 293, 299, 302, 306-307
Right whale, 3, 299, 406, 409
Right whale dolphin, 77

Index

Schools (*see also* Herds; Pods), 20, 74-76, 80, 152, 176, 182, 185, 219-220, 227-228, 230-232, 235-251, 254-260, 262-263, 269, 271-284
Seasonal occurrence, 176-181, 259
Sei whale, 307, 417
Sexual behavior, 194-196, 221-222, 251, 341
Sexual dimorphism, 180, 216, 233-234, 246, 266-267, 275, 277-279
Sexual maturity (*see* Reproduction)
Social interactions, 193-196, 201-202
Songs (*see also* Vocalizations), 302, 306, 403, 420-421, 424
Sotalia lentiginosa (*see also* Humpback dolphins), 172
Sound playback, 311-341, 426
Sound production (*see also* Communication; Vocalization), 218, 296, 298-299, 302, 306, 369-401, 403-427
Sousa, 165-226
  S. lentiginosa (*see also* Humpback dolphins), 172
  S. plumbea (*see also* Humpback dolphins), 172
Southern elephant seal, 282-283
Southern right whale, 208
Sperm whale, 55, 77, 149, 155, 157, 227-289
Spermaceti organ, 233, 278-279
Spinner porpoise, 51
Stellar sea lion, 266
Stenella, 5, 51, 53, 65, 78
  S. attenuata, 6, 18, 20-21, 51, 66
  S. coeruleoalba, 51, 55, 66, 213, 221
  S. graffmani, 51, 65-66
  S. longirostris (*see also* Whitebelly spinner), 28, 51
  S. plagiodon, 78

Stenella (*cont.*)
  S. styx, 65-66, 220
Strandings, 45, 57, 68, 77, 116-120, 144-146, 148, 230
Survey, ship, 100-102
Survey, aerial, 102, 105
Swimming, behavior, 112-114
Swimming speeds, 59, 70, 73-74, 80, 191, 303

Tagging, 34, 41, 102, 291, 301, 303, 306-309
Temperature, 17, 149-150, 178-181, 261, 267-268
Tidal cycles, 203-207, 214-215
True's porpoise, 51, 78
Tursiops, 6, 8, 28, 51, 53, 65, 70, 78, 105, 165, 208, 213, 215, 221, 369
  T. aduncus (*see also* Indian Ocean bottlenose dolphin), 194-195, 197, 212, 221, 277
  T. truncatus (*see also* Bottlenose dolphins), 3, 212, 214-215, 220-223, 311-312, 345, 354, 369-401

Vision, 77, 298, 303, 307
Vocalizations (*see also* Alarm call; Songs; Whistles), 114, 403-427, 369-401
  dialects, 333-334, 340, 413, 418, 421, 424
  functions of, 334, 336-337, 339-340, 407, 409, 412-413, 420-421
  signature, 345, 418, 425

Water depth, 148-149, 160
Water temperature, 106

Whaling, 34, 143-148, 150-151, 160, 227-229, 237, 258, 263, 280, 282-284
Whistles (*see also* Vocalizations), 77, 346-367, 369-401
Whitebelly spinner, 51

White-sided dolphin, 400

Zalophus californianus (*see also* California sea lion), 266
Ziphiids, 59